Troubleshooting Finite-Element Modeling with Abaqus

Raphael Jean Boulbes

Troubleshooting Finite-Element Modeling with Abaqus

With Application in Structural Engineering Analysis

Raphael Jean Boulbes
Lyon, France

ISBN 978-3-030-26742-1 ISBN 978-3-030-26740-7 (eBook)
https://doi.org/10.1007/978-3-030-26740-7

Mathematics Subject Classification (2010): 65Z05

This Springer imprint is published by the registered company Springer Nature Switzerland AG
The registered company address is: Gewerbestrasse 11, 6330 Cham, Switzerland

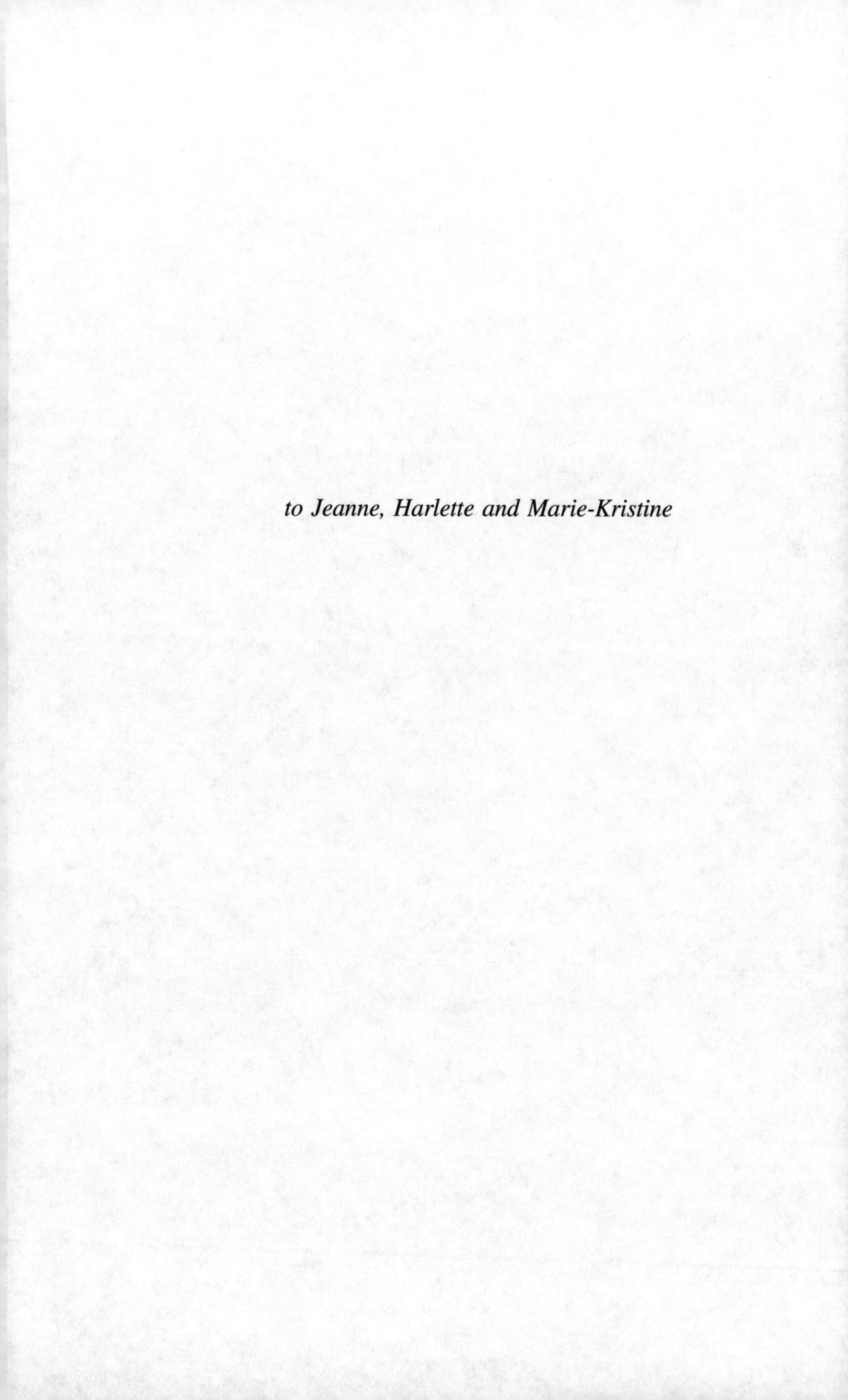

to Jeanne, Harlette and Marie-Kristine

Foreword by Dr. Sonell Shroff

I have been working with Finite Element Analysis software for the past decade and as our understanding of the subject and its utility has increased over time, so have the number of software and the analysis types. As a user and further, as a university teacher for master's students, I have spent a good amount of effort to get familiarized with multiple software platforms to get, in effect, the same result for structural problems. What concerns me most while imparting whatever knowledge I can to the students during the few hours of lessons is to be certain that once they leave that classroom, would they be able to become engineers who do not just work out the problems that face them, but solve them efficiently?

When Raphael contacted my boss at the university for a recommendation of a reviewer in the field for this book, fortunately, I was instantly put in contact with him. It is part of my courses to teach the students how to debug a model and what are the best ways to categorize the errors that they encounter. Coincidently, ABAQUS is one of the most popular softwares we use at the Aerospace faculty because of its wide usage in the industry. It comes with an expansive user guide, and expansive is an understatement; so much so that it is most of the times very difficult to find exactly what you are looking for in order to troubleshoot the program. As an advanced Finite Element course, my focus was so much on the science and the formulation of the finite-element theory that I lacked the organization of the troubleshooting techniques. Raphael's idea of this book came as a fabulous news for me. I would be able to rest easy with such a book out there for the interested students to find.

Ph.D. students, postdocs, M.Sc. graduate students, and even bachelor students use this software and do not necessarily have the time to go through the entire user guide. This book will help them understand how to categorize the error types and where to look for further information. It will teach us how to use the information we receive in the messages from the software itself. And the best part of it is that it is suited to all users, amateurs, and experienced. It contains enough background knowledge of important topics that you would need to get started with what the software is expected to do, and then holds your hand through the error resolution process. Raphael has managed to cover such important and difficult topics in a comprehensive manner. I hope my colleagues in the teaching world will recognize this book as a handy manual both for teaching and for use in their own modeling for research.

Brussels, Belgium Dr. Sonell Shroff
September 2018

Foreword by Gautam Puri

I first spoke with Raphael in 2011 when he was working for FMC Technologies in Norway. At the time, he was performing FEA simulations on various subsea technologies and contacted me after reading a book I had recently written on Python scripting for Abaqus. We had a conversation about running GUI scripts, and since I was about to graduate, he kindly offered to forward my resume to hiring managers at his company. I eventually ended up accepting a local offer from SIMULIA (the Dassault Systémes brand that develops Abaqus) here in the United States, but I must admit that Raphaels offer of an opportunity to work abroad was a tempting one.

Fortunately, a few years later Raphael found another opportunity where we could collaborate; he intended to write a book about troubleshooting/debugging finite-element models in Abaqus and requested me to be one of the reviewers. Anyone who has worked with Abaqus has experienced the frustration of spending many hours debugging an FE model and attempting to get it to converge, and so this sounded like a good subject for a book. I was excited to see a draft version a few months later.

In my opinion, this book will serve as a guide or a toolkit for engineers when they run into difficulties with their Abaqus simulations. The first few chapters provide general information on how an analyst should approach the process of troubleshooting FE models as well as an overview of some common errors and the potential causes. Later, chapters provide deeper coverage of specific topics such as materials and contact interactions. One could argue that much of what is covered in this book is already included in the Abaqus documentation; however, the documentation is vast and spans thousands of pages, and the shear amount of information often makes it very difficult to find tips that will help troubleshoot a model. This book attempts to alleviate this problem.

Many readers, particularly ones who are relatively new to Abaqus, would benefit by reading through at least the first few chapters. They do contain a lot of information, which may be difficult to digest in one sitting but they will help familiarize the reader with the types of issues they are likely to encounter over time. For more experienced Abaqus users, this book will be useful as a handy reference manual.

While it is not possible for simulation analysts to entirely escape the frustration of diagnosing and fixing Abaqus model errors and solver convergence issues, I do believe this book will help make these situations a little less painful for the reader.

Ohio, USA
September 2018

Gautam Puri

Foreword by Prof. David Bassir

First of all, thank you to Raphael Boulbes for providing such support for the mechanical community using the commercial finite-element software Abaqus to simulate complex problems. Indeed, this book will help users to deal with finite-element modeling under different aspects like boundary conditions and to solve troubleshooting issues with such commercial software. It will bring enormous knowledges to both intermediate and advanced users. I wish to get such support 20 years ago, when I was doing my Ph.D.

With concrete and real practical examples included in this book, it gives also a references for engineers at different mechanical industries. I am confident about the success of this work and I wish all the readers to enjoy reading this book as I did.

Shenzhen, China
September 2018

Prof. David Bassir

Preface

The finite-element method (FEM) has become a staple for predicting and simulating the physical bahavior of complex engineering systems. The commercial finite element analysis (FEA) software have gained common acceptance among engineers in industry and researchers at universities. Therefore, academic engineering departments include graduate or undergraduate senior-level courses that cover not only the theory of FEM but also its applications using the commercially available FEA programs.

The goal of this book is to provide students and engineers with an essential theoretical and deep practical knowledge of the finite-element method and the skills required to fix troubleshooting issues with Abaqus, a commercially available FEA software. This book designed for seniors and first-year graduate students, as well as practicing engineers, is introductory and self-contained in order to minimize the need for additional reference material.

In addition to the fundamental topics in finite-element method, it presents advanced topics concerning modeling and analysis procedures with Abaqus. These topics are introduced through some examples in a step by step fashion from various engineering disciplines mainly focusing on structural analysis. The book focuses on the methodology in analysis to structure debugging aspects in Part I then provides in Part II specific issues describing many understandings about potential troubleshooting in the model, and finally in Part III some practical toolbox protocols to resolve troubleshooting in the model. The content of this book is a collection of researches made from my different professional and academic experiences work with finite-element analyses using Abaqus, including my findings on the free web data, and solutions gathering from the Abaqus support, which I worked with.

In Part I, Chap. 1 provides an introduction to the principles about how to conduct a finite-element analysis. In Chap. 2, analysis convergence guidelines to obtain a converged solution are described. The methodology and control checks in order to have an efficient debugging diagnostic are given in Chap. 3. Chapter 4 gives an essential knowledge covering different analysis aspects including user subroutines.

In Part II, the main practical materials nonlinearities and customizations are discussed in Chap. 5. Chapter 6 provides an important field of expertise in order to understand how the mesher works in Abaqus and what are the good practices to mesh a structure, and how to create a finite element. The contact interaction module and feature options are explained in Chap. 7.

In Part III, Chap. 8 gives the solutions regarding most classics troubleshooting with Abaqus. The options to control the solution accuracy in different model cases are explained in Chap. 9. Chapter 10 provides some very specific issues, where the user will need to get help in modeling. Lastly, Chap. 11 gives solution about non-simulation issues, and a full cluster solution to implement.

Finally, the user will find some example solutions described in the book with the related code in Appendix, in order to have a better understandings about specific aspects. For instance, with coupling options, numerical convergences, and some meshing techniques.

Bourg-la-Reine, France Raphael Jean Boulbes
September 2018

Acknowledgements

This book would not have been published without the help and trust of a few people whom I wish to thank simply but sincerely for everything, because they made it happen.

First of all, I am thankful to my loving relatives who have supported and encouraged me. You have always given me the best, so it is natural that I dedicate this book to you.

Secondly, I thank Springer, especially Oliver Jackson, the editor of the engineering department; Mani Nareshkumar, the project coordinator; and Yasmin Brookes in administration. Thank you for your support and guidance during my project of writing this book.

Finally, I am hugely grateful to my scientific committee of three experts in structural analysis, whom I mention below. The time they spent reviewing the content, as well as their comments, feedback, and the contributions they made to this book, were priceless to me.

Dr. Sonell Shroff is a former assistant professor from the Department of Structural Integrity, Faculty of Aerospace Engineering at Delft University of Technology. Dr. Shroff was strongly recommended to me by Prof. Jan Hol (the Secretary of the Department of Aerospace Structures & Materials, Faculty of Aerospace Engineering, Delft University of Technology), and I was really pleased to work with her. Having completed high school in 2003, Dr. Shroff moved to Bangalore to study engineering. Dr. Shroff oriented herself with a bachelor's degree in Mechanical Engineering, with internships every semester break in reputable companies and research institutes. She graduated in 2007 and received a Royal Dutch Huygens Scholarship, which brought her to the Delft University of Technology for a master's course. Here, she graduated with a final thesis on the structural analysis of a retrofitted winglet on the Hercules C-130 at Marshall Aerospace in 2009. Subsequently she started her Ph.D. study, first with CleanEra and then on the ALaSCA project at Delft University of Technology, which she completed in 2014. Those who are interested in her work will find some papers she worked on cited as references.[1,2,3,4] She is a highly motivated online teacher in the field of Finite Element Method. As she says: "In this world of the future, education needs to be omnipresent. Therefore, I must be, too. Going online brings me closer to the farthest student and breaks the communication barrier. Online education will create a smarter and more learned world, which only means progress for more and a better tomorrow!"

Gautam Puri is a project manager, consultant, and long-time Abaqus user. He has degrees in engineering from the Georgia Institute of Technology and an MBA from Emory University. I was glad to have Gautam on board because I was really impressed after reading his book[5] about Python scripting for Abaqus. After earning his masters degree in Aerospace Engineering, he worked for Dassault Systémes, which develops Abaqus, CATIA, SolidWorks, and a range of other computer-aided engineering software tools. His book Python Scripts for Abaqus—Learn by Example is used by more than 40 engineering companies, including NASA, Boeing, Apple, and dozens of universities including MIT, Stanford, and Georgia Tech. He also runs a YouTube channel, AbaqusPython,[6] with training video material on running finite-element analyses with Abaqus. While at SIMULIA, Gautam provided consulting, training, and technical support services to various SIMULIA clients using Abaqus. He also created the Abaqus tutorial series and the Isight (workflow optimization software) training series for the SIMULIA Learning Community. Gautam is obsessed with cutting-edge technologies. Aside from automating simulations, he has coded and architected web pages, servers, databases, mobile apps, cloud services, and microcontrollers. His most recent role was as an R&D Project Manager at Amazon Web Services (AWS), developing PLC hardware and machine learning algorithms to help manage and scale out Amazons cloud infrastructure. He is now the chief technologist at a web-based startup that is currently in stealth mode.

---

[1]Y.L.M. van Dijk, T. Grtzl, M. Abouhamzeh, L. Kroll, S. Shroff, "Hygrothermal viscoelastic material characterisation of unidirectional continuous carbon-fibre reinforced polyamide 6", Composites Part B: Engineering, Volume 150, 2018, pp. 157–164.

[2]Sonell Shroff, Ertan Acar, Christos Kassapoglou, "Design, analysis, fabrication, and testing of composite grid-stiffened panels for aircraft structures", Thin-Walled Structures, Volume 119, 2017, pp. 235–246.

[3]Sonell Shroff and Christos Kassapoglou, "Designing Highly Loaded Connections in a Composite Fuselage", Journal of Aircraft, Vol. 51, No. 3 (2014), pp. 833–840.

[4]Sonell Shroff and Christos Kassapoglou, "Progressive failure modelling of impacted composite panels under compression", Journal of Reinforced Plastics and Composites, Vol. 34, No. 19 (2015), pp. 1603–1614.

[5]Python Scripts for Abaqus—Learn by Example, book website: http://www.abaquspython.com/.

[6]Gautam (Gary) Puris YouTube channel AbaqusPython: https://www.youtube.com/user/AbaqusPython.

I know Prof. David Bassir well, because he was the master responsible for overseeing my research studies.[7] Dr. Bassir is currently appointed as a Foreign expert at the IIT, Chinese Academy of Sciences (Guangzhou). Previously, he worked at the Ministry of Foreign Affairs (MAEE) to serve as an Attach for Science and Technology (AST) at the Consulate General of France in Guangzhou, China. Before moving to France, Prof. Bassir was General Director of Research at the special school of civil engineering (ESTP) in Paris. After a master's degree in Structural Optimization at ENS2M/UFC universities, in 1999 he obtained his Ph.D. from the same university, with distinction. Then, he worked as a spacecraft engineer for GECI Technology in different space agencies, such as Arianespace (France) and Matra Marconi Space (Astrium Group). In 2002, he joined the Mechanical Department of the Technical University of Belfort-Montbeliard as a professor assistant; here, he obtained his certification (HDR) to conduct research few years later within the framework related to some research activities at the FEMTO-ST Institute in Besancon. Before joining the board of Aerospace Structures at the Technical University of Delft in 2008, he was an invited professor at the Aircraft Manufacturing Department in Xian, China. Dr. Bassir is also the founder of the ASMDO association[8] and the International Journal IJSMDO,[9] which is published by EDP Sciences. He was involved in many international organizations as a member of expert committees. Furthermore, he has published more than 150 research papers in peer-reviewed scientific journals and peer-reviewed international scientific proceedings. He also acts as a reviewer in many international journals related to structural and material optimization. He has conducted more than 2 HDR, 12 Ph.D., 10 masters, and 20 final projects. His excels at project and team management, strategy, and business orientation, government liaison, research and development. His main scientific research is related to the design of composite materials (at macro-, micro-, or nanoscales) and optimization strategies.

[7]Parameter identification of a nonlinear model of a composite laminate shell (David Hicham Bassir, Raphael Boulbes, Lamine Boubakar) FEMTO-ST, University of Franche Comt, UTBM conference paper for the 5th International Conference on Computation of Shell and Spatial Structures, 2005, France. 2005, Salsburg, Austria, June 1–4.

[8]ASMDO association: http://www.asmdo.com/.

[9]International Journal IJSMDO: https://www.ijsmdo.org/.

Contents

Abbreviations

ALE	Arbitrary Lagrangian–Eulerian
CAD	Computer-Aided Design
CAE	Complete Abaqus Environment
CFD	Computational Fluid Dynamic
CPU	Central Processing Unit
DOF	Degrees of freedom
DSV	Dependent State Variables
FEA	Finite-Element Analysis
FSI	Fluid–Structure Interaction
GIGO	Garbage In, Garbage Out
HHT	Hilber–Hughes–Taylor
Mdb	Model database
MPC	Multi-point Constraints
XFEM	Extended Finite-Element Method

Part I
Methodology to Start Debugging Model Issues

All structural engineers—students or professors using FEA software distribution analysts—understand how it can be difficult to have a starting point to fix some issues in series from a finite-element model that cannot be computed as expected. In the present part of this book, the focus is first to obtain a practical methodology to follow as a guidance through troubleshooting met with an Abaqus model, and second, the section provides a quick overview of several solutions that can be applicable regarding some specific errors or warning messages.

Chapter 1
Introduction

1.1 Global Mindset

Before even describing a methodology for doing this or that, the most important thing is to have a preliminary global overview about the analysis purpose to perform and then to execute with a software, using here, for instance, the finite-element method with Abaqus. The mathematical way to solve any finite-element analysis model is unique, thus the method used to perform such task is unique too. Therefore, there is a single way to debug a model. This is the key point to have in mind from the beginning.

The essential idea is, therefore, to get a methodology to convert an engineering scope of work into an FEA model. To do so, the following diagram shows the different linear steps to discuss in order to determine what the FEA model could look like to compute.

The diagram shown in Fig. 1.1 starts with the engineering team or equivalent workgroup because the analyst is here to perform, to solve a structural analysis query made by other in orders to give a quantitative answer to a qualitative problem.

The different phases of the flowchart in Fig. 1.1 are described below:

1. The analyst must be sure to have the required expertise and knowledge to handle the submitted request from the work group; otherwise, specific training about the analysis case should be performed first.
2. The workgroup with the analyst must define the scope of work as a function of the different work directions are listed below:
   - Explain objectives and rephrase the main aim of the analysis in accordance with project requirements;
   - Locate and identify the component or system to analyze;
   - Clarify the function of each component or system to analyze;
   - Draw the border of structural analysis, which will help to determine the loading and boundary conditions;
   - Make a list of all documents to be used in order to perform the analysis;

R. J. Boulbes, *Troubleshooting Finite-Element Modeling with Abaqus*,
https://doi.org/10.1007/978-3-030-26740-7_1

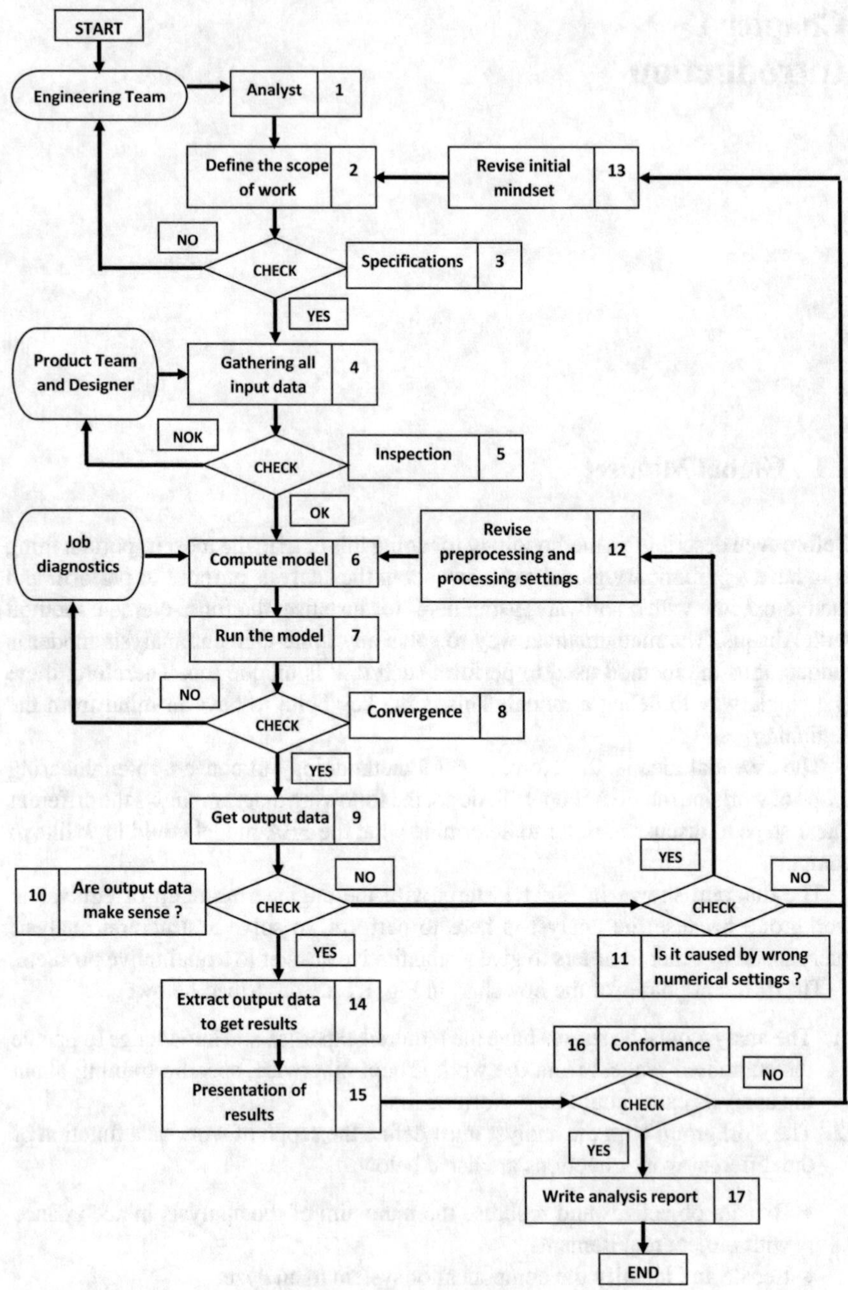

Fig. 1.1 Example of a global overview analysis flowchart methodology

- What standard(s) is/are applicable to the component or system regarding the structural analysis;
- Draw a sketch to summarize what will be analyzed;
- Make all assumptions to perform the analysis;
- If possible, make hand calculations to figure out local or global expected structural response(s) for some areas inside the analysis zone.

3. This phase essentially involves a preliminary contract agreement with the engineering team, first to ensure a certain quality of work and the structural integrity analysis of the component or system under analysis. The content of this specification document shall be understandable as best as possible for all engineering team members, even non-analysts.
4. Be sure to have all necessary documentations to make proper settings, for instance, get drawings or CAD files (e.g., a step file format can be used with Abaqus to import designs from CAD software).
   If CAD files are used then the analyst should have worked with the designer in order to clean the design to make it compatible with the analysis needs in accordance with the best and simplest level of design details as a function of the structural investigation areas.
   If there is no CAD file to import the design then the analyst will have to construct the FEA model from the drawing but still in accordance with the same level of design details to get the best approximation of geometry to represent the design structural response as accurately as possible.
5. The inspection phase is the moment where all specifications made in phase (2) have been transformed into a machine language regarding the design properties, the analyses criteria, and also the way of performing the structural analysis.
6. This is the modeling phase where the input data is implemented according to the work specifications already defined inside the scope of work descriptions, reviewed, and approved by using the proper Abaqus features with correct settings.
7. At this point, the FEA model is run by submitting the analysis job into the Abaqus solver.
8. It is important to check that the analysis job has been solved without error or critical warning messages. If error or critical warning messages have occurred then a job diagnostic must be investigated by the analyst, using all available support including this book to fix the troubleshooting issue causing incorrect termination of the Abaqus solver.
9. Getting output data is a very important phase because sometimes the FEA model has been solved but it does not mean the result returned is correct or even accurate enough to post-process the output data with a good confidence in the results. The point here is to identify a potential GIGO (Garbage In, Garbage Out) in the solution calculated by the Abaqus solver, meaning that the output is the result of an incorrect input.
   Therefore, the analyst must criticize the output data to ascertain how good the output results are and how accurate the output data is; trust first but always double

check! The best way of doing this is to return to phase (2) and check if whether the expected hand calculations make sense with the output data. Of course, a visual inspection can also be useful to determine incorrect settings, for instance, if the deflection of the structure goes in the wrong direction because the gravity load has been set in the wrong direction.

Most of time, it is very difficult to make this control check because hand calculations can be extremely difficult to translate into equations. Consequently, it is always a good practice to first take the time to think about what is a likely or unlikely solution expected from the structure analyzed under the loading and boundary conditions given. This can be rephrased as the following maxim: **"Think first, program after and trust second but always double check"**.

Being a thinker before being a builder in order to build right first time and having a bit of analysis philosophy, creates the ability to consider a problem with higher dimensions, which should be the core value of an expert analyst in that phase. Here this concept is known, as having the analysis mindset.

10. The output data should be inspected to ensure confidence according to phase (9).
11. The question asked here is to determine where is/are the source(s) of the numerical difficulties or unconverged solutions. If they are caused by wrong numerical settings, the analyst will have to revise the preprocessing and/or the processing settings made. Otherwise, the cause of the numerical issue is not caused by wrong settings made in the FEA model but instead the definition of the scope of work. Therefore, the analyst will have to look at the initial specifications to find out what has created those inconsistencies and instabilities within the FEA model.
12. In case of strange results behavior with the FEA model then a revision of the model settings will need to be made.
13. In case of unrealistic results behavior then a revision of the FEA model definitions and assumptions will be required.
14. While it is so obvious for analysts, it is perhaps less so for non-analysts, indeed the output data computed from the FEA solver is not generally the direct results given in the conclusion.

    For instance, if inside the scope of work the analyst needs to check a criterion given by a standard, the standard should outline the different operations to the analyst in order to make additional post-processing calculations with the output data following a specific method to determine the final results of the structural response to be checked in accordance with the standard criterion given inside. Once this standard criterion is checked, the analyst will be able to come to a conclusion regarding the structural integrity of the structure under analysis.
15. This is an important phase because the analyst will have to communicate the results to a group of people, some of whom will not have knowledge about FEA or the structure being analyzed. Communication is a difficult exercise consisting of telling a story about a specific topic using simple and clear explanations to be understandable for the audience.

    The best-recommended strategy is to give a short and clear explanation of results with images instead of having text boxes in order to convey the results and key points in an understandable way. It is a balance between a go straight to the point and a go to all analysis details approach, both extremes are not useful because, as the audience includes non-analysts, they will be lost due to the lack of detail or drowned in excessive detail, and therefore they will be unable to understand the results and conclusion.

So let's keep it as simple as possible and clear enough to understand the conclusion. The audience will then raise questions to dig more into detail about the results and conclusion.

16. After presenting the conclusion made by the analyst, the engineering team will have to decide the status of the structural analyses conformance. If more work is needed then the workgroup will have to go back to phase (13) in order to revise the mindset and make a new definition of the scope of work to redo.
17. This is the final phase, in which everything is reviewed and approved. The analyst will have to write a report about the work done for the engineering team.

    The format and the content section inside an FEA report can vary from one workplace to another, but essentially the most important point is to show the different phases as described in the flowchart shown in Fig. 1.1. A non-exhaustive list of section contents which can be expected inside an FEA report are listed below:

    a. write in a short introduction passage the aim of performing an analysis regarding the questions about the structural integrity of the model analyzed,
    b. write the scope of work tasks and subtasks,
    c. write all reference documents relating to the scope of work including drawings, standards, and so on,
    d. draw sketches showing in figures all the details to understand the analysis area with the loading and boundary conditions,
    e. write all relevant assumptions made with the FEA model,
    f. write all references regarding input data used to set the FEA model. For example, friction coefficient, materials data, and so on,
    g. put all figures needed to understand the FEA model assembly, material definitions, contact interactions, and so on.
    h. If applicable, write all hand calculations performed in accordance with the scope of work to predict a solution,
    i. put the most relevant figures about the structural responses to explain the results. For example, a stress, displacement, and strain plots.
    j. Write a report regarding the analysis results in a concise language and clear short text explanations.
    k. Write a short conclusion about the conformity of the product analyzed according to the questions raised in introduction.

In general, structural analysis cases with optimal conditions, if the analyst follows the methodology described above, the risk of having troubleshooting issues should be minimized and a good understanding about the structural analysis is provided.

The methodology can be readjusted in cases of research analysis about structure in development where some phases are not completely achieved.

The following methodology is therefore a guideline to ensure a good practice in structural analysis, and is naturally flexible enough to face different situations which are more or less specific to a work environment. As a result, the methodology described in the flowchart in Fig. 1.1 can be seen as a global mindset about the process and completion of structural analysis.

1.2 The Four Absolutes of Quality in Analysis

"Quality without tears" [1] gives the definition of four absolutes and can be rephrased for the needs of an analyst as a core focus in performing structural analysis. An absolute is a principle value defined and in accordance with a company standard or code in order to conduct a business in a way to control the quality of work performed at every competence level. To conduct analysis work, the absolutes are given below:

The first absolute: The definition of quality is conformance to requirements.
The second absolute: The system of quality is prevention against failure by having safety control measures.
The third absolute: The performance standard is zero defects.
The fourth absolute: The measurement of quality is the price of nonconformance, not indexes.

Initially, the quality improvement is built on getting everyone to do it right the first time. This is why quality in analysis has to be defined as conformance to work specification requirements, rather than correctness. After establishing a proper definition of the scope of work, the workgroup will need to prevent the risk of deviation in conformance.

Prevention is putting the engineering team in a state of knowledge about how to convert a design model into a finite-element model. Here, non-analysts will need to understand the basis of analysis process. The system for creating a global quality in the analysis methodology is prevention, not appraisal.

The zero defect is the confidence to reach a solid conclusion based on the calculated results. The performance standard must be a zero defect in the control of FEA, rather than state "this is similar enough".

The price of nonconformance is all expenses involved in doing things wrong. If the workgroup does not take the time to think about or to understand a small piece of work specification at the beginning, there might occur a huge deviation at the end. This will be the extra price to pay, or rather not to pay.

1.3 Checklist for Performing Analysis

A methodology mindset as shown in Fig. 1.1 will not be complete without having a control on material data, for instance, a checklist procedure, to summarize the most important milestones to translate the problem from the engineering team into structural analysis settings. Table 1.1 shows an example of a classic local FEA checklist.

Of course, it is up to the analyst group to make such checklist documentation in order to standardize a recurrent analysis task. It is important to take the time to make a checklist procedure as shown in Table 1.1 with the objective of standardizing analysis tasks as this, will help both the engineering team and the analyst, to minimize the risk of analysis troubleshooting and also to speed up the overall process. Indeed, having a mutual understanding about the specific needs between non-analyst who create the model, and analyst who get the model conformance, is essential to make a efficient and effective teamwork. This mutual trust will help improve the quality of work and help with troubleshooting issues that may occur during the analysis phases.

1.4 A Heuristic Analysis Confidence Ratio

Analysts must question the accuracy of their results when a solution to a model has been reached but comes from different directions; such as when comparing a finite element analysis (FEA) solution with a solution from hand calculation, or test data. Complex methods exist [2] or can be developed based on various heuristic techniques. Heuristic, from the Ancient Greek, means a technique to find or to discover something, and heuristics refers to any approach to problem-solving, learning, or discovery that employs a practical method: it is not guaranteed to be optimal, perfect, logical, or rational, but instead is sufficient for reaching an immediate goal. Where finding an optimal solution is impossible or impractical, heuristic methods can be used to accelerate the process of finding a satisfactory solution. A very simple heuristic method to correlate different solution sources is presented in this section.

There are only three directions by which to reach a solution for a given problem: first, by using a theoretical calculation model; second, through computer-aided modeling, using Abaqus software, for instance; and finally, with a measurement system. In most cases, analysts lack the measurement data required to correlate the model solution. Indeed, all three directions give different solution values for the same problem: this is possible because all three directions approach the problem in a different way, and therefore include some error deviations that arise from different sources. For example, a theoretical calculation model is a function of the assumptions made to present the problems solution as a system of equations: thus, if the assumptions do not relate as accurately as possible to the problem to be solved, then some error deviations will occur. Of course, having more assumptions will lead to more complex equations to solve; and, due to the limitations of hand calculations, this will not be a simple task.

Table 1.1 Example of a checklist document to standardize a structural analysis procedure

Phases	Tasks	Descriptions	Check
Specifications	Clauses	Do the clause rules comply with the required standard?	
	Documents	Are all component drawings, material data, project design basis received available and in compliance with the scope of work defined?	
FEA	Hand calculations	Have some basic hand calculation been performed to roughly estimate and validate the FEA results? (e.g., the reaction forces from FEA checked against applied loads in static equilibrium)	
	Model	Does it have sufficient level of detail?	
		Have symmetry planes or any other considerations to reduce model size been selected correctly?	
		Are the center of gravity and weight correct?	
		Does the model contain unwanted gaps? (clean operation on the CAD model)	
	Material properties	Are the material properties at the operating and design temperature correct?	
	Contact properties	Are the contact interaction surfaces correctly defined along with the friction properties?	
	Type of analysis	Is the chosen FEA appropriate for the loading conditions (static, dynamic, linear, nonlinear and so on)	
	Context	Have the loading, pre-described displacements and boundary conditions been applied correctly?	
		Have model sketches and color plots clearly showing these topics been included in the report?	
	Element type	Is the element type beam, truss, shell, solid, etc...) used for the FEA appropriate for the component or system being analyzed?	
	Meshing	Is the mesh density (element size) confirmed accurately by sensitivity analysis or other suitable method?	
		Are the stress precision values and/or mesh convergence studies acceptable?	
		Has the mesh been checked and found correct in critical areas? (especially around sharp corners or abrupt design transitions)	

(continued)

Table 1.1 (continued)

Phases	Tasks	Descriptions	Check
	Results	Are the stresses, strains, rotations, deflections and any output field within the design code allowed?	
		Is penetration of contact surfaces avoided? (for instance, a node penetration through a surface)	
		Has the methodology outlined in the governing standard been complied with?	
Documentations	FEA report	Does the FEA report contain the minimum information needed to clearly understand the analysis and the conclusions made?	
	Files	Create a zip a folder with all analysis files inside[a]	

[a]To reduce the size of the Abaqus CAE file, go to the menu bar and click on "-File". Then choose "-Compress Mdb". But remember to save the changes before using this solution

Moreover, it is not advisable to construct a very complex system of equations to be solved, because of the risk that the analyst will introduce some errors during the calculation procedure. Consequently, the second direction is the most commonly used by analysts: thanks to computer-aided modeling software such as Abaqus, the solution will be obtained by using a analysis solver; though even in this case, some error deviations will occur. For instance, the mesh function is only an approximation of the real geometry, and thus introduces a mesh error deviation; also, the material curve is an idealization of the real material behavior; furthermore, the contact interaction does not represent a real contact force, and so on. All of these issues will raise some questions about the accuracy of the computed solution, due to the contribution of all error deviations that emerged during the modeprocedure.

Finally, neither is the measurement system perfect. For example, a strain gauge has an error deviation, also called also tolerance, regarding the measurement data recorded: thus, the greater the complexity of a measurement system and the more it depends on a variety of equipment, the higher the error deviation value. Moreover, when considering the potential error deviations that a test team can make during the installation procedure or when running tests, it is clear that nothing is perfect.

This section describes a simple way to make a calibration that adapts the level of confidence in order to estimate the correlation of a given solution, based on the error deviations from one, two or three directions. In the case of having all three directions, it is possible to give a geometrical representation, using an orthogonal 3D space coordinates system that shows the problems solution at the center point. This geometrical representation is shown in Fig. 1.2, where S_1 is the solution provided by the theoretical hand calculations; S_2 is the solution from the computed calculation modeling using an FEA model, with Abaqus, for instance; and S_3 is the solution given by the measurement data systems. It is obvious that the model solution S cannot be reached perfectly, due to some error deviations caused by the theoretical

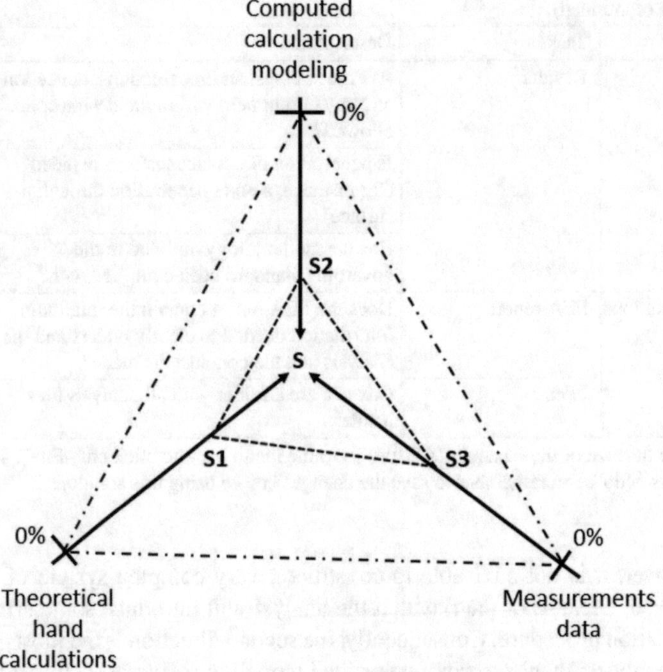

Fig. 1.2 Heuristic 3D space representation of the confidence ratio

models assumptions or by the computed model, or by the tolerance in measurement devices used.

According to Fig. 1.2, the heuristic analysis confidence ratio C_3 can be expressed as shown in Eq. 1.1, where V_{sol} is the tetrahedral volume containing the solution, and V_{unit} is the total tetrahedral volume per unit of solution.

$$C_3 = 1 - \frac{V_{sol}}{V_{unit}} \tag{1.1}$$

In this case, Eq. 1.1 is established as a function of the tetrahedral vertices (S_1, S), (S_2, S) and (S_3, S) in Eq. 1.2.

$$C_3 = 1 - \frac{(1/6)(S_1, S)(S_2, S)(S_3, S)}{(1/6)} \tag{1.2}$$

The difficulty is to determine the value of the vertices: each vertice is independent of the others, and the way to determine their values relies on the direction they represent. Thus, the vertex (S_1, S) in Fig. 1.2 represents the theoretical hand calculation direction. In order to use the value from solution S_1 to determine the real unknown solution S chosen as the origin of the coordinate system, only an error calculation can be determined. The error calculation to determine the vertex (S_1, S) will be established as a function of the list of assumptions and the method of quantifying each assumption. Consequently, an assumption which cannot be quantified will not be taken into account within the error calculation, or therefore in the confidence ratio. Once all quantifiable assumptions (a_i) are listed, the vertex (S_1, S) can be determined as shown in Eq. 1.3, because the vertex is the contribution of all the error assumptions made to determine the solution in the theoretical hand calculation direction.

$$(S_1, S) = a_1 * a_2 * \cdots * a_{n_1} = \prod_{i=1}^{n_1} a_i \tag{1.3}$$

Similarly, the vertex (S_2, S), representing the computed calculation modeling direction, can be determined after taking into consideration all possible quantifiable variations in the FEA model. These include variations between a fine or coarse mesh, the residual computed value with an element check or with an energy ratio, and so on. Once all quantifiable variations (v_i) are listed, the vertex (S_2, S) can be determined as shown in Eq. 1.4, because the vertex is the contribution of all error variations made to determine the solution in the computed calculation modeling direction.

$$(S_2, S) = v_1 * v_2 * \cdots * v_{n_2} = \prod_{i=1}^{n_2} v_i \tag{1.4}$$

Lastly, the vertex (S_3, S), which represents the measurement data direction, can be determined after taking into consideration all possible quantifiable device tolerances used with the measurement system. These include a strain gauge with its mechanical tolerance in the measurement, coupled with data acquisition equipment, an inspection of the quality of work done to prepare the specific zone for gluing the strain gauge, and so on. Once all quantifiable tolerances in measurement (m_i) are listed, the vertex (S_3, S) can be determined as shown in Eq. 1.5, because the vertex is the contribution of all errors in measurement made to determine the solution in the measurement data direction.

$$(S_3, S) = m_1 * m_2 * \cdots * m_{n_3} = \prod_{i=1}^{n_3} m_i \tag{1.5}$$

To better understand the error formula given in Eqs. 1.3–1.5, they can be seen as the product of all error contributions from each independent variable taken as many assumptions in the model, which are caused by variation due to different features used for modeling, or from device tolerances in the measurement data. It is assumed that x_1 and x_2 are two independent variables, with δx_1 and δx_2 as the error in variables x_1 and x_2, respectively. The total error contribution caused by these two variables is given in Eq. 1.6.

$$(x_1 + \delta x_1) * (x_2 + \delta x_2) = x_1 x_2 + x_1 \delta x_2 + x_2 \delta x_1 + \delta x_1 \delta x_2 \tag{1.6}$$

As both variables x_1 and x_2 are independent, it makes no sense to have in Eq. 1.6 a calculated term where one variable is a function of the error given by the other; therefore, Eq. 1.6 is simplified in Eq. 1.7.

$$(x_1 + \delta x_1) * (x_2 + \delta x_2) = x_1 x_2 + \delta x_1 \delta x_2 \tag{1.7}$$

Equation 1.7 can be expressed in a different way to show the total error correlation from variables x_1 and x_2, as determined in Eq. 1.8.

$$\frac{(x_1 + \delta x_1)}{x_1} * \frac{(x_2 + \delta x_2)}{x_2} = 1 + \frac{\delta x_1}{x_1}\frac{\delta x_2}{x_2} = 1 + \prod_{i=1}^{2} \frac{\delta x_i}{x_i} \tag{1.8}$$

The confidence ratio in Eq. 1.1, with a configuration shown in Fig. 1.2, can be calculated using the different error estimations in Eqs. 1.3–1.5, as shown below:

$$C_3 = 1 - \left(\prod_{i=1}^{n_1} a_i\right)\left(\prod_{i=1}^{n_2} v_i\right)\left(\prod_{i=1}^{n_3} m_i\right) \tag{1.9}$$

Following the same principle as C_3 determined in Eq. 1.9, a confidence ratio in two directions, C_2, can be also determined: as shown, for example, in Fig. 1.3, between the theoretical hand calculation and computed calculation modeling directions.

The transposing Eq. 1.9 in 2D, as shown, for example, in Fig. 1.3, is determined in Eq. 1.10.

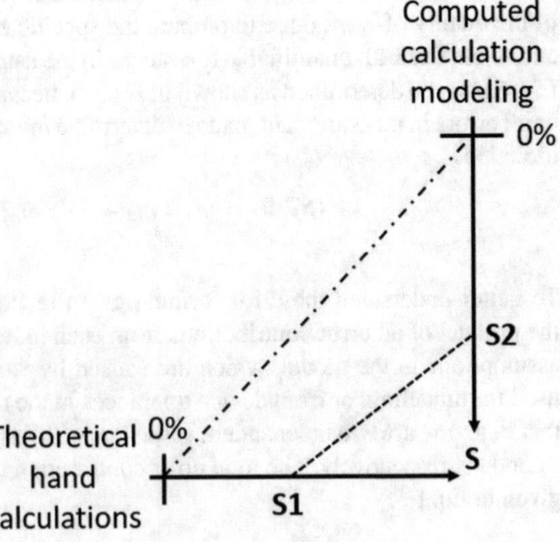

Fig. 1.3 Heuristic 2D space representation of the confidence ratio

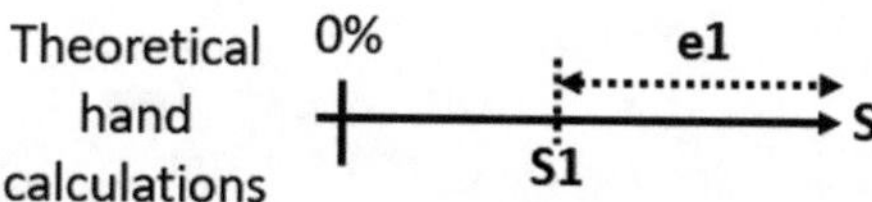

Fig. 1.4 Heuristic 1D space representation of the confidence ratio

$$C_2 = 1 - \left(\prod_{i=1}^{n_1} a_i\right)\left(\prod_{i=1}^{n_2} v_i\right) \tag{1.10}$$

Similarly, in 1D, as shown, for example, in Fig. 1.4, the confidence ratio in Eq. 1.10 becomes C_1 in Eq. 1.11.

$$C_1 = 1 - e_1 = 1 - \left(\prod_{i=1}^{n_1} a_i\right) \tag{1.11}$$

In conclusion, human beings regularly rely on heuristics, which can be described as various rough-and-ready problem-solving techniques; however, these techniques can lead us to make systematic reasoning errors called biases. This occurs, especially, when the analyst is not sufficiently critical of each step of the task performed. Therefore, to minimize biases, the error deviation criteria must first be discussed in an expert committee, which is a group of people with high skills and experience in the analysis field.

References

1. Crosby PB (1995) Quality without tears, the art of hassle-free management. McGraw-Hill, Inc., New York
2. Kaveh A (2014) Computational structural analysis and finite element methods. Springer, Berlin

Chapter 2
Analysis Convergence Guidelines

2.1 Symptoms of Convergence Problems

Convergence issue is a typical analyst issue related to engineering design involving the prediction of deflections, displacements, stresses, natural frequencies, temperature distributions, and so on. These parameters are used to iterate on material parameters and/or geometry to optimize their behavior. Traditional methods like hand calculations, involved idealization of physical models using simple equations to obtain solutions. However, these approximations oversimplify the problem, and an analytical solution can only provide conservative estimates. Alternatively, finite element method and other numerical methods are meant to provide an engineering analysis that takes into account much greater detail, something that would be impractical with hand calculations. Finite-element method divides the body into smaller pieces, enforcing continuity of displacements along these element boundaries. For those using finite-element analysis, the term convergence is often used. Most linear problems do not need an iterative solution procedure. Mesh convergence is an important issue that needs to be addressed. Additionally, in nonlinear problems, convergence in the iteration procedure also needs to be considered.

In this section, convergence issues will be investigated and address issues related to this term. First, to identify the symptoms of most convergence problems can be found in the message file (.msg) extension. In addition, the (.dat) and the (.sta) files may also contain symptoms of the problem. There are some common messages that may indicate convergence problems creating numerical difficulties in solving the finite-element model. Some examples are outlined as follows:

- WARNING: THE SOLUTION APPEARS TO BE DIVERGING
- WARNING: THE STRAIN INCREMENT HAS EXCEEDED FIFTY TIMES THE STRAIN TO CAUSE FIRST YIELD AT 7 POINTS
- WARNING: THE SYSTEM MATRIX HAS 3 NEGATIVE EIGENVALUES
- WARNING: ELEMENT 441 IS DISTORTING SO MUCH THAT IT TURNS INSIDE OUT

R. J. Boulbes, *Troubleshooting Finite-Element Modeling with Abaqus*,
https://doi.org/10.1007/978-3-030-26740-7_2

- NOTE: SUBDIVISION AFTER 12 ITERATIONS FOR SEVERE DISCONTINUITIES
- WARNING: OVERCLOSURE OF CONTACT SURFACES SLAVE_SURF AND MASTER_SURF IS TOO SEVERE CUTBACK WILL RESULT
- WARNING: SOLVER PROBLEM. ZERO PIVOT WHEN PROCESSING NODE 1 D.O.F. 1

2.2 Causes of Convergence Problems

Inadequate FE modeling is the most common cause of convergence problems in nonlinear simulations. Here are some examples:

- Defining conflicting constraints between boundary conditions, contact conditions, and/or multiple point constraints.
  - Not adequately constraining the model;
  - Having incomplete/inadequate material data;
  - Using an inappropriate element.
- Another common cause is a highly unstable physical system. These cases demand the correct element type and analysis techniques to be used.

2.3 Helping Abaqus Find a Converged Solution

A good method of identifying which symptom is the cause of the numerical difficulty is to isolate the maximum potential causes and rerun it to see what changes, thus fixing the symptoms one at a time. Here are some recommendations:

- The best way to help Abaqus is to build a light model test.
  - Do not put every detail into your first model.
  - Possibly start with contact, but no plasticity, friction, or nonlinear geometry to gain an understanding of how the model behaves.
  - Add one piece at a time to limit the number of sources of convergence issues.
- Give reasonable values for the initial increment, minimum increment size, and maximum increment size.
- Causes of convergence problems are reported in the .msg, .dat, .odb, and .sta files.
  - Do not limit the data written to the message file.
  - For contact issues, access the model input file "-.inp" and use the keyword command ***PRINT, CONTACT=YES** to get detailed contact information in the message file.

- For material issues use ***PRINT, PLASTICITY=YES** to get the output of element and integration point numbers for which the plasticity algorithms have failed to converge in the material routines.
- Request other additional information be written to these files to aid in locating the source of the convergence issue.

2.4 General Tools

A quick control with the global overview of warning messages above can give a reasonable guess for an analyst about what is going wrong and a rough idea about a corrective action to take. A list of some first logical operations to perform to fix the numerical issue is given here:

1. Use preferably a displacement control instead of a load control. For example, if the model is loaded in pure tension then apply an axial displacement value to simulate the tension load instead of using a concentrated force load, it will minimize convergence issue because displacement is a better controlled to iterate a solution. The iterated solution will be more stable. The same recommendation applied in case of a pure bending load by using a rotation displacement value instead of a concentrated moment load. Write the required nodal forces and displacements to the .dat file then extract data with the (-xydata) feature, thus generating an (x-y) data file of load versus displacement to plot in the Abaqus viewport.
2. Control increment sizes to prevent Abaqus from approaching a sudden stiffness change too aggressively. Set the initial increment size, the minimum step size, and the maximum step size using the command ***STATIC**. The initial increment size should normally be in the 0.01–0.1 range to start the analysis slowly (DEFAULT=1.0). The minimum step size can be decreased to allow the solver to cut back further, while (DEFAULT=0.00001) the maximum step size can be decreased to prevent Abaqus from overshooting a sudden stiffness change and can result in a more efficient run (NO DEFAULT). (e.g., ***STATIC 0.01, 1.0, 1.0000E-08, 0.1**)
3. Create an initial step that is very small for the purpose of initiating contact.
4. Use dashpot[1] or spring elements on specific nodes.
5. Use connector elements or beam elements instead of multi-point constraints.[2]

---

[1] Dashpots are used to model relative velocity-dependent force or torsional resistance. They can also provide viscous energy dissipation mechanisms. Dashpots are often useful in unstable, nonlinear, and static analyses where the modified Riks algorithm is not appropriate and where the automatic time-stepping algorithm is used because sudden shifts in configuration can be controlled by the forces that arise in the dashpots. In such cases the magnitude of the damping must be chosen in conjunction with the time period so that sufficient damping is available to control such difficulties but the damping forces are negligible when a stable static response is obtained.

[2] The MPC module provides the basic capability of modeling multi-point constraints. Multi-point constraints are a general way of relating degrees of freedom to one another within a model. They

6. If hourglassing [1][3] is a problem (usually only an issue with continuum elements rather than, shell elements), use fully integrated element types or hourglass control.
7. To help with problems with large rotations use parabolic extrapolation. (e.g., ***STEP, EXTRAPOLATION=PARABOLIC**)
8. Turn off extrapolation of the displacement correction so Abaqus does not approach a sudden stiffness change too aggressively. (e.g., ***STEP, EXTRAPOLATION=NO**)
9. For problems with follower loads or for finite sliding between highly curved deformable surfaces, the asymmetric matrix storage and solution scheme should be used. (e.g., ***STEP, UNSYMM=YES**)
10. For globally unstable problems like in global buckling, collapse, or snap-through where the nonlinear instability region is the region in which snap-through[4] occurs and in which the equilibrium path goes from one stable point A to another new stable point B, RIKS[5] can be used. If using RIKS, proceed without RIKS until needed, before creating an additional step that uses RIKS. It must be noted that using displacement control is more efficient than RIKS. For backtracking in RIKS analysis, specify a maximum arc length such as 1.5 under ***STATIC, RIKS**.
11. For locally unstable problems, use automated stabilization and monitor the damping energy. This cannot be used with RIKS, but can be combined with displacement control ***STATIC, STABILIZE *ENERGY OUTPUT**, ***ENERGY PRINT** or ***ENERGY FILE** to monitor the energies ELSD,[6] ESDDEN[7] and ALLSD.[8]

---

provide an extremely powerful tool which is useful for many modeling problems. An important example is transmitting load between nodes, which are separated in space or are attached to different degrees of freedom such as translations and rotations.

[3] It is essentially a spurious deformation mode of a finite-element mesh, resulting from the excitation of zero-energy degrees of freedom. It typically manifests as a patchwork of zigzag or an hourglass, where individual elements are severely deformed, while the overall mesh section is non-deformed. This happens on hexahedral 3D solid reduced integration elements and on the respective tetrahedral 3D shell elements and 2D solid elements.

[4] See Abaqus Example Problems Guide v6.14 in Sect. 1.2.1 Snap-through buckling analysis of circular arches.

[5] Geometrically nonlinear static problems sometimes involve buckling or collapse behavior, where the load–displacement response shows a negative stiffness and the structure must release strain energy to remain in equilibrium. Alternatively, static equilibrium states during the unstable phase of the response can be found by using the modified Riks method. This method is used for cases where the loading is proportional; that is, where the load magnitudes are governed by a single scalar parameter. The method can provide solutions even in cases of complex, unstable response.

[6] Total energy dissipated in the element resulting from automatic static stabilization. Not available for steady-state dynamic analysis.

[7] Total energy dissipated per unit volume in the element resulting from static stabilization. Not available for steady-state dynamic analysis.

[8] Energy dissipated by automatic stabilization. This includes both volumetric static stabilization and automatic approach of contact pairs (the latter part included only for the whole model).

12. Add a slightly increasing slope to the perfectly plastic region of the ***PLASTIC** material definition.
13. Use hybrid elements for highly incompressible elements (Poisson's ratio approaching 0.5) or for Anisotropic hyperelasticity formulations (large stiffness differences in elements such as bending versus axial stiffness).
14. Loosen the convergence criteria (avoid this if possible). The analyst might need to do this for an initial small step when contact is required, before then using default parameters for subsequent steps. ***CONTROLS, PARAMETERS=FIELD**.

2.5 Tools for Contact Stabilization

Here, some recommendations applicable in static equilibrium depending on contact interaction troubleshooting are provided.

1. Create an initial step that is very small for the purpose of initiating contact.
2. Use displacement control instead of load control. Write the required nodal forces and displacements to the .dat file and then employ the xydata features to generate an (x-y) data file of load versus displacement to plot in viewport.
3. Add springs that have a low stiffness compared to the total load to give some resistance to the contact pair until contact is established. If the spring force is too high, the second step can be established to remove the spring once contact is established.
4. To get contact established in an initial step without rigid body motions, the approach parameter should be utilized. Apply the structural load (or the vast majority of it) in a separate step and then monitor the energy levels of the contact pressure CPRESS and the energy ALLSD. ***CONTACT CONTROLS, APPROACH MASTER=master-name, SLAVE=slave-name**.

2.6 Tools for Contact Related Convergence Problems

In general, dealing with contact interaction definition must be used with caution, especially when using additional parameters to help with convergence (e.g., adjusting, approaching, shrinking, and automatic tolerance). A control check that can be performed afterward, makes sure the load flow or critical contact behavior is not affected.

1. Monitor the contact forces using ***CONTACT PRINT**. Forces will be written to the *.dat file, which will help determine which contact pairs are having difficulty establishing contact.
2. Choose master/slave surfaces and define the mesh accordingly to capture the desired contact behavior, the master surface should have the coarser mesh. More-

over, analysts can define the master surface such that it extends beyond the slave surface, but never the opposite.

3. Double check normals on contact surfaces. The contact normal direction is based on the master surface; thus, if the normal direction is critical then the master surface should be chosen accordingly. If large overclosures are observed, it could indicate that the contact normal directions are wrong.
4. Double check edges on contact surfaces and eliminate cracks on the master surface.
5. Do not have one node defined as a slave for two or more contact pairs or gap elements.
6. GAP elements should be used if possible to eliminate the contact. If gap elements defined as initial zero clearance are chattering, try changing to a very small nonzero clearance.
7. Add springs that have a low stiffness compared to the total load to provide some resistance to the contact pair until contact is established. If the spring force is too high, a second step can be created to remove the spring once contact is established (use the S11 option to monitor the reaction force).
8. Add dashpots.
9. If contacts are slightly over-penetrated in the initial stages, use adjust=0 but use with care if the initial contact force is critical.
10. Use softened contact to apply force relative to the amount of penetration (if chattering). ***SURFACE BEHAVIOR, PRESSURE-OVERCLOSE= EXPONENTIAL 0.1, 200**.
11. Increase the maximum number of allowed severe discontinuity iterations (DEFAULT=12) if the severe discontinuities are decreasing. ***CONTROLS, PARAMETERS=TIME INCREMENTATION, , , , , ,24,**
12. Turn on automatic tolerances so that Abaqus will compute an overclosure tolerance and a separation pressure tolerance. ***CONTACT CONTROLS, AUTOMATIC TOLERANCES**.
13. Eliminate friction in contacts unless it is absolutely necessary (such as in mechanisms/internals). Conversely, on rare occasions, a model will converge better with increased friction value. If any friction coefficient is greater than 0.2, Abaqus/Standard will use the asymmetric matrix storage and solution scheme automatically.
14. Turn on small sliding, but only where applicable. Small sliding creates an infinite master surface so use with caution. ***CONTACT PAIR, SMALL SLIDING**.
15. Use ***CONTACT DAMPING**[9] to damp relative motions of contact surfaces during approach or separation.
16. Increase the absolute penetration tolerance called HCRIT in ***CONTACT PAIR**. Although this rarely helps in some cases it will be better than nothing.
17. For finite sliding between highly curved deformable surfaces, use the asymmetric matrix storage and solution scheme. ***STEP, UNSYMM=YES**.

---

[9] See Abaqus Analysis User's Guide v6.14 in Sect. 37.1.3 Contact damping.

18. For severely discontinuous behavior such as frictional sliding, apply the discontinuous control. This could increase run time, especially for problems that are not severely discontinuous. ***CONTROLS, ANALYSIS=DISCONTINUOUS**.
19. Use an explicit solver in static to change the numerical scheme during the solving processing.[10] Indeed, explicit solvers are the best choice for simulating high energy, short-duration dynamic events such as impact, drop testing, and blast analysis.
    However, under certain circumstances they can also be useful for static analyses. Explicit solvers rely on the assumption that model properties are linear within each time step and matrices are updated at the end of each step. This assumption is considered to be accurate because only very small, conditionally stable time steps are used. The assumption is significant because it eliminates the need for convergence iterations, which can often prevent highly nonlinear implicit analyses from solving.
    This means that explicit solvers can be used to handle highly nonlinear statics problems that either will not solve with an implicit solver due to convergence problems or solve very slowly because many iterations are required.

A good practice with contact interaction settings in the finite-element model is to follow the basic rules listed below:

- As a rule, use ***PRINT, CONTACT=YES** to request detailed output of points that are contacting or separating in interface and gap problems.
- As a rule, do not use the ADJUST parameter in a ***CONTACT PAIR** that will be removed. The adjustment takes place prior to the removal and can distort elements if the surfaces are not in their final positions before contact is initiated.
- As a rule, if you are adding damping, use ***ENERGY PRINT** or ***ENERGY OUTPUT** or ***ENERGY FILE** to monitor ALLAE[11] and ALLSE.[12]

Reference

1. Belytschko T, Ong JSJ, Liu WK, Kennedy JM (1984) Hourglass control in linear and nonlinear problems. Comput Methods Appl Mech Eng **43**(3):251–276. http://www.sciencedirect.com/science/article/pii/0045782584900677. ISSN 0045-7825

[10]If the explicit solver is used for a static problem, the solution might take a very long time as Abaqus Explicit uses small increments, and then the user may need to carefully choose a loading rate or mass scaling in order to speed it up.

[11]"Artificial" strain energy associated with constraints used to remove singular modes (such as hourglass control), and with constraints used to make the drill rotation follow the in-plane rotation of the shell elements.

[12]Recoverable strain energy.

Chapter 3
Method to Debug a Model

3.1 Debugging Flowchart

The checking method of debugging a model described is a list of preventive actions before launching an analysis in order to identify all potential bugs within the model in advance. Indeed, a model can take a long time to run and, some bugs can occur during the processing of the solution sooner or later. If this happens sooner in the process then the model will exit with an error message but if the bug occurs later in the processing phase, then the analyst will have wasted some computation time for nothing.

The flowchart shown in Fig. 3.1 starts at the job diagnostics step shown in Fig. 1.1 to start debugging a model with convergence issues. The flowchart shown in Fig. 3.2 is the suite of Fig. 3.1.

3.2 Job Diagnostic

The job diagnostic is the starting box shown in Fig. 1.1 entitled "-Job diagnostics". It includes the different analysis techniques and procedures, which need to be checked when the model has some convergence issues.

3.2.1 *Making a Test Model*

For large model, it is strongly recommended to make a test model in order to speed up the debugging processes in order to identify the numerical troubleshooting causing the unconverged solution more quickly and more accurately. In general, it is never a waste of time and always a good practice to make a test model to start debugging numerical difficulties.

R. J. Boulbes, *Troubleshooting Finite-Element Modeling with Abaqus*,
https://doi.org/10.1007/978-3-030-26740-7_3

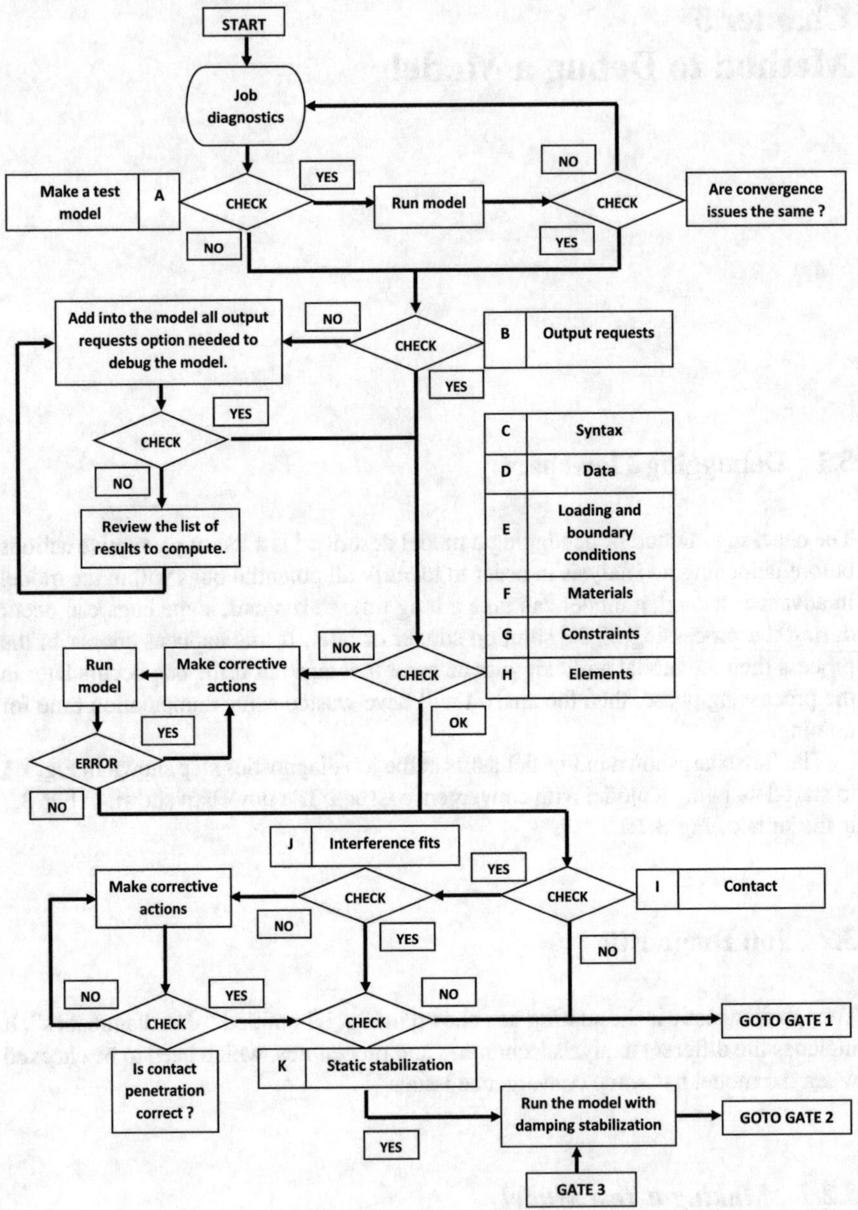

Fig. 3.1 Flowchart to debug a model—part one

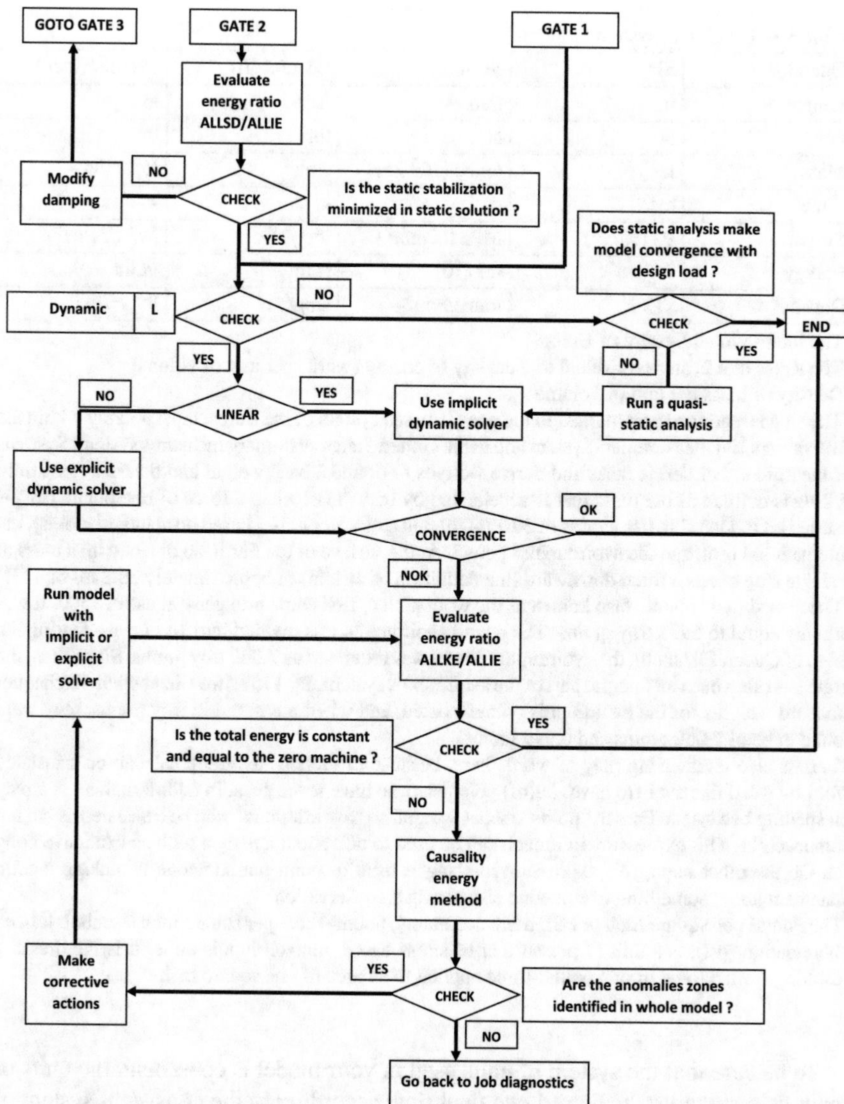

Fig. 3.2 Flowchart to debug a model—part two

Table 3.1 Consistent system of units

Quantity	SI[a]	SI (mm)	US unit (ft)	US unit (inch)
Length	m	mm	ft	in
Force	N	N	lbf	lbf
Mass	kg	tonne (10^3 kg)	slug[d]	$lbf.s^2$/in
Time	s	s	s	s
Stress[b]	Pa (N/m^2)	MPa (N/mm^2)	lbf/ft^2	psi[f] (lbf/in^2)
Energy	J	mJ (10^{-3} J)	ft.lbf	in.lbf
Density[c]	kg/m^3	tonne/mm^3	slug/ft^3	lbf.s^2/in^4

[a]The International System of Units
[b]The stress unit is also equivalent to a density of energy (work) per unit of volume
[c]Density of mass per unit of volume
[d]The slug is a derived unit of mass in the weight-based system of measures, most notably within the British Imperial measurement system and in the United States customary measures system. Systems of measure either define mass and derive weights or define a base weight and derive a mass unit. A slug is defined as the mass that is accelerated by one ft/s^2 when a force of one pound (lbf) is exerted on it. One slug has a mass of 32.174049 lbm or 14.593903 kg based on standard gravity, the international foot, and the avoirdupois[e] pound. At the surface of the Earth, an object with a mass of a single slug exerts a force downward (the definition of weight) of approximately 32.2 lbf or 143N
[e]The avoirdupois pound, also known as the wool pound, first came into general use c. 1300. It was initially equal to 6992 troy grains. The pound avoirdupois was divided into 16 ounces. During the reign of Queen Elizabeth, the avoirdupois pound was redefined as 7,000 troy grains. Since then, the grain has often been an integral part of the avoirdupois system. By 1758, two Elizabethan Exchequer standard weights for the avoirdupois pound existed, and when measured in troy grains, they were found to be of 7,002 grains and 6,999 grains
There is also an amusing play of words here, because in French "avoirdupois" can be translated word by word like avoir(to have) du(of) pois(peas), so have some peas. In addition, there is a play on spelling because in French "poi**d**s" means weight, so "avoirdupoids" can be translated as having some weight. This expression in French can be used to talk about a person with an excessive body fat. On the other hand, this expression in slang is used to point out someone is making a solid statement about something or someone else during a conservation
[f]The Pound per Square Inch or PSI, more accurately, pound-force per square inch (symbol: lbf/in^2 abbreviation: psi), is a unit of pressure or of stress based on avoirdupois units. It is the pressure resulting from a force of one pound-force applied to an area of one square inch

To be sure that the system of units used in your model is consistent, the units of input data should be reviewed one final time according to the consistent system of units shown in Table 3.1.

Once the system of units used in your model has been checked, then the analyst shall review the loading and boundary conditions again to ensure that the analysis did not converge because of an unrealistic scenario.

Although performing a test run is never a waste of time especially on heavy models, these reduced-size models should only be used for testing and debugging. The test model is tested for items on the checklist to detect a potential bug. If applicable, the test model should have the following settings:

- A linear tetrahedral, linear elastic and reduced integration version with reduced representative loading. Most preprocessors should be able to change element types to write input files with changes fairly quickly, like for instance: ***ELEMENT, TYPE=C3D4** instead of ***ELEMENT, TYPE=C3D10**.
- A coarse mesh version of the problem.
- A subset of the assembly subjected to anticipated load levels.
- Small test models to explore unfamiliar features.
- An explicit model run for a fraction of the intended step time to check behavior.
- A reduced dimensionality model (stick model, plane-stress, plane-strain, and axisymmetric).

3.2.2 Output Check

Before submitting an analysis job, it is always beneficial to request more calculations in **Field Output Requests** and/or in the **History Output Requests** module regarding selected outputs that will help to debug the model afterward. Of course, the computational time with these requests will take longer but it will help the analyst to identify the cause of unconverged solution. The following are some recommendations about what option requests to select:

1. Are the following quantities being output to the .dat file?
   - Energy histories using *ENERGY PRINT with no data lines
   - Reaction forces using *NODE PRINT
   - Contact forces using *CONTACT PRINT
   - Always save this .dat file as it will help identify load paths.
2. If running a heat transfer analysis, request the following output:
   - **NT** the nodal point temperatures,
   - **HFL** the magnitude and components of the heat flux vector (e.g., W/mm2),
   - **RFL** the reaction flux values due to the prescribed temperature (e.g., W),
   - **RFLE** the total flux value at a node (e.g., W) for interesting node sets using *NODE PRINT, TOTALS=YES,
   - **HFLA** the heat flux vector on a surface (multiplied by the nodal area) and **HTLA** the time integrated of **HFLA** for contact surfaces.
3. If a bolt connection with pretension is present in the analysis model, request the output **TF1** the total force to monitor the bolt preload versus the time increment. The direction 1 is the default direction where the bolt preload has been defined.
4. If using connector elements, is the output variable nodal force contact **NFORC** being output for global load path analysis if necessary?
   Remember that the contact force tangential **CFT**, the contact force normal **CFN** and the contact force surface **CFS** reported by Abaqus are the forces acting on the master surface. Here is an example of (.dat) file requests in history data for

an analysis with contact and connector elements for drawing free body diagrams using global axes:

```
*NODE PRINT,
GLOBAL=YES, NSET=PRESCRIBED-DISPLACEMENT-NODES
RF,
*CONTACT PRINT
CFT,
*EL PRINT, ELSET=MY-CONNECTOR-ELEMS
NFORC,
```

5. In addition to requesting output variables, the contact stress **CSTRESS** and the contact displacement **CDISP** output for contact analyses, it is sometimes useful to request output variables the contact force **CFORCE** and the contact status (open/close) **CSTATUS**. One can then symbol plot the resultant contact normal forces **CNORMF** and contact shear forces **CSHEARF** and contour plot the contact status **CSTATUS** for easier visualization of the contact conditions.
6. Do the file sizes indicate the need for less output or less frequent output for further runs?

3.2.3 Syntax Check

To run an analysis check on the current debugging directory, the first step is to open an MSDOS command prompt in the current directory. A simple way of doing this is to create a text file inside the current debugging directory and, enter the command line "cmd.exe" before saving and closing the file. Second, the file needs to be renamed and the extension changed from (.txt) to (.bat). Double click on, for example, "myMSDOS.bat" to open an MSDOS command prompt set into the current debugging directory. The next step is to open a command prompt corresponding to a specific version that user wants to use, for example, Abaqus v6.14-5 will need to use the command abq6145.[1] The following command can be used for running a syntax check analysis:

abq6145 syntaxcheck j=my-input-file-name-without-extension

[1] To identify the correct Abaqus version command a simple method is to click right on the shortcut icon to start Abaqus, go the "General" tab and, read the description. For Abaqus version 6.14-5, it shown as "abq6145.bat".

Some files have been generated from Abaqus with the following extensions (.com),[2] (.dat),[3] (.log),[4] (.odb)[5] and (.sim).[6] The log file should have shown a correct syntax check if the user receives the following return lines from Abaqus:

```
Abaqus JOB my-input-file-name
Abaqus 6.14-5
Begin Analysis Input File Processor
Run pre.exe
End Analysis Input File Processor
Abaqus JOB my-input-file-name COMPLETED
```

If the log file shows that Abaqus did not complete the analysis input file then both (.dat) and (.odb) files will have to be investigated regarding the test made using the syntax check. The analyst will therefore have to review and evaluate all error and warning messages in the data (.dat) file. At this point, the three main questions to ask, will be

1. Do these warning messages need to be addressed?
2. Does the model look correct in Abaqus/Viewer? (Display Group features can be used to control check if the model is as defined or if it is missing some settings.)
3. Are the surfaces for contact and coupling constraints defined correctly?

3.2.4 Data Check

Now that the Abaqus syntax has been checked, let's do the same for the Abaqus data check. The operation is identical to the syntax check so the following command line is entered into a command prompt:

abq6145 datacheck j=my-input-file-name-without-extension

[2]Command file, created by the Abaqus execution procedure.

[3]Printed output file. It is written by the analysis, syntax check, parameter check, and continue options. Abaqus/Explicit and Abaqus/CFD do not write analysis results to this file.

[4]Log file, which contains start and end times for modules run by the current Abaqus execution procedure.

[5]Output database. It is written by the analysis and continue options in Abaqus/Standard, Abaqus/-Explicit, and Abaqus/CFD. It is read by the Visualization module in Abaqus/CAE (Abaqus/Viewer) and by the convert=odb option. This file is required for restart.

[6]Model and results file, used by Abaqus/CFD. It is also used by Abaqus/Standard and Abaqus/-Explicit when the results format=sim or both option is specified. It is written by the syntax check option. It is read and can be written by the analysis and continue options. This file is required for restart.

The data check generates more files from Abaqus than the syntax check and therefore, the current debugging directory presents the following files extensions: (.023),[7] (.com), (.dat). (.fil),[8] (.log), (.mdl),[9] (.msg),[10] (.odb), (.prt),[11] (.res),[12] (.sim) and (.stt).[13]

A log file exited correctly should show the following returned lines:

```
Abaqus JOB my-input-file-name
Abaqus 6.14-5
Abaqus License Manager checked out the following
licenses:
Abaqus/Standard checked out 5 tokens from Flexnet
server localhost.
<1019 out of 1024 licenses remain available.>
Begin Analysis Input File Processor
Run pre.exe
End Analysis Input File Processor
Begin Abaqus/Standard Datacheck
Begin Abaqus/Standard Analysis
Run standard.exe
End Abaqus/Standard Analysis
Abaqus JOB my-input-file-name COMPLETED
```

[7]Communications file, used by Abaqus/Standard and Abaqus/Explicit. It is written by the analysis and data check options and is read by the analysis and continue options.

[8]Results file. This is written by the analysis and continue options in Abaqus/Standard and by the convert=select and convert=all options in Abaqus/Explicit.

[9]Model file, used by Abaqus/Standard and Abaqus/Explicit. It is written by the data check option. It is read and can be written by the analysis and continue options in Abaqus/Standard, while it is read by the analysis and continue options in Abaqus/Explicit. Multiple model files may exist if the element operations are executed in parallel in an Abaqus/Standard analysis. In such cases, a process identifier is attached to the file name. This file is required for restart.

[10]Message file. This is written by the analysis, data check, and continue options in Abaqus/Standard and Abaqus/Explicit. Multiple message files may exist if the element operations are executed in parallel in an Abaqus/Standard analysis. In such cases, a process identifier is attached to the file name.

[11]Part file, used by Abaqus/Standard and Abaqus/Explicit. This file is used to store part and assembly information and is created even if the input file does not contain an assembly definition. The part file is required for restart, import, sequentially coupled thermal stress analysis, symmetric model generation, and underwater shock analysis, even if the model is not defined in terms of an assembly of part instances. This file may also be needed for submodeling analysis.

[12]Restart file, which contains information necessary to continue a previous analysis and is used by Abaqus/Standard and Abaqus/Explicit. The restart file is written by the analysis, data check, and continue options. It is read by any restarted analysis.

[13]State file. This is written by the data check option in Abaqus/Standard and Abaqus/Explicit and is read and can be written by the analysis and continue options in Abaqus/Standard. It is read by the analysis and continue options in Abaqus/Explicit. Multiple state files may exist if the element operations are executed in parallel in an Abaqus/Standard analysis. In such cases, a process identifier is attached to the file name. This file is required for restart.

If the analysis data check exited with warning and error messages then all different files, especially, the text inside (.dat), (.msg) files and (.odb) file with Abaqus/Viewer will have to be inspected first.

At this point, the analyst needs to focus on several different analysis techniques below:

- out of contact definition:

1. Review and evaluate all warning messages in the data (.dat) and message (.msg) files.
2. Inspect the (.odb) file in Abaqus/Viewer: are there any anomalies present in the model?
3. Do these warning messages need to be addressed?
4. Search for the string memory in the .dat file.
   - Do the memory requirements seem reasonable?
   - Check MINIMUM MEMORY required and MEMORY REQUIRED TO MINIMIZE I/O (Input/Output). Are the appropriate servers being used? An example of memory estimates printed to the (.dat) file:

PROCESS	FLOATING POINT OPER-ATIONS PER ITERATION	MINMUM MEMORY REQUIRED (MBYTES)	MEMORY TO MINMIZE I/O (MBYTES)
1	$3.04E + 12$	2380	17418

   Therefore to consider running on the machine if FLOPS ~ 1E+10 or 1E+11 or less. The analyst will have to pay a close attention to MEMORY REQUIRED TO MINIMIZE I/O if running on desktop and evaluate if there is enough memory on desktop to run a majority of the job in core memory.
5. Can C3D10 elements, penalty contact, C3D8R elements with enhanced hourglass controls (moderate plasticity), C3D8R elements with default hourglassing (if lots of plasticity) or small sliding contact be used to reduce memory requirements? Check whether this is possible by running another data check.
   The simplest form of penalty contact looks like the following command lines inside the model input file (.inp):

```
*SURFACE INTERACTION, NAME=MY-SURF-INT
*SURFACE BEHAVIOR, PENALTY
```

   For an overall stiffer penalty response for interference fits, analyst can use non-linear penalty contact:

```
*SURFACE BEHAVIOR, PENALTY=NONLINEAR
```

The enhanced hourglass controls definition for C3D8R elements is as follows:

```
*SECTION CONTROLS, NAME=MY-CONTROLS,
HOURGLASS=ENHANCED
....
*SOLID SECTION, ELSET=MY-C3D8R-ELSET,
CONTROLS=MY-CONTROLS
```

6. If user is curve fitting a hyperelastic model, then is it stable for the strain ranges anticipated in the simulation?
   Make sure that user is using the following option to obtain this information:

```
*PREPRINT, MODEL=YES
```

- with contact interaction:

1. Do the initial contact opening "COPEN" values for each contact pair make sense? Please note an important command lines to add inside the model input file (.inp) in order to detect contact opening; the contact opening COPEN values are written to the .odb file at the start of the analysis if and only if the output variable contact displacement CDISP is requested, as shown below:

```
*OUTPUT, FIELD
*CONTACT OUTPUT
CSTRESS,
CDISP,
```

2. Are there any unwanted initial overclosures?
3. Can they be justifiably removed with the ADJUST parameter or using the keyword command ***CLEARANCE**?
4. Do certain components need to be repositioned?

3.2.5 Loading and Boundary Conditions Check

Now it is time to check the loading and boundary conditions. After the syntax and data check the analyst can be confident about the model input data gathered, which has been correctly entered into Abaqus to make proper settings according to the work specifications. The analyst needs to monitor the loading and boundary conditions to be sure that the applied load case(s) and boundaries have proper settings with appropriate Abaqus features.

The review of loading and boundary conditions is summarized with different analysis techniques below:

1. Do the loads and boundary conditions need to be applied in a local coordinate system?
2. Are they correctly defined in directions and magnitudes?
3. Can displacement control be applied to initially engage contact to make the analysis more robust? For example, a pretension load[14] used with bolt model.
   In Step 1, an example is given in the case of forcing a displacement to establish a proper initial contact interaction between both surfaces, with a boundary condition constraining a displacement of 0.01 mm for initial bolt pretension reference nodes:

```
*BOUNDARY
MY-PRETENSION-NODES, 1, 1, 0.01
OTHER-BC-NODES, 1, 6, 0.0
```

   Then, in Step 2, the displacement is changed in order to produce a concentrated load of 12 kN on the same node:

```
*BOUNDARY, OP=NEW
OTHER-BC-NODES, 1, 6, 0.0
*CLOAD
MY-PRETENSION-NODES,1, 12000.
```

4. Can symmetry boundary conditions be used to reduce the problem size? Sometimes, even if the structure is not perfectly symmetric, a conservative solution can be obtained by analyzing the weaker portion of the symmetric structure. Never use this simplification for dynamic problems!
5. Is the material density specified in the correct units for dynamic analyses or analyses with gravity **GRAV** and centrifugal **CENTRIF** loads?[15] (See the consistent system of units detailed in Table 3.1)
6. Do not distribute concentrated loads evenly between nodes on faces or edges of quadratic elements as, this leads to nonuniform pressures between corner and middle nodes as shown in Fig. 3.3.
7. Are the boundary conditions used to constrain tangential and/or radial degrees of freedom for nodes lying on the axis of a defined local cylindrical coordinate system? These directions are not uniquely defined! Use a rectangular coordinate system instead if needed and define appropriate boundary conditions (Fig. 3.4).

---

[14]In case of bolt preload, at least two analyses steps are needed. The first analysis step will only be used to set the applied bolt preload. The second analysis step will be used to apply the external load on the structure but with a propagated load from the previous step regarding the bolt preload in order to fix the bolt preload at a current length in the second step. Otherwise, the bolt preload

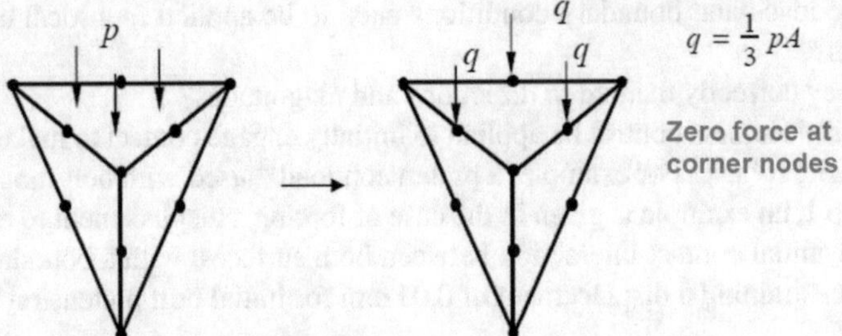

Fig. 3.3 3D quadratic tetrahedral element C3D10 with a pressure "p" applied on top surface and distributed on the "q" nodes. The equilibrium force on the top surface is equal to 3q. Figure used by permission ©Dassault Systemes Simulia Corp

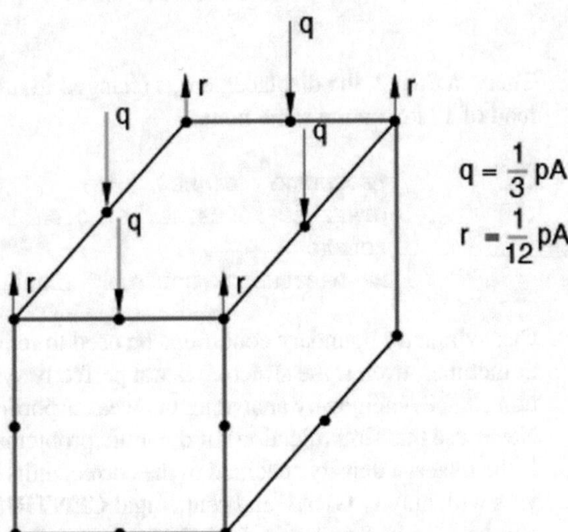

Fig. 3.4 Equivalent nodal loads produced by a constant pressure on the second-order element face in hard contact simulations for a quadratic brick element C3D20, with the pressure "p" on the top surface and spread in the "q" nodes and the reaction in "r" nodes. The equilibrium force on the top surface is equal to 4q − 4r. Figure used by permission ©Dassault Systemes Simulia Corp

3.2.6 Materials Check

At this point, the material properties need to be checked in order to correctly represent the structural responses of the model under the loading and boundary conditions. The function of the complexity of material data and the physical behavior law need to be implemented into the model, and it is recommended to follow the list of different analysis techniques in order to detect the potential risk of numerical difficulties:

---

at the beginning of the second analysis step will not be constrained as a compressive bolt load and consequently convergence difficulties with solver exit will occur.

[15]See the dynamic load keyword command ***DLOAD**. This option enables the specification of distributed loads. These include constant pressure loading on element faces and mass loading (load per unit mass) either by gravity forces or by centrifugal forces.

1. General considerations:

   - Is my material model selection aligned with the goals of my analysis?
   - What physics do I need to capture in my material model?
     - Is the elasticity linear or nonlinear?
     - Is time-dependent behavior such as viscoelasticity or creep important?
     - Do I need to capture plasticity?
     - Is my loading monotonic, load/unload, or cyclic?
     - Does rate dependence play a role?
     - Does my material display temperature-dependent behavior?
     - Are particular material models required to capture specific behaviors that will be important for downstream analyses, for instance, fatigue strength assessments?
   - Does the strain for purely linear elastic materials in the model exceed 5%? It might be necessary to use a different material model (elastic–plastic, hyper-elastic).
   - Can the plastic curve be extended to higher plastic strains with a small slope to avoid solving the exact plastic curve near singularity locations (point loads, couplings, sharp corners, etc.) that lead to artificial hot spots? An example of an extension would be

```
*MATERIAL, NAME=BRASS
*ELASTIC
97000., 0.31
*PLASTIC
310.99, 0.
612., 0.4189
** added point extending the curve to higher,
** plastic strains
620., 1.
```

     Another option is to use power law plasticity. An example would be as follows:

```
*PARAMETER
A=258.2523
B=838.4076
N=0.214319
E=200000.0
nu=0.3
*MATERIAL, NAME=STAINLESS
*ELASTIC
<E>, <nu>
*PLASTIC, HARDENING=JOHNSONCOOK
<A>, <B>, <n>, 1000, 500
```

- Is the hyperelastic material model being used stable for the range of strains anticipated in the simulation?
  For hyperelastic material models output non-default variable **NE**,[16] is the nominal strain. Although the maximum shear strain is an ambiguous concept in finite-strain problems, the quantity **NEP3**[17] minus **NEP1**[18] is a useful measure of the maximum shear strain. It is also important to look out for regions of negative pressure when analyzing components made out of hyperelastic materials as this indicates hydrostatic tension, which can be detrimental to durability in rubber components.
- Avoid using deformable elements with artificially stiff materials to enforce rigid constraints. Use rigid bodies or connector types like beam elements to do this. An example is given below regarding a beam connector model set with a rigid beam constraint:

```
*ELEMENT, TYPE=CONN3D2, ELSET=MY-BEAM
800001, 65001, 65002
.
*CONNECTOR SECTION, ELSET=MY-BEAM
BEAM,
```

3.2.7 Constraints Check

There are two methods available to couple the motion of the reference node to the average motion of the coupling nodes: the continuum coupling method and the structural coupling method. The continuum coupling method is used by default.

The default continuum coupling method couples the translation and rotation of the reference node to the average translation of the coupling nodes. The constraint distributes the forces and moments at the reference node only as a coupling nodes force distribution; no moments are distributed at the coupling nodes. The force distribution is equivalent to the classic bolt pattern force distribution when the weight factors are interpreted as bolt cross-sectional areas. The constraint enforces a rigid beam connection between the attachment point and a point located at the weighted center of position of the coupling nodes.

The structural coupling method couples the translation and rotation of the reference node to the translation and the rotation motion of the coupling nodes. This method is particularly suited for bending-like applications of shells when the coupling constraint spans small patches of nodes and the reference node is chosen to be on or very close to the constrained surface. The constraint distributes forces and

[16] All nominal strain components.

[17] Maximum principal nominal strains.

[18] Minimum principal nominal strains.

moments at the reference node as a coupling node-force and moment distribution. For this coupling method to be active, all rotation degrees of freedom at all coupling nodes must be active (as would be the case when the constraint is applied to a shell surface) and the constraints must be specified in all degrees of freedom (default). In addition, for the constraint to be meaningful, the local (or global) z-axis used in the constraint should be such that it is parallel to the average normal direction of the constrained surface.

With respect to translations, the constraint enforces a rigid beam connection between the reference node and a moving point that remains in the vicinity of the constrained surface at all time. The location of this moving point is determined by the approximate current curvature of the surface, the current location of the weighted center of position of the coupling nodes, and the z-axis used in the constraint. This choice avoids unrealistic contact interactions in situations where multiple distributed coupling constraints are used to fasten pairs of shell surfaces.

With respect to rotations, the constraint is different along different local directions. Along the z-axis (twist direction), the constraint is identical to the one enforced via the continuum coupling method. In contrast, the rotational constraint in the plane perpendicular to the z-axis relates the in-plane reference node rotations to the in-plane rotations of the coupling nodes in the immediate vicinity of the reference node. This choice provides a more realistic (compliant) response when the constrained surface is small and primarily deforms in a bending mode.

It is important to use surface-based coupling constraints whenever possible instead of the older style kinematic couplings or distributing coupling elements. Here is an example of a surface-based distributing coupling:

```
*COUPLING, CONSTRAINT NAME=C1, REF NODE=1000,
SURFACE=SURF-A
*DISTRIBUTING
1, 6
```

3.2.8 Elements Check

When elements are responsible for numerical difficulties in the model, different analysis techniques must be applied to make some corrective actions, as listed below:

1. Is excessive hourglassing shown in Abaqus/Explicit?
   One solution is to use enhanced hourglass controls, mesh refinement and penalty contact to distribute the contact to more nodes.
2. If using reduced integration elements, compare the artificial energy **ALLAE** to the internal energy **ALLIE** to ensure that the artificial energy used to constrain hourglassing modes is not excessive.

3. Will the more efficient orthogonal kinematic formulation be adequate for hexahedral solid elements in Abaqus/Explicit?
   An example of the definition of the orthogonal kinematic formulation is

```
*SOLID SECTION, ELSET=MY-C3D8R-ELSET,
MATERIAL=MY-MATL, CONTROLS=MY-CONTROLS
....
*SECTION CONTROLS, KINEMATIC-SPLIT=ORTHOGONAL,
NAME=MY-CONTROLS
```

   Do not use the orthogonal formulation for analyses with highly distorted elements, excessively coarse meshes or high confinement.
4. Are the nonlinear spring stiffnesses defined for sufficient range of deformation to avoid zero stiffness?
   Analysts can use the **EXTRAPOLATION=LINEAR** if appropriate with the *CONNECTOR ELASTICITY option to avoid zero stiffness.
5. Is the mesh locking in shear when bending fully integrated first-order elements?

***** Warning *****

Do not use C3D8 element type with a bending applied load problem.

6. Try fully integrated hybrid elements for incompressible hyperelastic materials if the reduced integration hybrid elements have trouble converging.
7. Reasonably shaped C3D8 elements (trapezoidal shapes in particular should be avoided) are probably the most cost-effective solid continuum elements for modeling bending and strain localization problems.

3.2.9 Interference Fits Check

This section covers how to use contact to resolve interferences, with a focus on interference fits. An interference or press fit is a common way for parts to be fastened together and used as a system in a global assembly model with several components in contact interaction. See different analysis techniques that can be used to simulate and effectively solve these issues with Abaqus.

1. Check to see what constraints the input file definition is imposing by relating it to the equation $h - v(t)$ is less than or equal to 0. For example, for the following example contact interference definition:

```
*AMPLITUDE, NAME=RAMP-DOWN
0., 1., 1., 0.
*CONTACT INTERFERENCE
```

```
MY-SLAVE, MY-MASTER, my_value
*STATIC
0.1, 1.
```

The amplitude **RAMP-DOWN** defines a function of time increment:

$$f(t) = 1 - t \tag{3.1}$$

The function v(t) is formed by taking the value on the ***CONTACT INTERFERENCE** data line and multiplying it with f(t).

$$v(t) = my\_value \times f(t) \tag{3.2}$$

The constraint that Abaqus will impose over the step is

$$h - v(t) \leq 0. \tag{3.3}$$

where h is the penetration.

2. For surface-to-surface contact, interference tends to be resolved via the slave facet normals. Is this correct or will node to surface contact with interference resolved along the master facet normals be more appropriate?
   At the same time, it is possible to establish surface to surface and a node to surface contact interactions applicable on the same master surface.

3.2.10 Contact Check

The contact interactions are relatively easy to define; however, they can cause instability in the global stiffness matrix by adding local contact stiffness and making the matrix nonsymmetric. Therefore, some numerical difficulties may appear.

A good method of setting the surface in contact is to ensure, the master surface is higher in surface area than the slave surface. The master surface should also have a coarse mesh pattern compared with the slave surface. To investigate the underlying causes of the main numerical difficulties in contact, different analysis techniques are listed as follows:

1. Is surface-to-surface penalty contact being used despite there being a strong case for not using it?
   This is necessary if the slave is defined on C3D10 elements and a Lagrange multiplier contact is employed. The simplest form of penalty contact looks like the following:

```
*SURFACE INTERACTION, NAME=MY-SURF-INT
```

```
*SURFACE BEHAVIOR, PENALTY
```

2. Is it appropriate to supplement the surface to surface contact with node-based slave surfaces and node to surface contact to capture edge contact?
   Here is an example of a node-based slave surface being used to capture edge contact for contact pairs:

```
*SURFACE, TYPE=NODE, NAME=MY-NODE-BASED-EDGE-SURF
NODE-SET-DEFINING-EDGE, 1.0
*CONTACT PAIR, TYPE=NODE TO SURFACE
MY-NODE-BASED-EDGE-SURF, MY-MASTER-SURF
```

   Here is an example of controlling which feature edges are activated in the general contact domain:

```
*CONTACT
*CONTACT INCLUSIONS, ALL EXTERIOR
*SURFACE PROPERTY ASSIGNMENT,
PROPERTY=FEATURE EDGE CRITERIA
, 20.
```

3. Can a small sliding contact be used to reduce the problem size?
   - Visualize the tangent planes created by the small sliding algorithm for your master and slave combination.
   - Will they be suitable?
4. Has the surface belonging to the stiffer component in the interaction been defined as the master surface?
   Stiffness is not just dependent on material properties; it also depends on the shape and the amount of restraint.
5. Does the fillet[19] parameter need to be used for the analytical rigid surfaces to smooth the surfaces?
6. Can contact chattering[20] be observed?
   Try softened contact or penalty contact to get more nodes involved in the contact.

---

[19] Regarding contact penetration, in some cases Eulerian material may penetrate through the Lagrangian contact surface near corners. This penetration should be limited to an area equal to the local Eulerian element size. Penetration can be minimized by refining the Eulerian mesh or adding a fillet to the Lagrangian mesh with a radius equal to the local Eulerian element size.

[20] A normal Lagrange contact chattering is an issue which often occurs with the normal Lagrange method. If no penetration is permitted, then the contact status is either open or closed (a step function). This can sometimes make convergence more difficult because contact points may oscillate between open/closed status and is called "chattering". If some slight penetration is allowed, it can make it easier to converge since contact is no longer a step change.

Moreover, ***CONTACT CONTROLS, STABILIZE** may help with chattering in some situations with the usual precautions.

7. Is the asymmetric[21] solver being used for 2D and 3D finite-sliding surface to surface contact?
8. Is the asymmetric solver being used for 3D finite-sliding node to surface contact with highly curved faceted master surfaces?
   Use the following option to indicate that nonsymmetric matrix storage and solution should be used:

```
*STEP, UNSYMM=YES
```

3.2.11 Initial Rigid Body Motion and Over Constraints Check

In Abaqus, a rigid body is a collection of nodes and elements whose motion is governed by the motion of a single node, known as the rigid body reference node. One cause of non-convergence is inadequate boundary conditions. Unreasonable boundary conditions can lead to extreme local deformations. A model can also be over or under constrained. With an under constraint, not all rigid body motion is suppressed. This leads to one or more degrees of freedom with zero stiffness and usually zero-pivot warnings. Over constraints also tend to cause zero-pivot warnings.

Although Abaqus checks for over constraints and tries to solve them are not always possible. For example, if the over constraint starts occurring after a period of time due to contact, it is recommended to check all warning messages related to over constraints, and it is important not to assume Abaqus will correctly resolve the over constraint, but to correctly define the constraint yourself.

Furthermore, look at the location of zero-pivot warnings: are there over or under constraints? Let's try to figure it out using the following different analysis techniques:

1. Are weak springs needed to prevent initial rigid body motion with numerical singularities?
   A useful technique is to define field variable dependent springs. The springs have non-negligible stiffness at the beginning of the step, but this stiffness is ramped off as contact is established:

```
*ELEMENT, TYPE=CONN3D2, ELSET=MY-SPRING
800001, 65001, 9965001
...
*CONNECTOR SECTION, ELSET=MY-SPRING,
BEHAVIOR=MY-SPRING-BEHAV
```

[21] It is strongly recommended, especially with a high coefficient of friction value set between both surfaces of, greater than 0.2.

```
CARTESIAN,
ORI-GLOBAL,
*CONNECTOR BEHAVIOR, NAME=MY-SPRING-BEHAV
*CONNECTOR ELASTICITY, COMP=1, ELSET=MY-SPRING,
DEPENDENCIES=1
50., , , 0.0
0.01, , , 1.0
...
*NSET, NAME=MY-SPRING-NODES
65001,
*NSET, NAME=MY-SPRING-GROUNDS
9965001,
...
*STEP, NLGEOM
*STATIC
0.1, 1.
...
*FIELD
MY-SPRING-NODES, 1.
MY-SPRING-GROUNDS, 1.
```

As demonstrated in the code shown above, the default initial value for field variables is 0. The spring stiffness is defined as a function of field variables (50 N/mm at a field variable value of 0 and 0.01 N/mm at a field variable value of 1). As the field variable at the node of the spring is ramped up over the step to 1, the stiffness of the spring is ramped down.

2. If using ***CONTACT CONTROLS, STABILIZE** to handle initial rigid body motion, is the stabilization energy **ALLSD** small compared to the global energy **ALLIE**? (Plot both to compare versus time increment or load magnitude.)
3. Is the contact damping pressure **CDPRESS** small compared to the contact pressure **CPRESS** at any increment where results are being used? (Plot both to compare versus time increment or load magnitude.)

***** To use with caution *****

Use this option only to handle initial rigid body motion issues. Improper use of this option can lead to extremely erroneous results therefore use all recommended checks and safeguard.

The usual form to use contact stabilization is the following:

```
*CONTACT CONTROLS, STABILIZE=0.01, TANGENT=0.01
, 0.001
```

A small amount of damping is left at the end of the step. In addition to writing whole energy quantities to the input file:

```
*OUTPUT, HISTORY
*ENERGY OUTPUT, VAR=PRESELECT
```

It is also recommended to add the single line that prints energies to the .dat file somewhere in the step definition. Data lines are not required:

```
*ENERGY PRINT
```

The contact damping pressure **CDPRESS** output can be obtained by requesting the **CDPRESS** output variable as demonstrated here:

```
*OUTPUT, FIELD, FREQUENCY=6
*CONTACT OUTPUT
CSTRESS,
CDSTRESS,
```

4. If using ***CONTACT CONTROLS, STABILIZE**, are the results being used before the step has ended?
   It is essential to check for equilibrium if these results are used before the step has ended by drawing free body diagrams using output quantities such as the total contact force **CFT**. The total contact force can be requested to the .dat file and .odb files, but it is important to remember that CFT as reported is the total contact force acting on the master surface.

```
*CONTACT PRINT
CFT,
*OUTPUT, HISTORY
*CONTACT OUTPUT
CFT,
```

5. Any zero-pivot messages in the message file indicating over constraints?
   Never accept a solution if the message file has zero-pivot warnings.
   These have to be resolved.
   If zero-pivot messages are caused by connector elements, try to change connector definitions to avoid redundant constraints while still modeling the kinematics appropriately.

Another strategy is to build the desired connector using basic connectors like CARTESIAN[22] and CARDAN[23] and using reasonably stiff elasticity definitions in components of relative motion that need to be constrained.

Here is an example of a TRANSLATOR[24] connector with the slot direction along the global X-axis:

```
*ELEMENT, TYPE=CONN3D2, ELSET=MY-TRANSLATOR-CONN
800001, 65000, 65001
...
*ORIENTATION, NAME=ORI-TRANSLATOR
1., 0., 0., 0., 1., 0.
*CONNECTOR SECTION, ELSET=MY-TRANSLATOR-CONN
TRANSLATOR,
ORI-TRANSLATOR,
```

Here is the same connector built with basic connectors and some flexibility to avoid over constraints:

```
*ELEMENT, TYPE=CONN3D2, ELSET=MY-TRANSLATOR-CONN
800001, 65000, 65001
...
*ORIENTATION, NAME=ORI-TRANSLATOR
```

[22]The CARTESIAN connection does not impose kinematic constraints. It defines three local directions at node "a" and measures the change in the position of node "b" along these local coordinate directions. The local directions at node "a" follow its rotation.

[23]Connection type CARDAN provides a rotational connection between two nodes where the relative rotation between the nodes is parameterized by Cardan (or Bryant) angles. A Cardan-angle parameterization of finite rotations is also called a 123 or yaw–pitch–roll parameterization. Connection type CARDAN cannot be used in two-dimensional or axisymmetric analysis. When connection type CARDAN is used with connector behavior, the relative rotation axis with the highest resistance to rotational motion should be assigned to the second component of relative rotation (component number 5) in order to avoid gimbal lock, a singularity in the rotation parameterization for relative rotation angles. The CARDAN connection does not impose kinematic constraints. A CARDAN connection is a finite rotation connection where the local directions at node "b" are parameterized in terms of Cardan (or Bryant) angles relative to the local directions at node "a". Local directions are positioned relative to by three successive finite rotations.

[24]Connection type TRANSLATOR imposes kinematic constraints and uses local orientation definitions equivalent to combining connection types SLOT and ALIGN. The connector constraint forces and moments reported as the connector output depend strongly on the order and location of the nodes in the connector. Since the kinematic constraints are enforced at node "b" (the second node of the connector element), the reported forces and moments are the constraint forces and moments applied at node "b" to enforce the TRANSLATOR constraint. Thus, in most cases, the connector output associated with a TRANSLATOR connection is best interpreted when node "b" is located at the center of the device enforcing the constraint. This choice is essential when moment-based friction is modeled in the connector since the contact forces are derived from the connector forces and moments, as illustrated below. Proper enforcement of the kinematic constraints is independent of the order or location of the nodes.

```
1., 0., 0., 0., 1., 0.
*CONNECTOR SECTION, ELSET=MY-TRANSLATOR-CONN,
BEHAV=FLEX
CARTESIAN, CARDAN
ORI-TRANSLATOR,
...
*CONNECTOR BEHAVIOR, NAME=FLEX
*CONNECTOR ELASTICITY, COMP=2
1.e6,
*CONNECTOR ELASTICITY, COMP=3
1.e6,
*CONNECTOR ELASTICITY, COMP=4
1.e7,
*CONNECTOR ELASTICITY, COMP=5
1.e7,
*CONNECTOR ELASTICITY, COMP=6
1.e7,
```

3.2.12 Static Stabilization Check

A stabilization into a static solution is a stabilized equation put into the static equation to help the solver calculate a solution; however, this solution is no longer purely static. To understand this part, it is possible to see this as the static equation representing the model to solve, where K is the global stiffness matrix, x is the nodal displacement vector and f_{ext} is the nodal external force vector. For a better understanding, the static equation will be written as a scalar equation in one dimension as shown in Eq. 3.4.

$$Kx = f_{ext} \tag{3.4}$$

For some reasons, the static Eq. 3.4 cannot be solved because of an instability in the model built which generates some numerical difficulties. A trick to solve such equations is to add an stabilized equation, which is the damping force component with a damping coefficient D. The Eq. 3.4 becomes Eq. 3.5

$$D\dot{x} + Kx = f_{ext} \tag{3.5}$$

The damping coefficient D is the stabilization value to add in order to solve a quasi-static solution. Of course, the closer this damping coefficient value to zero, the more the solution will be a pure static solution. However, in cases where stabilization is needed, this stabilized force will be added into the static solution from Eq. 3.4. This transforms the static solution in a quasi-static solution as shown in Eq. 3.5. Once the model is solved according to the equilibrium equation shown in Eq. 3.5, the next step is to determine the proportion of this stabilized force compared with the real static force ($K \times x$), to determine how the quasi-static solution can move closer to the static solution that cannot be solved using a classic static method.

The best way to figure it out is to plot both energies **ALLSD** and **ALLIE** versus the time increment. Indeed, the energy **ALLSD** represents this quantity of artificial force added into the whole model in question. The internal energy **ALLIE** represents the summation of energies, for instance the sum of the stored strain energy **ALLSE** plus the inelastic dissipated energy **ALLPD** plus the energy dissipated by viscoelasticity **ALLCD** plus an artificial strain energy **ALLAE**.[25]

The energy ratio to check is given in Eq. 3.6 and should be as low as possible (**ALLSD** should be close to zero) to be confident regarding the quasi-static solution; as a result, the stabilization force introduced to stabilize and therefore to solve the whole model does not have a significant effect on the calculated solution.

$$E_{ratio} = \frac{ALLSD}{ALLIE} \tag{3.6}$$

Now let's explore different analysis techniques which can be used in such cases:

1. Is the ***STATIC, STABILIZE** option being used only for either instability due to loss of contact or problems involving local or global buckling?
   Note that both problem classes need a way of handling the release of stored internal energy **ALLIE**. All other uses of this option must be questioned very strongly!
2. If **ALLSD** is large compared to **ALLIE**, can it be explained?
3. Is **ALLSD** more or less constant toward the end of the simulation indicating that no significant viscous forces are present at that time?
4. Contour plot nodal output quantity **VF** (stabilization forces). Do they make sense in terms of the regions in which the forces are predominant?
   Here is an example requesting energy output and viscous forces **VF** to the (.odb) file:

```
*OUTPUT, HISTORY
*ENERGY OUTPUT, VAR=PRESELECT
*OUTPUT, FIELD
*NODE OUTPUT
U,
RF,
VF,
```

[25] It is important to be careful with vocabulary here as, this energy is an artificial energy caused by the hourglassing effect. An energy analysis will have to minimize this effect as much as possible with meshing techniques. When reduced integration is used in the first-order elements (the 4-node quadrilateral and the 8-node brick), hourglassing can often make the elements unusable unless it is controlled. In Abaqus the artificial stiffness method and the artificial damping method given in Flanagan and Belytschko [1] are used to control the hourglass modes in these elements.

It is also helpful to add the single line that prints energies to the (.dat) file somewhere in the step definition, with no data lines needed:

```
*ENERGY PRINT
```

5. Will a physical situation involving a release of energy be helped by using implicit dynamics as an alternative to static stabilization?
   To solve the class of static problems involving instability due to loss of contact or local buckling, you can use the following:

```
*DYNAMIC, APPLICATION=QUASI-STATIC, NOHAF
```

6. Track **ALLIE** if an instability is encountered due to loss of contact and see whether the energy lost appears as **ALLKE** the kinetic energy, **ALLFD** the frictional dissipated energy or the energy dissipated by automatic stabilization **ALLSD**.

3.2.13 Dynamics Check

The differences between implicit and explicit dynamic analysis are such as that in static analysis, there is no effect of mass (inertia) or damping. In dynamic analysis, nodal forces associated with mass/inertia and damping are included. Static analysis is done using an implicit solver. Dynamic analysis can be done via the explicit solver or the implicit solver.

In nonlinear implicit analysis, the solution for each step requires a series of trial solutions (iterations) to establish equilibrium within a certain tolerance. On the other hand, in explicit analysis, no iteration is required as the nodal accelerations are solved directly. The time step in explicit analysis must be less than the current time step (the time it takes a sound wave to travel across an element). Implicit transientanalysis has no inherent limit in terms of the size of the time step. As such, implicit time steps are generally several orders of magnitude larger than explicit time steps. Implicit analysis requires a numerical solver to invert the stiffness matrix once or even several times over the course of a load/time step. This matrix inversion is an expensive operation, especially for large models. Explicit analysis doesn't require this step.

Explicit analysis handles nonlinearities with relative ease in comparison to implicit analysis. This would include treatment of contact and material nonlinearities. In explicit dynamic analysis, nodal accelerations are solved directly (not iteratively) as the inverse of the diagonal mass matrix times the net nodal force vector, where net nodal force includes contributions from exterior sources (body forces, applied pressure, contact, etc.), element stress, damping, bulk viscosity, and hourglass control. Once accelerations are known at time n, velocities are calculated at time $n + \frac{1}{2}$, and displacements at time $n + 1$. From displacements comes strain, and from strain comes stress. The cycle is then repeated.

The choice of operator used to integrate the equations of motion in a dynamic analysis is influenced by many factors. Abaqus/Standard is designed to analyze structural components, by which we mean that the overall dynamic response of a structure is sought in contrast to wave propagation solutions associated with relatively local responses in continua. Belytschko [2] labels these inertial problems and classifies them by stating that wave effects such as focusing, reflection, and diffraction are not important.Structural problems are considered inertial because the response time sought is long compared to the time required for waves to traverse the structure.

3.2.13.1 Linear Dynamics Check

In static, the time has no physical sense for an applied load because the applied load in static should act fully immediately on the structure rather than with a time increment. Indeed, when a gravity load is applied on a structure in reality, the gravity load does not need time to act fully on the structure; gravity acts immediately on the structure. However, numerical schemes in all FEA solvers need a time increment to perform calculations. In contrast, dynamic analysis has a physical sense regarding time versus applied load. Therefore, the applied load versus time increment should be checked carefully to ensure realistic solution. Some different analysis techniques are listed to establish a control check for linear dynamic analysis:

1. Is the mesh sufficiently fine to capture the mode shapes corresponding to the highest frequency of interest?
2. Are enough modes included to capture the dynamic response?
   As a rule of thumb, modes with frequencies from 1.5 to 3 times the highest frequency of the loading must be included for accuracy.
3. Is there at least some damping to prevent unbounded responses at excitation frequencies corresponding to the fundamental frequencies?
4. Are residual modes needed to capture shapes that are not captured by the dynamic modes of the system?
5. Is the SIM[26] parameter being specified for the *FREQUENCY option with the Lanczos solver to use the high-performance SIM software architecture for subsequent mode-based linear dynamics analysis steps?

```
*FREQUENCY, EIGENSOLVER=LANCZOS, SIM
```

6. For transient dynamics, do the time increments taken capture sufficient data to describe the highest frequencies of interest in the response?

[26] SIM is a high-performance software architecture available in Abaqus that can be used to perform modal superposition dynamic analyses. The SIM architecture is much more efficient than the traditional architecture for large-scale linear dynamic analyses (both model size and number of modes) with minimal output requests. SIM is used with the eigensolver to calculate modal frequencies using Lanczos or AMS techniques but is not applicable for subspace iteration techniques.

A rule of thumb is that 8 or more points are needed to capture a cycle of the highest frequency of interest.

3.2.13.2 Implicit Dynamics Check

In Abaqus/Standard, the time step for implicit integration can be chosen automatically on the basis of the half-increment residual, a concept introduced by Hibbitt and Karlsson [3]. By monitoring the values of equilibrium residuals at $t + \frac{\Delta t}{2}$ once the solution $t + \Delta t$ has been obtained, the accuracy of the solution can be assessed and the time step adjusted appropriately. The following analysis recommendations are useful if numerical difficulties are encountered:

1. Use the following guidelines to select the appropriate implicit dynamics setting:
   - For dynamic applications where accurate resolution of high-frequency vibrations is not of interest, use:

     ```
     *DYNAMIC, APPLICATION=MODERATE DISSIPATION
     ```

   - For cases in which a static solution is desired but the stabilizing effects of inertia are beneficial, use:

     ```
     *DYNAMIC, APPLICATION=QUASI-STATIC
     ```

   - For dynamic applications where accurate resolution of high-frequency vibrations is of interest, use:

     ```
     *DYNAMIC, APPLICATION=TRANSIENT FIDELITY
     ```

   The default amplitude for the APPLICATION=MODERATE DISSIPATION and the APPLICATION=TRANSIENT FIDELITY setting is STEP and the default amplitude for the APPLICATION=QUASI-STATIC setting is RAMP.
2. For transient implicit dynamics, do the time increments taken capture sufficient data to describe the highest frequencies of interest in the response?
   A rule of thumb is that 8 or more points are needed to capture a cycle of the highest frequency of interest.
3. ***CONTACT CONTROLS, STABILIZE** may help with chattering when quasi-static implicit dynamics are being used. If this option is employed, the usual precautions should be taken, for instance, a solution stabilized by damping coefficient D on the structure K.
   Although an example in static stabilization has been given according to Eq. 3.5, a dynamic stabilization now follows the same principle according to Eq. 3.7 with

a mass M. Therefore, energy ratios must be checked between the internal energy **ALLIE**, the kinetic energy **ALLKE** and the stabilized damping energy **ALLSD**.

$$M\ddot{x} + D\dot{x} + Kx = f_{ext} \tag{3.7}$$

3.2.13.3 Explicit Dynamics Check

The explicit dynamics procedure performs a large number of small time increments efficiently. An explicit central-difference time integration rule is used; each increment is relatively inexpensive (compared to the direct-integration dynamic analysis procedure available in Abaqus/Standard) because there is no solution for a set of simultaneous equations. The explicit central-difference operator satisfies the dynamic equilibrium equations at the beginning of the increment t; the accelerations calculated at time t are used to advance the velocity solution to time $t + \frac{\Delta t}{2}$ and the displacement solution to time $t + \Delta t$. Different analysis recommendations are listed below for cases of numerical difficulties:

1. Are smooth-step amplitudes being used to apply loads or boundary conditions for quasi-static analyses?
   An example of a smooth-step amplitude curve definition is

```
*AMPLITUDE, NAME=MY-SMOOTH-AMP, DEF=SMOOTH STEP
0., 0., 0.005, 1.
```

2. Can time scaling or mass scaling be used to speed up a quasi-static analysis of interest?
   If time or mass scaling, perform an energy balance check and verify that the kinetic energy, **ALLKE**, is a small fraction of the internal energy **ALLIE**. An example of variable mass scaling is

```
*VARIABLE MASS SCALING, DT=1.e-7, TYPE=BELOW MIN,
FREQUENCY=5
```

   Remember to verify that **ALLKE** is a small fraction of **ALLIE** when mass scaling or time scaling a quasi-static analysis. Users can contour plot the element mass scaling factor and the element stable time increment by requesting:

```
*OUTPUT, FIELD, NUMBER INTERVAL=15
*ELEMENT OUTPUT
EMSF, EDT
```

3. Can a general contact, with geometric feature edge specification if needed, be used?
An example of a general contact definition in Abaqus/Explicit is given as follows:

```
*SURFACE INTERACTION, NAME=GLOBAL-PROPERTY
*FRICTION
0.0,
*SURFACE INTERACTION, NAME=GRIP-ANVIL
*FRICTION
0.3,
...
*CONTACT
*CONTACT INCLUSIONS, ALL EXTERIOR
*CONTACT EXCLUSIONS
ANVIL, PUNCH
*CONTACT PROPERTY ASSIGNMENT
, , GLOBAL-PROPERTY
GRIP, ANVIL, GRIP-ANVIL
```

4. Check the initial contact state using data check analysis. Initial over closures for surfaces based on shell, membrane, and rigid elements can be reduced by modifying the surface thickness used for general contact if it is acceptable to use the adjusted geometry. An example of reducing the contact thickness by a scaling factor is given here:

```
*CONTACT
...
*SURFACE PROPERTY ASSIGNMENT, PROPERTY=THICKNESS
, ORIGINAL, 0.65
```

5. When using general contact and especially when modeling rigid to rigid body contact, look at a plot of the work done by contact penalties as the simulation progresses with contact penalty work **ALLPW** and make sure it is small compared to physically meaningful energy quantities.
Energy output is particularly important in checking the accuracy of the solution in an explicit dynamic analysis. In general, the total energy **ETOTAL** should be a constant or close to a constant; the "artificial" energies, such as the artificial strain energy **ALLAE**, the damping dissipation **ALLVD**, and the mass scaling work **ALLMW** should be negligible compared to "real" energies such as the strain energy **ALLSE** and the kinetic energy **ALLKE**.
In a quasi-static analysis, the value of the kinetic energy **ALLKE** should not exceed a small fraction of the value of the strain energy **ALLIE**.
It is a good practice to output the constraint penalty work **ALLCW** and the

contact penalty work **ALLPW** in analyses involving constraints (such as ties and fasteners) and contact. The value of these energies should be close to zero.

6. Use the built-in anti-aliasing filtering for history data output. An example of the built-in anti-aliasing filter is given below:

```
*OUTPUT, HISTORY, FILTER=ANTIALIASING,
TIME INTERVAL=7.e-5
*NODE OUTPUT, NSET=COMPONENT-REFS
U, V, A
```

7. To obtain better stress results for quasi-static analysis, filter the field output using a Butterworth filter with an appropriate cutoff frequency. An example of a field output filter definition is given here:

```
*FILTER, NAME=MY-FILTER, TYPE=BUTTERWORTH
2000.,
...
*OUTPUT, FIELD, FILTER=MY-FILTER, NUMBER INTERVAL=16
*ELEMENT OUTPUT
LE,
```

3.3 Causality Energy Method

To understand the causality energy method, it is first important to understand the difference between causality, causation, and correlation. Correlation refers to two or more factors behaving in the same manner, such as both increasing or decreasing simultaneously. However, this behavior is not necessarily due to the factors influencing each other. Causation, on the other hand, refers to a factor having a direct influence on the behavior of another factor. It is important to distinguish between the two because if the user makes a prediction based on the assumption of causation and the factors turn out to be only correlated then the user is likely to arrive at incorrect insights. Furthermore, the causality is more important than causation because causality is a relation, and not only a factor, between cause and effect. Therefore by using the causality energy method, the user will outline the response behavior inside a local area of a global model or the whole model response, in order to determine what is the aspect of the numerical solution which was computed not as expected or failed to compute. Once the causality will be known then the user will figure out what causation or correlation might be the reason for having this causality result. For instance, what is the factor or what might be a correlation of factors influencing the numerical solution difficulties.

The conservation of energy implied by the first law of thermodynamics states that the time rate of change of kinetic energy and internal energy for a fixed body of material is equal to the sum of the rate of work done by the surface and body forces. In contrast to dynamic change a static time increment has no physical sense, therefore the energy method is a method to estimate linear or nonlinear dynamic behaviors from quasi-static, implicit or explicit analyses. Causality and energy derivatives from the energy method can be used to correct or adjust a model by removing undesirable distortions and also to provide a first prediction from the output data computed.

The conservation of energy and the energy balance are general methods combined applicable for both static or dynamic problems, but in dynamic issues, there is a physical sense versus time. The energy balance is formulated in general, for example, the applied external work **W** is the sum of the internal strain energy **IE** plus the kinetic energy **KE** plus the dissipation effects made with other types of energy.

Abaqus standard has defined the total energy **ETOTAL**[27] available only for the whole model equal to

$$\begin{aligned}\mathbf{ETOTAL} =& \mathbf{ALLKE} + \mathbf{ALLIE} + \mathbf{ALLVD} + \mathbf{ALLSD} + \mathbf{ALLKL} + \mathbf{ALLFD} \\ &+ \mathbf{ALLJD} + \mathbf{ALLCCE} - \mathbf{ALLWK} - \mathbf{ALLCCDW}\end{aligned} \tag{3.8}$$

Abaqus explicit has defined the total energy **ETOTAL**[28] available only for the whole model equal to

$$\begin{aligned}\mathbf{ETOTAL} =& \mathbf{ALLKE} + \mathbf{ALLIE} + \mathbf{ALLVD} + \mathbf{ALLFD} + \mathbf{ALLIHE} - \mathbf{ALLWK} \\ &- \mathbf{ALLPW} - \mathbf{ALLCW} - \mathbf{ALLMW} - \mathbf{ALLHF}\end{aligned} \tag{3.9}$$

3.3.1 Basic Energy Approaches, Assumptions and Limitations

The energy methods are used to estimated impact behavior with the following preliminary considerations:

- According to the energy conservation principle of an isolated system, the energy is always conserved; thus, it is possible to combine static or dynamic analyses with energy methods to predict impact results.
- The internal energy contains elastic and inelastic energies, the summation of the internal energy **ALLIE** plus the dissipated through frictional energy **ALLFD** at

[27]As defined in Abaqus Analysis User's Guide v6.14 Sect. 4.2.1 Abaqus/Standard output variable identifiers paragraph Total energy output quantities.

[28]As defined in Abaqus Analysis User's Guide v6.14 Sect. 4.2.2 Abaqus/Explicit output variable identifiers paragraph Total energy output.

the peak event time, which should balance the kinetic energy **ALLKE** just before the impact.

- Apply static loading and boundary conditions until the energy balance has been reached (**ALLKE=ALLIE+ALLFD**) prior to impact. This statically deforms the structure into the expected dynamic deformation.
- The energy transfer assumption is typically a conservative assumption and is only effective for certain load cases. In general drop test problems, the structure will always have a mix of kinetic and stored internal energy throughout the impact event time.
- Higher frequency modes are ignored to focus only on the first deformation mode.
- Material strain rate sensitivity is ignored because using material properties data "at rate" in quasi-static simulations is not possible in a physical test.
- The use of inertial relief can be considered in certain simulations to avoid over constraints as a function of displacement with boundary conditions. Analysts must define a sufficient step of boundary conditions in static displacement to avoid rigid body motion.
- Understanding the correct static deformation mode to apply is not too difficult in some load cases and nearly impossible in other cases.

3.3.2 The Energy Method

Using energy plots in an analysis is a particularly helpful control check in the whole model to determine if an energy contribution is significant into the computed solution. For example, in Abaqus explicit, if after a certain time increment the **ALLKE** becomes significant compared with **ALLIE**, it can be a sign that the model is running too fast if the quasi-static method was used.

Checking the energy balance with **ETOTAL** in Abaqus explicit from Eq. 3.9 can be rewritten as

$$\mathbf{ETOTAL} = \sum(-\mathbf{ALLWK} + \mathbf{ALLIE} + \mathbf{ALLKE} + \mathbf{ALLFD} + \mathbf{Others}) \tag{3.10}$$

where the work energy **ALLWK** is moved to the same side as other quantities in **ETOTAL** shown in Eq. 3.10 are introduced to catch all quantities and should therefore remain constant throughout the entire solution equal to the zero machine. The model with initial kinetic energy will have large nonzero **ETOTAL** but it should still be constant over the solution time increments.

Total energy is used to find a problem inside an FEA model, demonstrating at what time increment the whole model solution begins to look suspicious. Second, the analyst will have to apply different analysis techniques to determine the anomaly areas inside the model causing this suspicious computed solution.

The time increment representing the applied load magnitude that looks suspicious can be easily identified in the model by plotting all energies according to the model physics and by checking where the total energy starts having a nonconstant zero machine value. This plot allows the analyst to ultimately find a bug within the model. If the total energy remains constant and equal to the zero machine during the entire solution time increment, then the solution has passed this sanity check and the analyst can be confident in the computed solution.

3.3.3 Energy Method Example to Scale Analyses

The energy method can also be powerful in some simple load cases to scale a static response from a model to obtain a dynamic result afterward. For example, for a simple drop test, a simple scaling equation to estimate impact behavior applicable for a linear analysis can be done, first by performing a static analysis on the elastic structure including the definition of some contact interactions if necessary. From static analysis, it is possible to extract static displacement computed at an impact point and get output data energies with Abaqus standard, for instance, the recoverable strain energy **ALLSE** (in steady-state dynamic analysis this is the cycle mean value) and the total energy dissipated through frictional effects **ALLFD**, which is only available for the whole model.

In linear analysis the total work is equal to

$$W = \frac{1}{2} F u \tag{3.11}$$

where F is the external force and u the displacement.

According to Eq. 3.11 and the static equilibrium Eq. 3.4 the static work is therefore equal to

$$W_s = \frac{1}{2} K u_s^2 = \mathbf{ALLSE} + \mathbf{ALLFD} \tag{3.12}$$

where u_s is the displacement in static solution and K is the global stiffness assembly of the whole model.

To scale the static response and thus obtain a dynamic result for a drop test, the equivalent total work in dynamic needs to be hand calculated as the sum of the potential energy function of the structure mass (m), the gravity acceleration (g) and the drop height (h) plus the kinetic energy function of the same structure mass (m) and the velocity (v) applied to the structure.

The equation of dynamic work seen as a dynamic energy prior to impact can be written as

$$W_d = \frac{1}{2} K u_d^2 = |\mathbf{PE}| + \mathbf{KE} = mgh + \frac{1}{2} m v^2 \tag{3.13}$$

where u_d is the displacement in dynamic solution.

Assuming that the structural stiffness is independent in displacement for both static and dynamic cases then it is possible to combine Eqs. 3.11 and 3.13, as follows:

$$\frac{1}{2}K = \frac{W_s}{u_s^2} = \frac{W_d}{u_d^2} \tag{3.14}$$

According to Eq. 3.14, it is now simple to link static and dynamic thanks to the energy method in order to scale a static displacement response to get a dynamic displacement result:

$$u_d = u_s\sqrt{\frac{\mathbf{|PE| + KE}}{\mathbf{ALLSE + ALLFD}}} \tag{3.15}$$

Of course, from the displacement value in linear analysis, it will be possible to return to the deformation and stress results in dynamic analysis.

In conclusion, the energy method is a powerful tool for analysts to ensure accurate simple analysis only applicable to a limited number of load case conditions to switch from static to dynamic responses.

3.3.4 Causality and Energy Derivatives

During a debugging process using the energy method, the analyst will need to determine how the energy will show the largest cause of the force while the structure is loaded and in corollary what materials, components, and so on. They have little influence on the structural response.

As defined in Eq. 3.16, the derivatives of external complementary work with respect to forces produce displacements like quantities. With respect to rotation, derivatives produce moments.

$$\frac{\partial \mathbf{ALLWK}}{\partial u} = F_{applied} \tag{3.16}$$

Let's assume model physics such as the work energy will be written only with:

$$\mathbf{ALLWK = ALLIE + ALLKE + ALLFD} \tag{3.17}$$

Consequently, the derivatives of energy relate directly to the force balance like for springs in parallel; indeed, as shown in Eq. 3.18 the force balance is the sum of all forces but for the same displacement given:

$$\frac{\partial \mathbf{ALLIE}}{\partial u} + \frac{\partial \mathbf{ALLKE}}{\partial u} + \frac{\partial \mathbf{ALLFD}}{\partial u} = \Phi_{IE} + \Phi_{KE} + \Phi_{FD} \tag{3.18}$$

Φ_{IE}, Φ_{KE} and Φ_{FD} are so-called the causality coefficients. These terms assess contributions to the overall structure response. They are used to determine what need to be changed to have the most influence on the structural response. They can also help to make estimated prediction equations for changes in quantities without the need to run the entire model again.

Energy derivatives are useful to investigate changes independently to other parameters and have a focus on specific expertise in physics modeling. An estimation of changes in the structure responses from energy derivatives can be made with the following interpretations:

- The derivative of internal energy **ALLIE** will indicate stiffness changes including changes in material or geometry; this derivative can be investigated to determine how component stiffness changes affect the structural response.
- The derivative of kinetic energy **ALLKE** can be investigated to determine how mass or velocity changes affect the structural response.
- The derivative of frictional dissipation energy **ALLFD** can be investigated to determine how the coefficient of friction changes affects the structural response.

These estimates of change or influence are most accurate if the structure's components behave as springs in parallel. Assessing energy derivatives for structures with components that act like spring in series (for instance, instead of making a derivation of energy a function of force rather than displacement) is also helpful, but more difficult and typically less accurate since obtaining complementary energy is often impossible and this factor must be estimated.

The energy balance must be satisfied, confirming the total energy is equal to a constant, which is the key sanity check for all models.

Assessing kinetic energy relative to work and internal energies is important for quasi-static and explicit dynamic models.

Energy methods can be effectively used to estimate impact behavior with quasi-static models or experiments. They are powerful techniques with some limitations.

Causality is derived from energy derivatives, with most accuracy for structural behavior similar to springs in parallel. This enables key components and materials to be computed from a single run and also allows prediction equations in both linear and nonlinear problems.

References

1. Flanagan DP, Belytschko T (1981) A uniform strain hexahedron and quadrilateral with orthogonal hourglass control. Int J Numer Methods Eng 17:679–706
2. Belytschko T (1976) Survey of numerical methods and computer programs for dynamic structural analysis. Nucl Eng Des 37:23–34
3. Hibbitt HD, Karlsson BI (1979) Analysis of pipe whip, EPRI, Report NP-1208

Chapter 4
General Prerequisites

4.1 Vocabularies

The first step in debugging analysis is to understand the meaning of error and warning messages, which are preprogrammed and therefore can be referenced. Tables 4.1 and 4.2 give a list of error and warning causes respectively, with some clues about the potential causes of troubleshooting seen as numerical issues or numerical difficulties. Definitions of the main causes of these error and warning messages are given below. Troubleshooting can be the sign of:

- **Excessive strain** increments means the current strain increment is so large that convergence of material point calculations is judged to be unlikely. Therefore, Abaqus will reduce the load and attempt to perform the increment again.
- **Large strain** increments mean the Abaqus strain criterion for the last increment exceeds "fifty" times the strain which caused the first yield. Thus, Abaqus will attempt to perform the material point calculations but convergence problems may result.
- **Negative eigenvalues** are often associated with a loss of stiffness or solution uniqueness such as might happen when a structure begins to buckle or a material becomes unstable.
  - Negative eigenvalues can also be associated with modeling techniques that make use of Lagrange multipliers to enforce constraints.
  - Negative eigenvalues warnings that pop up during iterations that do not converge can generally be ignored. If negative eigenvalues warnings appear during iterations that converge then the computed solution must be carefully evaluated.
- **Numerical singularities** typically result from rigid body motion, in which a portion of the model offers no resistance to the applied loads. A numerical singularity may indicate the need for additional boundary conditions or constraints in a portion of the model.

R. J. Boulbes, *Troubleshooting Finite-Element Modeling with Abaqus*,
https://doi.org/10.1007/978-3-030-26740-7_4

Table 4.1 List of error messages

Messages	Descriptions	Numerical issues
1	Too many increment	ZERO PIVOT warnings in the message file of my analysis
		Diverge with NUMERICAL SINGULARITY warnings
2	Too many attempts	The mesh of my Abaqus model has gone into a nonphysical shape with a regular pattern
		Source of convergence difficulty in the first increment of a contact analysis
		Correct convergence difficulties caused by local instabilities
		Using follower loads (including distributed pressures) in nonlinear analyses
3	Time increment	The mesh of my Abaqus model has gone into a nonphysical shape with a regular pattern
		Source of convergence difficulty in the first increment of a contact analysis
		Correct convergence difficulties caused by local instabilities
		Using follower loads (including distributed pressures) in nonlinear analyses
		Analysis ends prematurely, even though all the increments converged
		Explicit or quasi-static scheme stabilized convergence

Table 4.2 List of warning messages

Messages	Descriptions	Numerical issues
1	Strain increment	The current strain increment exceeds the strain to first yield mean
2	Element distortion	Excessive element distortion
		Modeling contact with second-order tetrahedral elements
		Understanding finite-sliding surface-to-surface contact
		How much hourglass energy is acceptable?
3	Negative eigenvalue	Using stiffness-based contact stabilization
		Model a structure undergoing a global instability
		Understanding negative eigenvalue messages
		Debugging divergence with too many cutbacks in the last attempted increment
		Explicit or quasi-static scheme stabilized convergence
		Job has completed, but the results look suspicious
4	Numerical singularity	Diverge with NUMERICAL SINGULARITY warnings
5	Zero pivot	ZERO PIVOT warnings in the message file of my analysis
6	Unconnected region	Identified unconnected region in your model

- **Zero-pivots** generally indicate an over constraint in the model, typically due to redundant boundary conditions or constraints. An over constrained node may still behave appropriately, but the presence of redundant constraints could be a modeling problem that leads to undesirable behavior in other portions of the model. A zero-pivot also sometimes arises due to rigid body motion.

For certain warning messages, error messages, and contact diagnostics, the nodes or elements which are involved in each diagnostic message can be seen using the Highlight selection in the viewport option. For warning and error messages, the nodes or elements causing the warning or error are highlighted in the model. For contact diagnostics, the nodes that are overclosing or opening are highlighted in the model.

4.1.1 Interpreting Error Messages

The interpretations of error messages listed in Table 4.1 are given below, with some recommendations and suggestions for corrective actions which can be made to the model.

1. **ERROR: TOO MANY INCREMENTS NEEDED TO COMPLETE THE STEP**
   - Check the message file for any warning messages, such as numerical singularity or zero-pivot warnings, that may be causing slow convergence.
   - If there appear to be no convergence issues, it may be necessary to increase the limit to the number of increments for that step.
2. **ERROR: TOO MANY ATTEMPTS MADE FOR THIS INCREMENT—ANALYSIS TERMINATED**
   - This is just an error message which explains why Abaqus finally aborted.
   - Do not modify the solver controls to increase the number of allowable attempts per increment.
   - Check the message file for warning messages that could cause convergence difficulties.
   - Check the model definition, and make sure that the model can actually withstand the applied loads.
3. **ERROR: TIME INCREMENT REQUIRED IS LESS THAN MINIMUM SPECIFIED-ANALYSIS ENDS**
   - Again, this is just an error message which gives a reason for why Abaqus finally aborted, it is not a suggestion by Abaqus to reduce the minimum allowable increment size.
   - Check the message file for warning messages that could cause convergence difficulties.
   - Check the model definition, and make sure that the model can actually withstand the applied loads.

4.1.2 Interpreting Warning Messages

The interpretations of error messages listed in Table 4.2 are given below, with some recommendations and suggestions for corrective actions which can be made to the model.

1. **WARNING: THE STRAIN INCREMENT HAS EXCEEDED FIFTY[1] TIMES THE STRAIN TO CAUSE FIRST YIELD AT 500 POINTS**
   - Indicates excessive plastic yielding in the given increment.
   - Possible causes would be:
     - Excessive or unrealistic loading (for example inconsistent units).
     - Incorrect or insufficient stress–strain plastic data.
     - Insufficient mesh refinement.
     - Unstable deformation, such as buckling.
2. **WARNING: ELEMENT 441 IS DISTORTING SO MUCH THAT IT TURNS INSIDE OUT**
   - Insufficient mesh refinement.
   - Hourglassing in first-order reduced integration elements.
   - Inconsistent units for material properties and/ or loads.
   - Excessive or unrealistic loads.
   - Adjusting slave nodes of severely overclosed surfaces for contact pairs or tie constraints (initial over closures).
3. **WARNING: THE SYSTEM MATRIX HAS 9 NEGATIVE EIGENVALUES[2]**
   - Some form of loss of stiffness suggesting that the stiffness matrix is assembled around a nonequilibrium state:
     - Geometrical instability: buckling, compression.
     - Material instability: inappropriate hyperelastic material models, onset of perfect plasticity.
   - Numerically, the use of Lagrange multipliers in certain cases may also lead to these warning messages.
   - Use of 3D second-order elements as contact (slave) surfaces.
   - Typically, these warning messages do not appear in converged iterations. If they do, ensure that the solution is physically acceptable.
4. **WARNING: SOLVER PROBLEM. NUMERICAL SIGULARITY WHEN PROCESSING NODE 1 D.O.F. 3 RATIO=3.141E+15[3]**
   - Typically suggests an unconstrained rigid body motion.
   - Even if the analysis runs to completion with these warnings, the results may not be accurate.

---

[1] Please note that the factor "fifty" mentioned in the warning message cannot be modified.

[2] See the Sect. 8.9 inside this book to get further explanations.

[3] See the Sect. 8.10 inside this book to get further explanations.

5. **WARNING: SOLVER PROBLEM. ZERO PIVOT WHEN PROCESSING NODE 1 D.O.F.1**[4]

   - Typically suggests an over constraint.
   - Even if the analysis runs to completion with these warnings, the results will not be accurate.

6. **WARNING: THERE ARE 2 UNCONNECTED REGIONS IN THE MODEL**

   - The message concerning unconnected regions simply means that there are isolated portions of the model that are not connected through common nodes, initially closed contact conditions, or constraint equations. If the unconnected regions have been introduced unintentionally, numerical problems such as unconstrained rigid body modes may result. Successful execution of the analysis might be prevented.

4.2 An Identified Unconnected Region in the Model

First, find the file named "abaqus_v6.env", normally installed in the work directory for Abaqus release, here for example with version v14-5 in the installation file C:\Simulia \Abaqus \6.14-5 \SMA \site.

Second, copy the file "abaqus_v6.env" in a backup folder to keep the original Abaqus environment settings with default mapping.

Then open the file "abaqus_v6.env" in the installation directory for modification and add the code lines below at the end line of the file.

unconnected_regions=ON

An example of piece of code is given below:

```
#       System-Wide Abaqus Environment File
#-------------------------------------------------
# DO NOT MODIFY THE CODE
#-------------------------------------------------
standard_parallel = ALL
mp_mode = MPI
mp_file_system = (DETECT,DETECT)
mp_num_parallel_ftps = (4, 4)
mp_environment_export = ('MPI_PROPAGATE_TSTP',
'ABA_CM_BUFFERING',
...
del driverUtils, os, graphicsEnv
license_server_type=FLEXNET
```

[4]See the Sect. 8.11 inside this book to get further explanations.

```
abaquslm_license_file="xxxx@xxxxxxx.net.xxx.com"
doc_root="http://xxxxxxx:xxxx/v6.14"
doc_root_type="html"
#-------------------------------------------------
# AT THIS POINT THE CODE CAN BE MODIFIED
#-------------------------------------------------
# Customized settings
unconnected_regions=ON
#
```

Run the model with the modified Abaqus environment file as shown above. Once the output data in (.odb) file is completed, the output file for visualization should be read in Abaqus viewport. Open the **Create display group** module before selecting the item to identify unconnected region nodes or elements and select the "Highlighted item in viewport" button to identify the unconnected region in viewport. Typically, the unconnected regions will have names such as "STEP 1 MESH COMPONENT 1".

- A technique used to identify unconnected regions is to perform a frequency extraction analysis on the model. This requires mass to be associated with all degrees of freedom. The unconnected regions will be easily identified by visualizing the rigid body mode shapes of the model in Abaqus/Viewer.
- An alternative approach using Abaqus/CAE is available. The first step is to import the model via the menu tab:
  File → Import → Model
  then select the input model file to be imported using the "Import Model" dialog box. If the model is not written as an assembly of part instances, i.e., if the input file is "flat", all flexible bodies in the model are collected into one part. The disconnected regions can then be identified by copying the single part containing the deformable elements. To do so click:
  Part → Copy
  once in the "Part" module, check the separate disconnected regions into the parts box in the "Part Copy" dialog box. The original part containing multiple disconnected regions will be split into separate parts, which makes it easy to identify the disconnected regions.
  The parts created during the copy process are listed in the "Part Manager": click **Part → Manager** and select the component inside the "Part" module. For example, if PART-1 is copied to PART-2, and PART-1 contains two unconnected regions, the Part Manager will list PART-2-1 and PART-2-2 as the newly created parts.
  If the model contains an assembly of part instances, importing it will create a model that has separate parts, one for each defined inside the input data. The technique of copying parts outlined above can still be used. For verification of the interaction between the different part instances, the "Interaction" module in Abaqus/CAE can be used to verify constraint and contact definitions.

Note that if the model contains regions joined by ***TIE** constraints, this will not be considered by Abaqus/CAE and these will subsequently be split by the "Part Copy" command as well.

- An unconnected region check is performed by Abaqus during preprocessing and the number of unconnected regions is printed as a warning in the message file. During the analysis, the use of the ***MODEL CHANGE** option may create unconnected regions. Unconnected regions are not necessarily troublesome, as long as they are constrained properly. If unconnected regions are unconstrained, then a static analysis will cause numerical singularity warnings to be printed in the message file.

  In a frequency analysis, the unconstrained, unconnected region of the model will be associated with zero stiffness rigid body modes, which may produce unexpected results. Abaqus presently issues warnings for unconnected, unconstrained regions for ***STATIC** analysis, but not for ***FREQUENCY** analysis. This is why it is not a bad idea for a dynamic modal model using ***FREQUENCY** option to be tested in static analysis first in order to detect unconnected regions.

4.3 Correction of Errors During the Data Check Phase of an Abaqus/Standard Analysis

The data check phase of an analysis ascertains whether or not the model is set up correctly. Abaqus cannot make sure that the physics of the model are correct; instead, it will check as much as possible that the model makes sense. When trying to correct data check errors, consider the following points.

Make sure the system of units is consistent with the set of units in the model under analysis. Abaqus does not use any built-in units.

A common error is to use the incorrect mass units for density in dynamic analyses, especially when using Imperial units. On the inside cover of every printed Abaqus manual and at the beginning of the online manuals there are tables that can be used for unit conversions. Table 3.1 shows the correct consistent system of units to use for both metric and Imperial systems. Using units that make numbers very large or very small can cause round-off problems during the analysis in rare cases, therefore the unit system and input values need to be monitored very carefully before processing.

Check the geometry, boundary conditions, loadings, surface definitions, and material properties to make sure they are correct and reasonable.

- The printout to the data (.dat) file in the data check phase is switched off by default. It is wise to switch this printout on, at least until it is certain that the model has been set up correctly. If the model has not been set up correctly, it is useful to check how Abaqus has interpreted the input data by looking at the data file. Switch on this printout by using:

– Job Module: **Job** → **Create...** → **Continue...** → **General** → **Preprocessor** then **Printout** and request printout of the input data, model definition data and history data.
– Keyword:

```
*PREPRINT, MODEL=YES, HISTORY=YES
```

- Material data and section property data are most easily checked in the .dat file. There is an environment file setting, **printed_output=ON**, that can be used to make this printout the default. Note however that this can sometimes cause the memory estimates for the problem to become erroneously large.
- If user has contact defined in the model to check the initial state of contact (open or closed and by how much), use:

  – Job Module: **Job** → **Create...** → **Continue...** → **General** → **Preprocessor** then **Printout** and request printout of the contact constraint data

- Keyword:

```
*PREPRINT, CONTACT=YES
```

- Unless there is a serious error in the input file, Abaqus will write information to the output database (.odb) file and the restart (.res) file (if restart output option is requested). Use Abaqus/Viewer to read the output database file in order to check the model graphically.
  Look at the model to make sure that the geometry is correct, the mesh is reasonable, the boundary conditions are applied correctly, the loads are in the correct places, and the surfaces are defined correctly. If contact has been defined in the model, checking the contact pairs in Abaqus/Viewer is fairly straightforward. In addition, check the normals to contact surfaces; if these are pointing in the wrong direction, the contact algorithm will try to apply this incorrect definition and probably cause severe distortion in the mesh.

If the error message is **THERE IS NOT ENOUGH MEMORY ALLOCATED TO PROCESS THE INPUT DATA** then read the Abaqus documentation inside appropriates sections, for instance, the "Using the Abaqus environment settings" section inside the Analysis User's Guide and the "Memory and disk management parameters" section inside the Installation and Licensing Guide, in order to understand memory usage in Abaqus/Standard functions of the machine configuration.

Particular analyses can use large amounts of memory in the data check phase as well as the analysis phase even if the number of degrees of freedom does not seem especially large. These analyses include:

- Finite-sliding contact problems with many nodes on the contact surfaces and a large SLIDE DISTANCE setting (refer to the ***CONTACT CONTROLS** option): a reasonable slide distance is automatically computed for three-dimensional finite-sliding problems.

- Cavity radiation problems with cavities that are finely meshed. It is strongly recommended to mesh the cavity more coarsely than the surrounding mesh. This will involve tying the heat transfer elements defining the cavity to the finer surrounding mesh using tie constraints or MPC.
- Problems in which equations or multi-point constraints are used to constrain a large number of nodes to a single node. If possible, use another technique to tie the nodes together.
- Substructures with large numbers of retained degrees of freedom. If possible, reduce the number of retained degrees of freedom.

Only free format is allowed, so this will not be a problem. If a line has only one data item on it, this data item must be followed by a comma. For more information see the "Input syntax rules" section inside the Analysis User's Guide.

It is easy to convert the input file from a version to version. The first step is to run the Abaqus free utility before running the Abaqus upgrade utility. For more information see the "Fixed format conversion utility" and the "Input file and output database upgrade utility" sections inside the Analysis User's Guide.

When the data check phase has been performed, review the data file for any warning messages. Make sure there is a good reason to ignore any warnings before continuing. The warnings are helpful for users to create good models and often they indicate serious errors. Sometimes they can be ignored. It is vital to understand the implications of ignoring any warning messages.

4.4 Tips and Tricks for Diagnostic Error Messages

The main tips and tricks for bypassing some numerical difficulties during the diagnostic phase are summarized as follows:

- Make sure that all assumptions made will represent the physics covered by the model and make sure all settings have been correctly defined with the proper values.
- To find the causes of an instability requires a step by step procedure in order to isolate the factor which has caused the unconverged solution by first removing all types of nonlinearity in the model (Large displacement, material properties, contact properties). After this, all nonlinearity functions of the settings and constraints of the model are individually investigated.
- In general, the large displacement **NLGEOM** does not help to stabilize the solution but it makes the iteration compute faster with fewer increments.
- Switch in module step, the analysis step where a numerical problem is returned from "Static" to "Dynamic implicit" can help the iteration to converge.
- If a beam element is used into the model as the bolt, first mesh it with three elements in order to define the bolt pretension at the middle element. Second, select the beam formulation with the hybrid formulation called B31H which is a better element for

a long and slender model, otherwise, the risk of getting a warning for a numerical singularity will be high.

- If the beam elements used to model a bolt preload are not working when there are large rotations in the model, it is recommended to use a strategy to model the bolt with MPC connectors or a mixed elements MPC/beam instead.
- To avoid wrong contact interactions be aware that the option “Finite sliding” is used in contact interaction in order to update the contact during the calculation. That is not the case for a “Small sliding”. Essentially, “Small sliding” can be used if the contact is not sliding more than one third of the typical element in contact size, and the contact status is done but has not updated during the calculation.
- Unconverged solutions regarding material plasticity can be avoided by extending material physic linearly up to the final calculated slope of the Stress–Strain curve, even to the extent reaching 100% strain in the material curve in order to allow solver computation to calculate the final increment. The analyst will be responsible for knowing where to make a stop sense with the computed solution.
- Quadratic element formulation should not be used for contact interaction to avoid numerical difficulties at the middle node.

4.5 Trying to Recover a Corrupted Database

Two procedures are described in this section in order to recover a corrupted Abaqus file. Choose procedure 1 or procedure 2 depending on what is the Abaqus message.

4.5.1 Procedure 1

Abaqus CAE has returned the following error message “**SMACkmCaeKerMod** has stopped working”, as shown in Fig. 4.1 which is generic and does not say anything specific about the causes of the crash. It is likely that the CAE file was corrupted and is impossible to recover. If Abaqus/CAE has crashed with the following message, follow procedure 1.

1. Make a backup: first copy and paste the (.cae) and (.jnl) files into a backup directory.
2. Open the Abaqus/CAE file inside the current working directory and do a: “**File → Compress mdb**”. If it is not possible to work again normally with the (.cae) file afterward then stop the procedure at this point and try procedure 2; otherwise, proceed to the next step.
3. Once the (.cae) file is opened, import the model “**File → Import model**” and select and import the model from the old (.cae) model. If this is not possible then stop this procedure and attempt procedure 2.

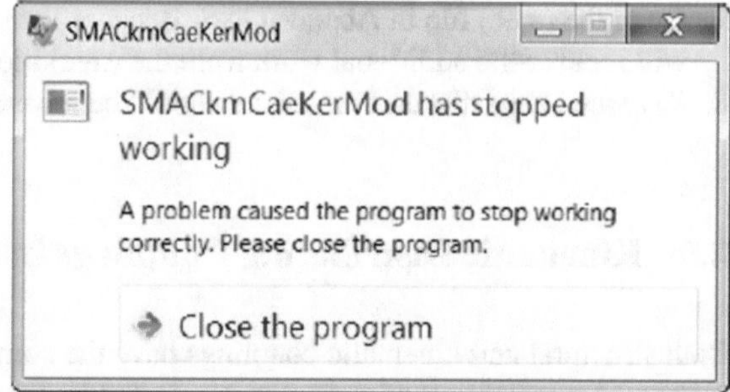

Fig. 4.1 Crash message **SMACkmCaeKerMod**; this is an executable file running in Task Manager as the process "abqcaek.exe"

4. At this point, if this procedure has failed to get back the model in the Abaqus/CAE file, perhaps the (.cae) database is corrupted at a higher level. Therefore to recover a model from a highly corrupted (.cae) file follow procedure 2.

4.5.2 Procedure 2

The following procedure can be used to recover a highly corrupted Abaqus/CAE database.

1. Copy all the files from the current working directory to a backup directory. The working directory is the directory from which Abaqus/CAE was running, and this directory will contain the following file extensions: (.cae), (.jnl), (.rpy), and (.rec).
2. Operate safely now in the current working directory to delete all the files except the journal file (.jnl). The (.jnl) file may have references (imported files used to create the model, etc...) which need to be at the same location as earlier. If the location of these files is uncertain then an error message will pop up during recovery to guide the user. Alternately, the location of the journal file can be identified with a manual scanning operation.
3. Run the following command inside the current working directory with the command prompt **abaqus cae recover=model-name.jnl**, where Abaqus is the command to run Abaqus and model-name is the name of the Abaqus/CAE database.
4. Once the model has been recovered, save the file as model-name.cae and quit the Abaqus/CAE session.
5. There may also be a recover file inside the backup directory. This will contain operations that were carried out after the journal file was created. If the user wants to recover this work too, copy the (.rec) file from the backup directory to the current working directory and it should have the name model-name.rec.
6. Edit the (.rec) file to remove the last set of commands (this is model dependent and by removing these commands it is assumed that the last set of commands caused the database corruption).

7. Open the (.cae) file in Abaqus/CAE. Because there is a (.rec) file, Abaqus/CAE will recover the additional work from the (.rec) file.
8. Save the (.cae) file, and the model should be recovered.

4.6 Kinematic Distributing Couplings in Abaqus

Both structural and kinematic couplings have the common purpose of coupling the motion of a collection of nodes on a selected surface (the coupling nodes) to the motion of a reference node. Some differences between the two methods are outlined in this section.

4.6.1 Nature of the Constraint Enforcement

- The kinematic coupling is enforced in a strict master–slave approach. Degrees of freedom (DOFs) at the coupling nodes are eliminated, and the coupling nodes will be constrained to move with the rigid body motion of the reference node.
- Distributing coupling is enforced in an average sense. DOFs at the coupling nodes are not eliminated. Rather, the constraint is enforced by distributing loads such that:
  - the resultants of the forces at the coupling nodes are equivalent to the forces and moments at the reference node, and the force and moment equilibrium of the distributed loads around the reference point is maintained.

A kinematic coupling constraint does not allow relative motion among the constrained DOFs. However, it does allow relative motion among the unconstrained DOFs.

A distributing coupling allows relative motion among the constrained and unconstrained DOFs. The relative motion of the coupling nodes will be such that the equilibrium condition on the distributed loads is maintained.

As an example, consider the cantilever beam shown in Fig. 4.2. It is meshed with second-order brick elements and is built-in at the right end. A coupling constraint is defined at the free end. Degrees of freedom 1 through 6 of the end surface nodes are included in the constraint. At the reference node, a displacement is applied in the vertical direction, while all other displacement and rotation components are held at zero.

The model is analyzed using the kinematic and distributing constraint methods; the deformed shape of the beam for both constraint types is similar, and is demonstrated in Fig. 4.3.

A closer inspection of the displacements at the coupled end of the beam reveals the difference between the results of the two constraint methods. Figures 4.4 and 4.5

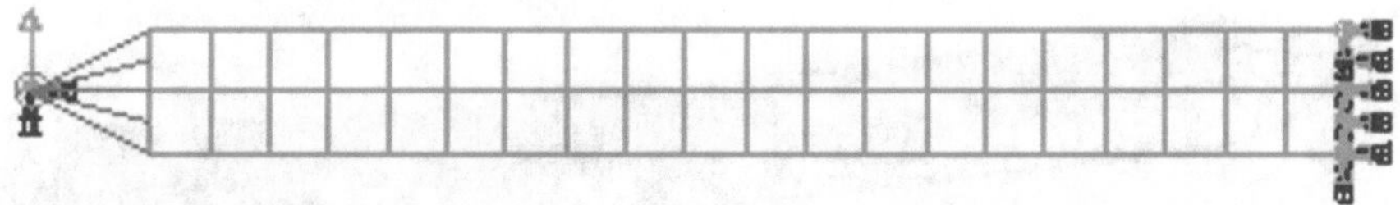

Fig. 4.2 Cantilever beam, undeformed, loaded with a vertical displacement

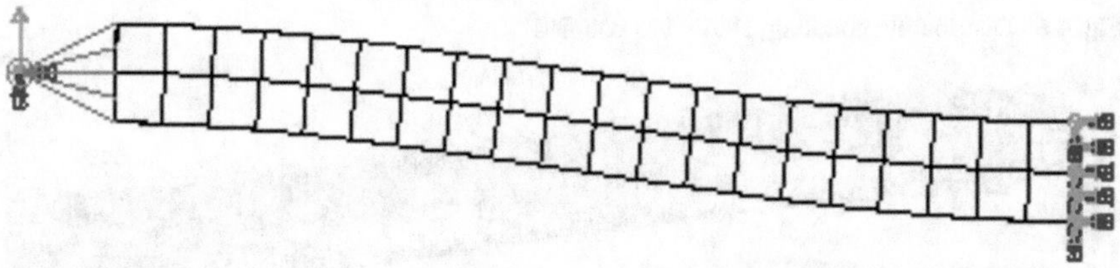

Fig. 4.3 Cantilever beam, deformed, loaded with a vertical displacement

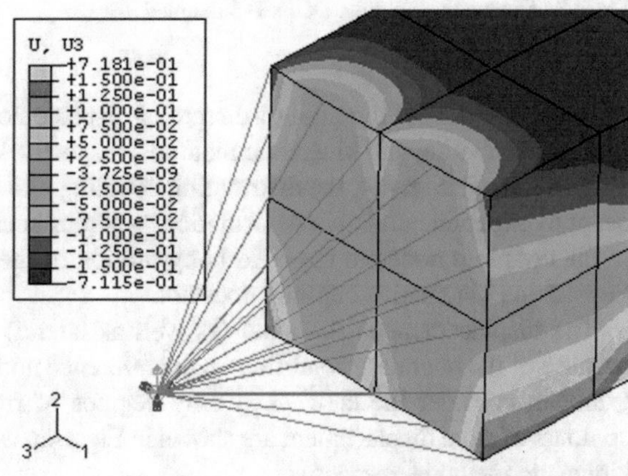

Fig. 4.4 Axial displacement, distributing coupling

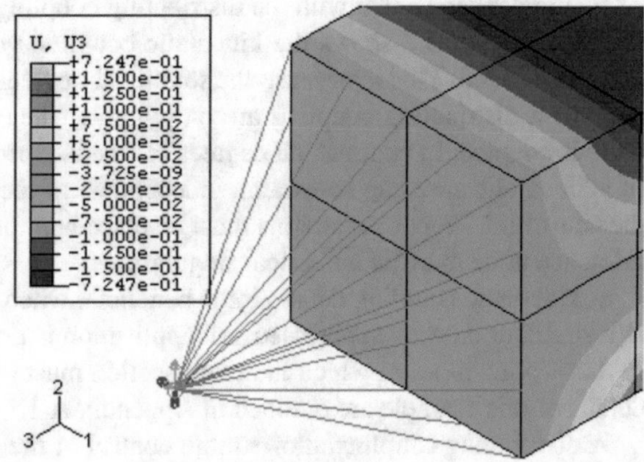

Fig. 4.5 Axial displacement, kinematic coupling

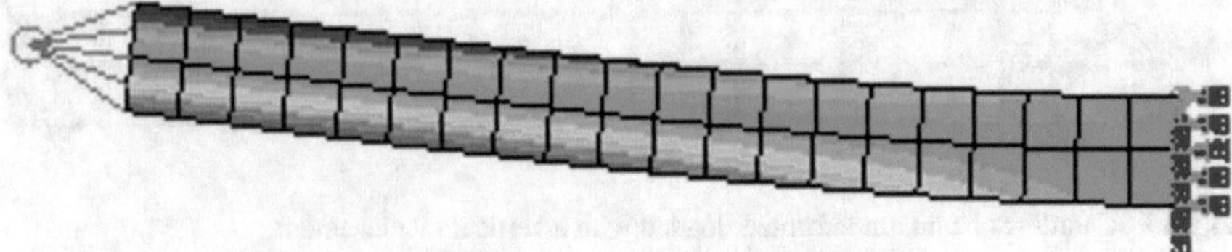

Fig. 4.6 Distributing coupling, DOFs 1–3 coupled

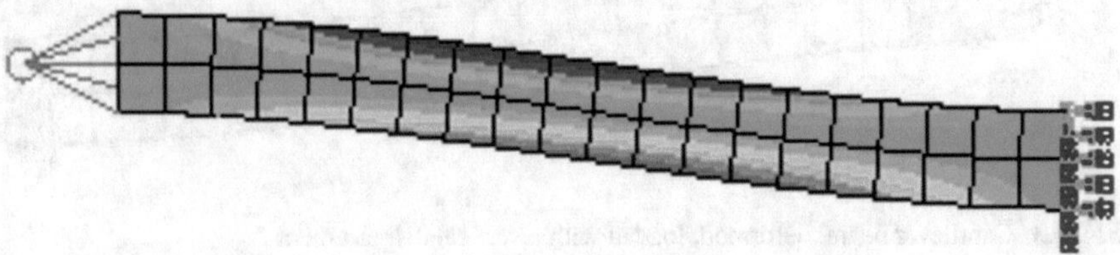

Fig. 4.7 Kinematic coupling, DOFs 1–3 coupled

show the contours of the axial (3-direction) displacement at the free end of the beam for the distributing and kinematic methods, respectively.

As shown in Fig. 4.4, the distributing coupling allows the nodes at the end of the beam to experience relative deformation, whereas because it constrains the motion of the coupling nodes to the rigid body motion of the reference node as shown in Fig. 4.5 the kinematic coupling, does not.

The displacement in the axial (as well as lateral) direction is identically zero because of the boundary condition on the reference node. Continuing with the above example, consider the case when only degrees of freedom 1–3 are coupled: the contours of axial displacement are shown in Figs. 4.6 and 4.7 for the distributing and kinematic methods, respectively.

Figure 4.6 shows that with the distributing coupling, the end of the beam is free to rotate. Figure 4.7 shows the kinematic coupling with the rotation of the beam end constrained. This is because the rotational degrees of freedom at the reference node of a distributing coupling are only active when at least one slave rotational DOF is coupled. In contrast, all degrees of freedom are active at the reference node of a kinematic coupling constraint, independent of the slave DOFs participating in the constraint. Proper constraints must be placed on the unconstrained DOFs of the reference node to avoid numerical singularities.

A kinematic coupling constraint is beneficial when a particular kinematic mode in a structure must be suppressed. An application is the simulation of pure bending in thin-walled pipes in which the cross section must ovalize but remain plane. The details of this example are outlined in Appendix A.1.

A distributing coupling allows more control of the distribution of load from the reference node to the coupling nodes. In addition to a uniform distribution, distributions can be made to decrease linearly, quadratically, or cubically in terms of distance from the reference node. No control is provided with the kinematic method.

Distributing coupling must always constrain all available translational degrees of freedom at the coupling nodes. This is not necessary with the kinematic coupling constraint.

Once any combination of degrees of freedom at the coupling nodes is specified in a kinematic coupling constraint, none of the remaining degrees of freedom are available for further constraint. This is not the case for the distributing coupling. Additional details are provided in Appendix A.2.

For either type of constraint, concentrated loads or displacements can be applied at the reference node. A large number of coupling nodes in a distributing coupling definition can cause excessive memory usage and long run times; this is a result of the large wavefront that is produced when forming the constraint.

4.6.2 Defining Constraints in Abaqus/CAE

Coupling constraints are defined in the Interaction module of Abaqus/CAE. First define the surface to be coupled. Then, if the reference node is not part of the existing geometry, a separate reference node must be created using **Tools → Reference Point**. Then select:

Constraint → **Create...** → **Coupling** → select the reference node → select the coupling surface.

The Edit Constraint dialog box will appear, from which the constraint type and coupled degrees of freedom can be selected.

4.7 Abaqus Geometric Nonlinearity

The option in the "Step" module to select nonlinear geometry is **NLGEOM**; put simply, the nonlinear geometry is an option to calculate the response to a changing model shape when large deformations exists and provide nonlinear changes in the components' stiffness. But what does a large deformation mean exactly?

In continuum mechanics, stress is a physical quantity that expresses the internal forces that neighboring particles of a continuous material exert on each other, while strain is the measure of the deformation of the material. Therefore, the strain is a consequence of a deformation, and the deformation is a consequence of the structure loaded. The relationship between mechanical stress, deformation, and the rate of change of deformation can be quite complicated, although a linear approximation may be adequate in practice if the quantities are small enough (known as small deformation). Here, there is a choice to make between small deformation **NLGEOM=NO** or large deformation **NLGEOM=YES**. But how can this choice be made and what effect will it have on the results?

The previous section has highlighted how the different kinematic distribution coupling options affect a structure; however, the kinematics are a function of the type of element used in the model. The kinematic or so-called the defined degree of freedom. For example, a solid 3D element has no rotation and just three displacements in three directions, whereas a beam element has three rotations and three displacements in a 3D space and thus a total of six degrees of freedom in its kinematics. Consequently, all are linked with the kinematics, for a small or large deformation, the material response as a function of deformation and therefore the structural stress response.

According to equilibrium and virtual work for both Abaqus standard and explicit solver, the Cauchy stress matrix must be symmetric and therefore the strain matrix too because according to continuum solid mechanics theory there is a relationship between stress and strain.

According to matrix algebra, Eq. 4.1 shows that any square matrix (here a matrix called "A") can be broken into the sum of a symmetric matrix (the strain matrix) and an antisymmetric matrix (the skew matrix) respectively.

$$A = \frac{1}{2}[A + A^T] + \frac{1}{2}[A - A^T] \tag{4.1}$$

According to Eq. 4.1, the strain matrix in a small deformation used with the option **NLGEOM=NO** can be immediately written like the sum of a small deformation matrix and a small skew matrix:

$$\varepsilon_{small} = \frac{1}{2}\left[\frac{\partial \mathbf{u}}{\partial \mathbf{x}} + \left(\frac{\partial \mathbf{u}}{\partial \mathbf{x}}\right)^T\right] + \frac{1}{2}\left[\frac{\partial \mathbf{u}}{\partial \mathbf{x}} - \left(\frac{\partial \mathbf{u}}{\partial \mathbf{x}}\right)^T\right] \tag{4.2}$$

When a large deformation is activated with Abaqus **NLGEOM=YES** then Eq. 4.2 becomes a large strain matrix in a large deformation equal to the sum of a large deformation matrix, also called the Green Lagrangian strain tensor and still a small spin matrix:

$$\varepsilon_{large} = \frac{1}{2}\left[\frac{\partial \mathbf{u}}{\partial \mathbf{x}} + \left(\frac{\partial \mathbf{u}}{\partial \mathbf{x}}\right)^T\right] - \frac{1}{2}\frac{\partial \mathbf{u}}{\partial \mathbf{x}}\left(\frac{\partial \mathbf{u}}{\partial \mathbf{x}}\right)^T + \frac{1}{2}\left[\frac{\partial \mathbf{u}}{\partial \mathbf{x}} - \left(\frac{\partial \mathbf{u}}{\partial \mathbf{x}}\right)^T\right] \tag{4.3}$$

From Eq. 4.3 the large deformation includes a second-order corrective term added into Eq. 4.2 with small deformation, making the large deformation result less conservative than with the small deformation.

Another observation from Eq. 4.3 is that a large deformation does not necessarily mean having a large rotation since it can be limited to rigid body rotations. For components undergoing finite rotations, large deflection effects must be included in order to correctly calculate and combine the rigid and flexible deformations to accurately predict the impact of the updated geometric stiffness on the solution. As a general rule of thumb, if the angular rotation is greater than 10 degrees, large deformation effects should be included.

Large strain implies the change in shape at the element level, such that individual elements are stretched, squeezed, or sheared in such a way that the final element shape is significantly different from the initial shape. From a practical perspective, large strain effects almost always require the use of a nonlinear material representation which can accurately model the material behavior in the finite strain regime, simply the use of a nonlinear structure deformation representation, or the use of both forms of representation.

A brief description of stress stiffening is that it is technically not a geometric nonlinearity. The effects of stress stiffening can be significant, and often it is lumped together with geometric nonlinearities for convenience. Generally speaking, stress stiffening is the increase (or decrease) in transverse stiffness when a long, thin structure is loaded in tension (or compression) along an axial direction, producing a membrane stress. A thin sheet metal part, a spinning blade or a violin string are some examples. Even when strains and rigid-body motions are small, significant stiffening (or softening) of the structure can still occur due to membrane stresses, which may play an important role in the accuracy of the analysis. Stress stiffening is typically important for structures that are small in at least one dimension and/or subject to significant axial or membrane stress. For these types of problems, if deformations and rotations are small, nonlinear geometric effects can be ignored, and stress stiffening is all that is required to reach a correct solution. However, from a practical perspective, since large deformation effects also include the effects of prestress, most analysts simply turn large deformation on and do not consider stress stiffening independently.

In some large displacement problems, the decision to also include stress stiffening with large deflection active becomes a convergence tool. In many cases, a large deflection solution with stress stiffening turned on converges faster than when it is not activated. Although the resulting answers are identical, turning on prestress helps the solution converge more effectively. This works both ways with sometimes where deactivating stress stiffening helped to get a nonlinear dynamic buckling beam model to converge more effectively.

The following are some directions to determine whether the analysis should run with small or large deformations:

- Even if the nonlinear effect is very small, invoking the nonlinear solver will give to user a significantly longer solution time. This is not an issue for small models, but when the model contains several million degrees of freedom, a reduction of the solution time by a factor of two can be hugely beneficial.
- Sometimes the analyst wants to be able to compare with an analytical solution and such solutions are often based on linear theory, so use a small deformation in such case.
- The analyst needs to follow a standard or analysis procedure where a linear or nonlinear approach is assumed.
- In a geometrically nonlinear problem, it is necessary to use the actual load. If the analyst wants to do a conceptual study of a structural response, the solution may not converge if the estimated load was too large.

4.8 Differences Between Implicit and Explicit Schemes

The main difference is regarding the numerical scheme pattern [1] used to calculate a derivative term. In static analysis, there is no effect of mass (inertia) or of damping. In dynamic analysis, nodal forces associated with mass/inertia and damping are included. Static analysis is conducted using an implicit solver in Abaqus standard. Dynamic analysis can be done via the explicit solver or the implicit solver in Abaqus explicit.

In nonlinear implicit analysis, the solution for each step requires a series of trial solutions (iterations) to establish equilibrium within a certain tolerance. In explicit analysis, no iteration is required as the nodal accelerations are solved directly.

The time step in explicit analysis must be less than the current time step, because the time in explicit analysis has a true physics meaning related with the smallest length meshed which is the time takes a sound wave to travel across an element. At this point, the analyst must pay serious attention to the mesh size and identify the smaller element size in the model. Implicit transient analysis has no inherent limit on the size of the time step. As such, implicit time steps are generally several orders of magnitude larger than explicit time steps.

Implicit analysis requires a numerical solver to invert the stiffness matrix once or even several times over the course of a load/time step. This matrix inversion is an expensive operation, especially for large models. Explicit analysis does not require this step.

Explicit analysis handles nonlinearities with relative ease compared to implicit analysis. This would include treatment of contact and material nonlinearities.

In explicit dynamic analysis, nodal accelerations are solved directly (not iteratively) as the inverse of the diagonal mass matrix times the net nodal force vector, where the net nodal force includes contributions from exterior sources (body forces, applied pressure, contact, etc.), element stress, damping, bulk viscosity, and hourglass control. Once accelerations are known at time n, velocities are calculated at time $n + \frac{1}{2}$, and displacements at time $n + 1$. From displacements comes strain, and from strain comes stress. The cycle is then repeated.

The performances of both solvers in Abaqus standard versus Abaqus explicit can be listed as follows:

1. Abaqus/Standard
    - Uses implicit time integration to calculate the transient dynamic or quasi-static response of a system.
    - Three application types:
        - dynamic responses requiring transient fidelity and involving minimal energy dissipation;
        - dynamic responses involving nonlinearity, contact, and moderate energy dissipation;

  - quasi-static responses in which considerable energy dissipation provides stability and improved convergence behavior for determining an essentially static solution.

2. Abaqus/Explicit
   - Uses an explicit time integration scheme to calculate the transient dynamic or quasi-static response of a system.

4.8.1 Equations for Dynamic Problems

Abaqus rewrites Eq. 3.7 into the equation form Eq. 4.4, i.e., the finite-element (time continuous) approximation to the equilibrium equations is written with the inertia forces isolated, where "M" is the mass matrix (assumed constant in time), ($\ddot{u}$) is the nodal accelerations component vector, "I" are the internal forces at the nodes (the distribution of stress in the elements defines the equivalent nodal forces), and "P" are the external nodal forces.

$$M\ddot{u} + I - P = 0 \tag{4.4}$$

"I" and "P" are assumed to depend only on (u) and ($\dot{u}$) which are the nodal displacements and velocities component vectors, respectively, so the system is second order in time.

These equilibrium equations are very general. They apply to the behavior of any mechanical system. And/or when, the inertial or dynamic force is small enough during the first term, the equations reduce to the static form of equilibrium.

Viscous effects, such as damping, viscoplasticity, or viscoelasticity, are included in the internal force vector "I", which is a function of displacements (u) and velocities ($\dot{u}$) components vectors.

4.8.2 Time Integration of the Equations of Motion

Nonlinear dynamics problems require direct integration of the equations of motion. Integration is computed first with the spatial discretization (finite-element approximation), which turns the partial differential equations describing the dynamic equilibrium into a set of coupled, nonlinear, ordinary differential equations in time. Then, the time integration is needed to solve this system of ordinary differential equations. Finally, the methods used to integrate these equations through time distinguish

Abaqus/Explicit and Abaqus/Standard. The difference between both solvers are listed below, for instance, the time integration is computed in accordance with[5]:

1. Abaqus/Standard
   - Uses a second-order accurate, implicit scheme called the Hilber–Hughes–Taylor (HHT) [2] rule unless the application type is quasi-static.
   - This is a generalization of the Newmark method.
   - Second-order accurate means the scheme integrates a constant acceleration exactly.
   - The method is unconditionally stable: any size time increment can be used and the solution will remain bounded.
2. Abaqus/Explicit
   - Uses a second-order accurate, explicit integration scheme to calculate the transient dynamic or quasi-static response of a system.
   - The method is conditionally stable: it gives a bounded solution only when the time increment is less than a critical value.

4.8.2.1 Implicit Dynamics Time Integration

As shown with Eq. 4.5, the march forward in time with implicit time integration is computed by utilizing a known nodal solution of displacement, velocities, acceleration component vectors iterated into the next time increment (Δt) to find the solution.

$$\begin{pmatrix} u_n \\ v_n \\ a_n \end{pmatrix} \rightarrow \Delta t \rightarrow \begin{pmatrix} u_{n+1} \\ v_{n+1} \\ a_{n+1} \end{pmatrix} \tag{4.5}$$

As a consequence of the numerical logic in Eq. 4.5, implicit schemes solve a nonlinear system of equations for each time increment using equation solver and Newton iterations (as is the case for statics) and a consistent mass matrix.

The time integrators used by Abaqus/Standard are unconditional stability, meaning that the time increment size is governed by convergence rate and accuracy.

Compared to explicit time integration, there is a higher cost per increment but fewer increments (larger Δt) and possibility a lack of convergence (residual tolerances in effect).

[5] More details about the time integration scheme are presented inside Abaqus documentation Theory Manual, in Sect. 2.2.1 Nonlinear solution methods in Abaqus/Standard and in Sect. 2.4 Nonlinear dynamics

4.8.2.2 Explicit Dynamics Time Integration

The march forward in time using the central difference method is computed from a known nodal solution of displacement and acceleration component vectors at time n with the velocity components vector at time $n - \frac{1}{2}$ this solution vector is going to be iterated with the next time increment Δt. Note how the order sequence of calculation components vector has changed to find the solution after the time increment Δt; the primary focus of the increment is to find the new net force $f\,(t_{n+1}; u_{n+1})$ as shown in Eq. 4.6.

$$\begin{pmatrix} u_n \\ v_{n-\frac{1}{2}} \\ a_n \end{pmatrix} \rightarrow \Delta t \rightarrow \begin{pmatrix} v_{n+\frac{1}{2}} = v_{n-\frac{1}{2}} + a_n \Delta t \\ u_{n+1} = u_n + v_{n+\frac{1}{2}} \Delta t \\ a_{n+1} = \frac{1}{m} f\,(t_{n+1}; u_{n+1}) \end{pmatrix} \tag{4.6}$$

As a consequence of the numerical logic in Eq. 4.6, there is no matrix inversion (lumped mass) reducing the computational time to ensure each increment calculation is very fast. For example, a one second analysis time/increment for a model with two million elements can be performed.

On the other hand, the conditional stability required for a small-time increment "Δt" carry out many increments calculation. For example, 100 000 increments for a 0.1 second event can be required. The gain regarding computational time is real but need to be compensated with a huge allocation of RAM memory and hard drive disk space available.

4.8.3 Automatic Time Incrementation with Abaqus Standard

The incrementation scheme depends on the dynamic application type. Abaqus standard contains three different options:

1. `*DYNAMIC, APPLICATION=TRANSIENT FIDELITY`

    - Transient fidelity applications (default for models without contact).
    - Require minimal energy dissipation.
    - Small-time increments required to accurately resolve the vibrational response of the structure, and numerical energy dissipation is kept at a minimum.

2. `*DYNAMIC, APPLICATION= MODERATE DISSIPATION`

    - Moderate dissipation applications (default for models with contact).
    - A moderate amount of energy is dissipated by plasticity, viscous damping, or other effects.
    - Accurate resolution of high-frequency vibrations is usually not of interest.
    - Improved convergence for analyses involving contact.

3. *DYNAMIC, APPLICATION=QUASI-STATIC

- For quasi-static applications, these problems typically show monotonic behavior and inertia effects are introduced primarily to regularize unstable static behavior.

The default parameter settings applied on the Abaqus standard solver or to be modified in accordance with numerical difficulties regarding the numerical implicit options used are shown in Table 4.3. These parameters α, β and γ, can be adjusted or modified individually if the Hilber–Hughes–Taylor operator is employed.

If the default settings of these parameters correspond to the transient fidelity settings shown in Table 4.3 and need to be explicitly modified by the α parameter alone, the other parameters will be adjusted automatically to $\beta = \frac{1}{4}(1-\alpha)^2$ and $\gamma = \frac{1}{2} - \alpha$.

This relation provides control over the numerical damping ξ associated with the time integrator while preserving desirable characteristics of the integrator. The numerical damping grows with the ratio of the time increment to the period of vibration of a mode. Negative values of α provide damping, whereas $\alpha = 0$ results in no damping (energy preserving) and demonstrates the trapezoidal rule[6] exactly (sometimes called the Newmark β-method, with $\beta = \frac{1}{4}$ and $\gamma = \frac{1}{2}$).

The setting $\alpha = -\frac{1}{3}$ provides the maximum numerical damping. It gives a damping ratio of about 6% when the time increment is 40% of the period of oscillation of the mode being studied.

Allowable values of α, β and γ are: $\left(-\frac{1}{2} \leq \alpha \leq 0\right)$, $(\beta > 0)$ and $\left(\gamma \geq \frac{1}{2}\right)$.

The implicit dynamic procedure is invoked with the ***DYNAMIC** option.

```
*DYNAMIC,
APPLICATION=...,
TIME INTEGRATOR=...,
IMPACT=...,
INCREMENTATION=...,
HAFTOL=..., HALFINC SCALE FACTOR=..., NOHAF,
ALPHA=..., BETA=..., GAMMA=...,
DIRECT
```

According to Table 4.3 the **APPLICATION=TRANSIENT FIDELITY** option should be used with the following settings:

- **TIME INTEGRATOR=HHT-TF**
- **IMPACT=AVERAGE TIME**
- **INCREMENTATION=CONSERVATIVE**

According to Table 4.3 the **APPLICATION=MODERATE DISSIPATION** option should be used with the following settings:

[6] An example of stability and accuracy of the trapezoidal rule is given in Appendix A.3.

Table 4.3 Automatic time increment used with Abaqus standard solver

Application	Time incrementation method	Default incrementation scheme	Default half-increment residual tolerance[a]	Integration parameters[b]
Transient fidelity	HHT	Same rules as for static analyses for a Δt_{max} equals to $0.01 \times T_{step}$ with a limit on half-increment residual	Based on 1000 × time of the average force without contact or 10000 × time of the average force with contact	$\alpha = -0.05$ $\beta = 0.275625$ $\gamma = 0.55$
Moderate dissipation	HHT	Same rules as for static analyses for a Δt_{max} equals to $0.1 \times T_{step}$	Not considered unless use conservative incrementation	$\alpha = -0.41421$ $\beta = 0.5$ $\gamma = 0.91421$
Quasi-static	Backward Euler	Same rules as for static analyses	Not considered	Not applicable

[a]The half-increment residual is the out-of-balance force that exists halfway through a time increment
[b]The integration parameters called in ***DYNAMIC** option ALPHA, BETA and GAMMA respectively

- **TIME INTEGRATOR=HHT-MD**
- **IMPACT=NO**
- **INCREMENTATION=AGRESSIVE**

According to Table 4.3 the **APPLICATION=QUASI-STATIC** option should be used with the following settings:

- **TIME INTEGRATOR=BWE**
- **IMPACT=NO**
- **INCREMENTATION=AGGRESSIVE**
- Default step amplitude is set to RAMP instead of STEP.

The four **TIME INTEGRATOR** methods are:

1. **HHT-TF**, a time integrator with slight numerical damping.
2. **HHT-MD**, a time integrator with moderate numerical damping.
3. **BWE**, the Backward Euler time integrator.
4. **HYBRID**, which closely resembles the HHT-TF except that it has fully implicit treatment of contact.

In an implicit dynamic scheme as shown in Eq. 4.5 the **HHT** time integrator works as described below in order to calculate the solution after the time increment "Δt" with the displacement "$u_{t+\Delta t}$", the velocity "$v_{t+\Delta t}$" and the residual force "$R_{t+\Delta t}$" making the zero machine at equilibrium Eq. 4.4:

$$\begin{pmatrix} u_{t+\Delta t} \\ v_{t+\Delta t} \\ -R_{t+\Delta t} \end{pmatrix} = \begin{pmatrix} u_t + \Delta t \times v_t + \Delta t^2 \left[\left(\frac{1}{2} - \beta \right) a_t + \beta a_{t+\Delta t} \right] \\ v_t + \Delta t \left[(1-\gamma) a_t + \gamma a_{t+\Delta t} \right] \\ M a_{t+\Delta t} + (1+\alpha)(I-P)_{t+\Delta t} - \alpha (I-P)_t \end{pmatrix} \tag{4.7}$$

There are three methods of treating impact releases:

1. **IMPACT=AVERAGE TIME** employs average time of impact release cut backs to enforce an energy balance and maintains compatible velocities and accelerations on the active contact interface.
2. **IMPACT=CURRENT TIME** "Marches through" increment without impact release cut backs. The velocities and accelerations are compatible with the active contact interface.
3. **IMPACT=NO** "Marches through" increment without impact release cut backs and without velocity/acceleration compatibility computations.

There are two general incrementation types:

1. **INCREMENTATION=CONSERVATIVE** maximizes solution accuracy.
2. **INCREMENTATION=AGGRESSIVE** incrementation based on convergence history, similar to the scheme typically used in static problems.

The half-increment residual tolerance utilizes three different options:

- **HAFTOL** Use this parameter to directly specify tolerance.
- **HALFINC SCALE FACTOR** is the default tolerance calculated automatically when CONSERVATIVE time incrementation is used:
  - 10000 × time average force for models with contact,
  - 1000 × time average force for models without contact.
  - This approach is preferable to specifying HAFTOL directly.
- **NOHAF** Suppresses calculation of the half-increment residuals.

The time incrementation data can be set using the **DIRECT** option with a fixed time incrementation or by setting the initial time increment "Δt_{init}", the time period of the step analysis "t_{total}", the minimum time increment "Δt_{min}" and the maximum time increment "Δt_{max}".

The main differences between the "Moderate Dissipation" setting versus the "Transient Fidelity" setting are that there are some additional numerical dissipations with a better convergence behavior for contact applications, thus requiring fewer solver passes. The three main reasons for using a "Moderate Dissipation" option are:

1. No direct enforcement of velocity and acceleration compatibility across contact interfaces.
2. No half-increment residual tolerance.
3. Different parameter settings for the HHT time integrator.

The "Quasi-Static" option is mainly used for cases in which a static solution is desired but stabilizing effects of inertia are beneficial. Upon convergence, there can be difficulties with a static procedure. The performances versus Abaqus explicit is problem dependent but can also be applicable to some dynamic events. Like in the general static procedure, the default loading amplitude type is "ramp" instead of "step". There is a high numerical dissipation and the Backward Euler time integrator is used instead of the HHT time integrator, in that case the Eq. 4.7 becomes:

$$\begin{pmatrix} u_{t+\Delta t} \\ v_{t+\Delta t} \\ -R_{t+\Delta t} \end{pmatrix} = \begin{pmatrix} u_t + \Delta t \times v_{t+\Delta t} \\ v_t + \Delta t \times a_{t+\Delta t} \\ Ma_{t+\Delta t} + (I - P)_{t+\Delta t} \end{pmatrix} \tag{4.8}$$

4.8.3.1 Algorithmic Details

It is important to know the main calculation steps used in implicit solver, in order to get a better insight about algorithm procedure, and to make appropriate settings adjustments regarding numerical parameter values to fix troubleshooting issue during the processing phase. The main calculation steps are the following:

1. Each increment iteration is very similar to statics.
2. Each Newton–Raphson iteration considers a system of equations of the matrix form $\bar{K} \times \Delta u = \bar{R}$, where the coefficient matrix and the residual vector $\bar{K}$ and $\bar{R}$, incorporate static terms plus inertia with damping effects and relationships implied by the time integrator between nodal vector displacement **u**, the velocity **v** and acceleration **a**.
3. Optional extra elements can be added, for instance:
   - Half-increment residual tolerance.
   - Direct enforcement of velocity and acceleration compatibility across the contact interface.
4. The effects of inertia on the effective "stiffness" matrix K can be referenced:
   - For the trapezoidal rule of time integration with interval parameters set equals to $\alpha = 0$, $\beta = \frac{1}{4}$ and $\gamma = \frac{1}{2}$. The system to solve becomes like $\bar{K} = K + \left(\frac{4}{\Delta t^2}\right) M$ similar for other time integrators in order to recall the system of equations per iteration $\bar{K} \times \Delta u = \bar{R}$ afterward.
   - Some singular modes of static stiffness K are not singular for $\bar{K}$, for example with unconstrained rigid body modes.
   - The inertia contribution to $\bar{K}$ increases as the time increment size is reduced because now the variation of time increment Δt^2 is in the denominator and has with the following effects:
     - The stabilizing effects of inertia become more significant after a cutback in the time increment size.
     - Snap-through examples like including inertia effects should stabilize the negative eigenvalue of K if the time increment is sufficiently small.
     - In practice, typical entries of mass matrix M are often orders of magnitude smaller than those of K.

5. Use of supplementary stabilization methods to stabilize K can enable a larger increment time Δt for dynamics.

The comparison to statics is directly obvious as pure static analysis is usually more efficient than quasi-static analysis if a model is statically stable. Furthermore, when quasi-static analysis is used with the dynamic procedure it should be more robust than pure static analysis, however, it is also good to supplement the solver with other stabilization methods if necessary.

The comparison to explicit dynamics is mainly related as a function of cost for increments or iterations versus the number of increments or iterations required to solve the problem. Relative overall performance is also problem dependent. There is only an easy satisfaction of residual tolerances in implicit. Moreover, the effects of "mass scaling" which is the only way to scale the mass in Abaqus standard is to adjust the density: this increases the stable time increment in Abaqus explicit and increase inertia effects in both solvers.

4.8.4 Automatic Time Incrementation with Abaqus Explicit

With Abaqus explicit the time increment size is controlled by the stable time increment. Therefore, the explicit dynamics procedure only provides a bounded solution when the time increment is less than a critical or stable time increment. The stability limit is given in terms of the highest eigenvalue in the model "ω_{max}" and the fraction of critical damping "ξ" in the highest mode as shown below:

$$\Delta t_{min} \leq \frac{2}{\omega_{max}} \left(\sqrt{1+\xi^2} - \xi \right) \tag{4.9}$$

From Eq. 4.9, it is evident that the damping reduces the stable time increment. As a result, it is not feasible to compute "ω_{max}", so easy to compute, conservative estimates are used instead. Indeed, the concept of a stable time increment is explained easily by considering a one-dimensional problem. In such scenarios, the stable time increment is the minimum time that a dilatation wave takes to move across any element in the model. A dilatation wave consists of volume expansion and contraction. The dilatation wave speed "c_d" in Eq. 4.10, can be expressed for a one-dimensional problem, where "E" is Young's modulus and "ρ" shows the current material density.

$$c_d = \sqrt{\frac{E}{\rho}} \tag{4.10}$$

Based on the current geometry each element in the model has a characteristic length L^e. Thus, the stable time increment can be expressed as

$$\Delta t = \frac{L^e}{c_d} \tag{4.11}$$

The time increment can be linked with the element length by combining Eqs. 4.11 and 4.10 as follows:

$$\Delta t = L^e \sqrt{\frac{\rho}{E}} \tag{4.12}$$

From Eq. 4.9 into Eq. 4.12, it is now possible to have a condition criterion regarding the minimum element length meshed in the whole model to ascertain numerical stability as a function of the highest eigenvalue "ω_{max}", the critical damping "ξ" and the material properties (E, ρ):

$$L^e_{min} \leq \frac{2}{\omega_{max}} \sqrt{\frac{E}{\rho}} \left(\sqrt{1+\xi^2} - \xi \right) \tag{4.13}$$

Decreasing "L^e" and/or increasing "c_d" will reduce the size of the stable time increment, such as

- Decreasing element dimensions reduces L^e.
- Increasing material stiffness increases c_d.
- Decreasing material compressibility increases c_d.
- Decreasing material density increases c_d.

Abaqus explicit monitors the finite-element model throughout the analysis to determine a stable time increment. The stability limit has major implications for large-scale finite-element analysis. Except for wave propagation or high-velocity impact problems, the stable time increment is typically much smaller than the time associated with the dynamic event; hence, a very large number of time increments will be required.

For example consider a typical car crash model with a structure including steel members for which the wave speed is 5000 m/s. A typical model may include elements as small as 20 mm. Therefore, the maximum stable time increment is $\Delta t_{max} = 20 \times 10^{-3}/5000 = 4 \times 10^{-6}$ s. A typical crash event lasts 400 ms, so the number of increments required is about 100 000! At the first glance, this stability limit may seem to be very restrictive to make the explicit method useful for low-speed dynamic problems.

However, a number of factors make this method attractive in certain situations:

- It only requires the inversion of the mass matrix.
- If the mass matrix can be diagonalized (lumped), which is the case for lower order elements, the inversion and storage of the matrix is trivial and has negligible computational costs. This is particularly advantageous for very large three-dimensional problems.
- The element operations do not require the formation of a stiffness matrix. The method is non-iterative and, therefore offers considerable additional benefits for problems with severe nonlinearities due to contact, material behavior, or (localized) buckling.

4.8.5 Dynamic Contact

Discontinuous nonlinearity impact occurs if contact conditions change. Contact in dynamic cases requires careful consideration because momentum exchange is involved and energy is lost instantaneously through certain mechanisms which are not modeled. While both Abaqus standard and Abaqus explicit have robust contact analysis capabilities, their implementations are quite different for impact problems. However, Abaqus explicit offers a more efficient solution method.

In Abaqus standard, basic contact is set by default with contact constraints to enforce contact interaction by using Lagrange Multipliers (hard contact with direct enforcement). Iteration is therefore required to establish a valid contact state, which means that very large over closures are possible.

Dynamic contact for impact problems in Abaqus standard is even more complex because, by default, the dynamic application type is set to "moderate dissipation" in cases where contact is present in the model. In this case, compatibility of velocities and accelerations across active contact interfaces is **not** enforced by default.

Moreover, the compatibility of velocities and accelerations across active contact interfaces can be enforced by setting the application type explicitly to "transient fidelity" or the impact parameter within the ***DYNAMIC** options. This implies that the penetration (or separation) of a node (or set of nodes) is detected, then the increment is split (re-analyzed) to find the point at which the average penetration (or separation) is zero, the momentum is transferred (in a "zero time" increment), and finally the remainder of original increment is re-analyzed.

In conclusion, with either approach, there is some level of inaccuracy due to incompatible velocities and accelerations using the default dynamic application type or inefficiency due to increment splitting and re-analysis with the transient fidelity application type. This is why Abaqus standard is not generally recommended for impact problems and why Abaqus explicit will be much more effective and accurate.

Indeed, dynamic contact is much simpler in Abaqus explicit because, first the penetration is detected and subsequently the node is moved back onto the surface; finally, the velocities are equalized. Here, over closure is usually very small because the time increments are insignificant.

4.8.6 Material Damping

The mass M and stiffness K matrices are proportional in what is known as the Rayleigh damping, which can be introduced into any material definition in an Abaqus model. With this form of damping, a damping matrix C is added to the system and is defined as

$$C = \alpha M + \beta K \tag{4.14}$$

The mass damping parameter α and the stiffness damping parameter β are unrelated to the HHT alpha parameter or the beta parameter that appears in the Newmark formulae. The mass and stiffness matrices proportional damping are introduced with the **ALPHA** and **BETA** parameters on the ***DAMPING** option. For each natural frequency of the system "ω_a", the effective damping ratio is equal to:

$$\xi\left(\omega_a\right)=\frac{\alpha}{2\omega_a}+\frac{\beta\omega_a}{2} \tag{4.15}$$

Thus, mass proportional damping dominates when the frequency is low, and stiffness proportional damping dominates when the frequency is high.

When stiffness damping is used in nonlinear dynamics, the most straightforward implementation would be to make the damping proportional to the tangent stiffness. However, this may cause numerical difficulties since the tangent stiffness may develop negative eigenvalues during deformation; this produces negative damping, thereby creating energy. Therefore, Abaqus uses the elastic stiffness matrix to define the damping matrix.

4.8.6.1 Numerical Damping Versus Material Damping

Since the stiffness damping is stronger for high frequencies, it can be used as a numerical damping technique. The damping coefficient can be tuned to damp frequencies with a period close to the time increment. For example, if 10% of critical damping is desired for vibrations with period T, the stiffness damping parameter β should be equal to:

$$\beta=\frac{2\xi}{\omega}=\frac{\xi T}{\pi}=\frac{0.1}{\pi}T \tag{4.16}$$

The stiffness damping does not diminish as quickly with decreasing frequency "ω" as the numerical damping provided by the α parameter of the HHT algorithm. Therefore, stiffness damping also has a strong dampening effect on low-frequency modes.

4.8.7 Half-Increment Residual Tolerance

The half-increment residual tolerance is the solution residual (or out of balance force) at the midpoint of an increment. The residual at $\frac{\Delta t}{2}$ is evaluated from solutions at the beginning and the end of the increment (inexpensive calculation) as shown in the Fig. 4.8.

The half-increment residual tolerance is used to control the accuracy of the HHT integration method. The tolerance can be set directly using the HAFTOL parameter or calculated automatically by Abaqus (by default). When calculated automatically, the value is based on a scaling of the time average force in the model with a default scale

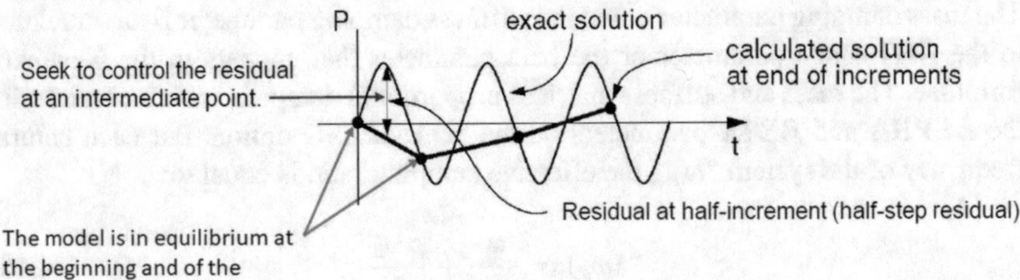

Fig. 4.8 Progression of the half-increment residual tolerance versus the exact solution

factors of 10000 for models with contact or 1000 for models without contact. It can specify non-default scale factor using the **HALFINC SCALE FACTOR** parameter, but as a good practice it is preferable to specify **HAFTOL** directly.[7] For a better understanding, a complement is given in Appendix A.4.

4.8.8 Comparing Abaqus/Standard and Abaqus/Explicit

Table 4.4 provides a summary of the benefits obtained from both Abaqus solvers when assessing analysis problems.

The techniques and logic in Abaqus are the same for dynamic analysis as for static analysis; most Abaqus features and techniques can still be used and, the model essentially looks the same and Abaqus logic generally functions the same way.

The hardest part about running dynamic analysis in Abaqus is not Abaqus but the dynamics themselves. Therefore, analysts must check the following points to ensure a realistic dynamic model:

- What happens when inertial forces are included?
- What happens when damping forces are included?
- How does your structure respond?
- What happens during impact?
- How does Abaqus handle these issues?

[7] Users will be able to find an example of a double cantilever elastic beam under point load in Abaqus Benchmark Problem 1.3.2 in the Abaqus Benchmarks Manual. After comparing the Abaqus/Standard and Abaqus/Explicit results, the results obtained with the default incrementation schemes show excellent agreement. Using a tighter half-increment residual tolerance for the implicit analysis further improves the agreement.

Table 4.4 Difference between both Abaqus standard and explicit solvers

Abaqus standard	Abaqus explicit
Time increment size is not limited, generally fewer time increments required to complete a given simulation	Time increment size is limited: generally, many more time increments are required to complete a given simulation
Each time increment is expensive since each requires the solution for a set of simultaneous equations	Each time increment is relatively inexpensive because not required to solve a set of simultaneous equations. Most of the computational expense is associated with element calculations
Convergence criteria and iteration are still needed. Non-convergence may still occur!	Non-convergence is not a concern
Ideal for problems where the response period of interest is long compared to the vibration frequency of the model. Slow dynamics problems difficult to solve effectively using explicit dynamics because of the limit on the time increment size	Ideal for high-speed dynamic simulations require very small-time increments; implicit dynamics is inefficient especially for large models
Use for problems that are mildly nonlinear and where the nonlinearities are smooth (e.g., plasticity). With a smooth nonlinear response Abaqus standard will need very few iterations to find a converged solution. In impact problems, Abaqus standard has to perform very expensive momentum transfer calculations for each impact	Usually more reliable for problems involving discontinuous nonlinearities. Contact behavior is discontinuous and involves impacts, both of which cause problems for implicit time integration. Other sources of discontinuous behavior include buckling and material failure

4.9 Unstable Collapse and Post-buckling Analysis

Abaqus standard uses the Riks method [3] to computed a post-buckling solution ***STATIC, RIKS**.[8] This method is generally used to predict unstable, geometrically nonlinear collapse of a structure; it can include nonlinear materials and boundary conditions. It often follows an eigenvalue buckling analysis to provide complete information about a structure's collapse and can be used to improve the convergence of ill-conditioned or snap-through problems that do not exhibit instability.

An unstable response can occur in the geometrically nonlinear static problems sometimes involving buckling or collapse behavior, where the load–displacement response shows a negative stiffness and the structure must release strain energy to remain in equilibrium. Several approaches are possible for modeling such behavior. One is to treat the buckling response dynamically, thus actually modeling the response with inertia effects included as the structure snaps. This approach is eas-

[8]Using the ***STATIC, RIKS, DIRECT** command gives a user direct control over the increment size; in this case, the incremental arc length Δl, is kept constant. This method is not recommended for a Riks analysis since it prevents Abaqus standard from reducing the arc length when a severe nonlinearity is encountered.

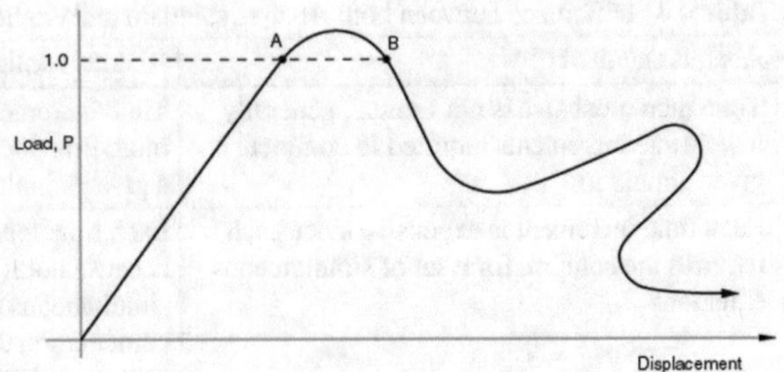

Fig. 4.9 Proportional loading with unstable response. Figure used by permission ©Dassault Systemes Simulia Corp

ily accomplished by restarting the terminated static procedure and switching to a dynamic implicit procedure when the static solution becomes unstable. In some simple cases, displacement control can provide a solution, even when the conjugate load (the reaction force) is decreasing as the displacement increases. Another approach would be to use dashpots to stabilize the structure during a static analysis.

Alternatively, static equilibrium states during the unstable phase of the response can be found by using the modified Riks method. This method is used for cases where the loading is proportional; that is, where the load magnitudes are governed by a single scalar parameter. The method can provide solutions even in cases of complex, unstable responses such as the one demonstrated in Fig. 4.9. The Riks method is also useful for solving ill-conditioned problems such as limit load problems or almost unstable problems that exhibit softening.

The Riks method uses the current load magnitude P_{total} as an additional unknown as shown in Eq. 4.17; it solves loads and displacements simultaneously. Therefore, another quantity must be used to measure the progress of the solution; Abaqus standard uses the "arc length" l along the static equilibrium path in load–displacement space. This approach provides solutions regardless of whether the response is stable or unstable.

$$P_{total} = P_0 + \lambda \left(P_{ref} - P_0\right) \tag{4.17}$$

where P_0 is the "dead load", P_{ref} is the reference load vector, and λ is the "load proportionality factor". The load proportionality factor is found as part of the solution. Abaqus standard prints out the current value of the load proportionality factor at each increment.

Abaqus standard uses Newton's method to solve the nonlinear equilibrium equations. The Riks procedure uses only a 1% extrapolation of the strain increment. Analysts need to provide an initial increment in arc length along the static equilibrium path Δl_{in}, when user defined the step. The initial load proportionality factor $\Delta\lambda_{in}$ is computed as shown below:

$$\Delta\lambda_{in} = \frac{\Delta l_{in}}{l_{period}} \tag{4.18}$$

where l_{period} is a user-specified total arc length scale factor (typically set equal to one). This value of $\Delta\lambda_{in}$ is used during the first iteration of a Riks step. For subsequent iterations and increments the value of λ is computed automatically, so users have no control over the load magnitude. The value of is part of the solution. Minimum and maximum arc length increments Δl_{min} and Δl_{max}, can be used to control the automatic incrementation.

There are some restrictions involved in using the Riks method in analysis such as

- a Riks step cannot be followed by another step in the same analysis. Subsequent steps must be analyzed by using the restart capability.
- If a Riks analysis includes irreversible deformation such as plasticity and a restart using another Riks step is attempted while the magnitude of the load on the structure is decreasing, Abaqus standard will find the elastic unloading solution. Therefore, if plasticity is present a restart should occur at a point in the analysis where the load magnitude is increasing.
- For post-buckling problems involving loss of contact, the Riks method will usually fail; inertia or viscous damping forces (such as those provided by dashpots) must be introduced in a dynamic or static analysis to stabilize the solution.

4.10 Low-Cycle Fatigue Analysis Using the Direct Cyclic Approach

The traditional approach for determining the fatigue limit of a structure is to establish the S-N curves (load versus the number of cycles to failure) $\Delta\sigma = f\,(\Delta N)$ for the materials in the structure. Such an approach is still used as a design tool in many cases to predict the fatigue resistance of engineering structures. However, this technique is generally conservative, and it does not define a relationship between the cycle number and the degree of damage or crack length.

From the S-N curve function $N = aS^{-m}$,[9] it is possible to find the T-N curve (pure tension load versus the number of cycles to failure) or the M-N curve (pure bending load versus the number of cycles to failure). Indeed the pure tension or bending moment stresses are given in Eqs. 4.19 and 4.20 respectively. The geometrical parameters are A for the cross section in tension, I for the second moment of area and y for the distance from the neutral fiber in the bending load case.

$$\sigma = \frac{T}{A} = S \tag{4.19}$$

$$\sigma = \frac{M}{I}y = S \tag{4.20}$$

[9] This equation can be linearized in a log_{10} base such as $Log(N) = Log(a) - mLog(S)$, with the couple of parameters $(Log(a); m)$ depending of the construction details of the crack, the loading conditions and the number of cycles.

The direct cyclic[10] analysis capability in Abaqus standard provides a computationally effective modeling technique to obtain the stabilized response of a structure subjected to periodic loading and is ideally suited to performing low-cycle fatigue calculations on a large structure. The capability uses a combination of Fourier series and time integration of the nonlinear material behavior to directly obtain the stabilized response of the structure.

A low-cycle fatigue step using the direct cyclic approach[11] can be the only step in an analysis, can follow a general or linear perturbation step, or can be followed by a general or linear perturbation step. Multiple low-cycle fatigue analysis steps can be included in a single analysis. In such situations, the Fourier series coefficients obtained in the previous step can be used as starting values in the current step. By default, the Fourier coefficients are reset to zero, thus allowing the application of cyclic loading conditions that are markedly different from those defined in the previous low-cycle fatigue step.

Use the following option to specify that the current step is a continuation of the previous low-cycle fatigue step using the direct cyclic approach:

```
*DIRECT CYCLIC,  FATIGUE, CONTINUE=YES
```

Use the following option (default) to reset the Fourier series coefficients to zero:

```
*DIRECT CYCLIC,  FATIGUE, CONTINUE=NO
```

4.11 Steady-State Transport Analysis

It is cumbersome to model rolling and sliding contact, such as a tire rolling along a rigid surface or a disk rotating relative to a brake assembly, using a traditional Lagrangian formulation since the frame of reference in which motion is described is attached to the material. For an observer in this reference frame, even steady-state rolling is a time-dependent process since each point undergoes a repeated history of deformation. Such an analysis is computationally expensive since a transient analysis must be performed and fine meshing is required along the entire surface of the cylinder.

This description can be viewed as a mixed Lagrangian/Eulerian method, where rigid body rotation is described in a spatial or Eulerian manner and deformation, which is now measured relative to the rotating rigid body, is described in a material or Lagrangian manner. It is this kinematic description that converts the steady-

[10] A direct cyclic analysis can be used to directly calculate the cyclic response of a structure subjected to a number of repetitive loading cycles. The direct cyclic algorithm uses a modified Newton method in conjunction with a Fourier representation of the solution and the residual vector to directly obtain the stabilized cyclic response.

[11] An example is given inside the Abaqus Benchmarks Guide Sect. 1.19.2 Crack propagation in a plate with a hole simulated using XFEM.

state moving contact problem into a purely spatially dependent simulation. The steady-state rolling and sliding analysis capability provides solutions that include frictional effects, inertia effects, and material convection for most rate-independent, rate-dependent, and history-dependent material models.

The steady-state transport analysis capability has several limitations. First, the deformable structure must be a full 360° cylindrical body of revolution. Convective boundary conditions are not available for modeling segments of a cylinder. Second, the capability is not available in two dimensions. Third, only one deformable spinning body is permitted; the symmetric model generation capability must be used to generate the deformable body. Finally, the material model definitions for models defined using the parallel rheological framework cannot include plasticity.

4.11.1 Convergence Issues in a Steady-State Transport Analysis

The steady-state transport procedure may experience convergence difficulties in certain situations that are described below:

with friction: The frictional forces that develop on the contact surface as a result of steady-state rolling are functions of the spinning angular velocity ω and the traveling straight line velocity c, or cornering velocity Ω. When these frictional forces are large, convergence of Newton's method becomes difficult. Convergence problems in Abaqus standard are usually resolved by taking a smaller load increment. However, contact forces caused by steady-state rolling do not usually abate when the magnitudes of the velocities are reduced.

For example, if a spinning object is prevented from moving ($c = 0$), full slipping conditions will develop over the entire contact zone for all values of spinning angular velocity $\omega > 0$. Consequently, the frictional force remains constant for all $\omega > 0$ (provided that the normal force remains constant), so that smaller increments in the velocities (ω, Ω, c) do not reduce the magnitude of the frictional forces and, hence, do not overcome convergence difficulties.

To provide convergence through the use of smaller increments in such cases, the friction coefficient can be increased from zero to the desired value over the analysis step. This is accomplished by setting the initial friction coefficient for the model to zero and then increasing the friction coefficient to its final value during the steady-state transport analysis step.

with the Mullins effect material model: If the Mullins effect material model is included in the material definition, there could be a strong discontinuity in the response of a structure in transitioning from a static (non-rolling) state to a steady-state rolling state. This discontinuity is due to the damage that occurs during the transient response (such as the damage that occurs as the structure undergoes its first revolution after static preloading). Since the transient response is not modeled during a steady-state transport analysis, the resulting discontinuity in the response can lead to convergence problems.

The damage associated with the Mullins effect is independent of the angular speed of rotation; as a result, time increment cutbacks fail to resolve the convergence problems. The Mullins effect can be ramped up over the time period of the step in these situations to obtain a converged solution. In such cases, the change in response due to damage is applied gradually over the step. The solution at the end of the step corresponds to the fully damaged material; solutions during the step correspond to a partially damaged material and are, therefore, physically meaningless.

Thus, it is recommended that prior to going from a static to a steady-state rolling solution, a do-nothing step at a low angular speed of rotation should be carried out with the Mullins effect ramped on. This facilitates resolution of the discontinuity in a gradual manner. The do-nothing step can then be followed by the regular steady-state transport step with the Mullins effect applied instantaneously at the beginning of the process.

with streamline integration in plasticity/creep models: In principle any material point along a streamline can be used as a starting point for the streamline integration when material convective calculations are performed. Abaqus standard always uses the material points in the original sector or the material points in the original cross section as starting points for the streamline integration in a model with periodic geometry or axisymmetric geometry, respectively.

If the pass-by-pass solution technique is used after an increment has been performed for all the streamlines, Abaqus standard will automatically use the state obtained at the end of the streamline as the starting state for the streamline integration in the subsequent increment. This iterative process is repeated for each increment until a steady-state solution is reached.

If the direct steady-state solution technique is used several local iterations are usually required for each streamline, with a local iteration corresponding to an integration over a closed-loop streamline. After a local iteration has been performed for a streamline, Abaqus standard will check to see whether the steady-state condition is satisfied for the streamline. This is best measured by ensuring the differences between the stresses/strains at the starting point of the streamline obtained before and after the iteration are sufficiently small. If the steady-state condition is not satisfied for the streamline, Abaqus standard will automatically use the state obtained at the end of the previous local iteration as the starting state for the streamline integration in the subsequent local iteration. This iterative process is repeated until a steady-state solution is reached for all the streamlines.

To improve the rate of convergence, the analyst should apply the loads on elements or nodes away from the starting points of the streamlines.

with unconstrained mesh motion: Unconstrained rigid body modes of the mesh motion will cause convergence problems for a steady-state transport analysis, similar to convergence problems for unconstrained rigid body modes in a static analysis. Friction cannot be relied upon to restrict rigid body modes in a steady-state transport analysis because frictional stresses depend on relative material velocities rather than relative nodal displacements for steady-state transport. Restricting the (steady-state) material velocity does not restrict nodal displacements for steady-state transport analyses. The material velocity includes the effects of the material flowing through the mesh and is governed by the spinning motion, reference frame motions, and nodal positions relative to the spinning axis.

4.12 Heat Transfer Analysis

Abaqus can solve the following types of heat transfer problems:

Uncoupled heat transfer analysis: Heat transfer problems involving conduction, forced convection, and boundary radiation can be analyzed in either Abaqus standard or Abaqus CFD.

Sequentially coupled thermal stress analysis: If the stress/displacement solution is dependent on a temperature field but there is no inverse dependency, a sequentially coupled thermal stress analysis can be conducted in Abaqus standard. Sequentially coupled thermal stress analysis is performed by first solving the pure heat transfer problem and subsequently reading the temperature solution into a stress analysis as a predefined field. In the stress analysis, the temperature can vary with time and position but is not altered by the stress analysis solution. Abaqus allows for dissimilar meshes between the heat transfer analysis model and the thermal stress analysis model. Temperature values will be interpolated based on element interpolators evaluated at nodes of the thermal stress model.

Fully coupled thermal stress analysis: A coupled temperature–displacement procedure is used to solve the stress/displacement and the temperature fields simultaneously. A coupled analysis is used when the thermal and mechanical solutions affect each other strongly. Although both Abaqus standard and Abaqus explicit provide coupled temperature–displacement analysis procedures, the algorithms used by each program differ considerably. In Abaqus standard, the heat transfer equations are integrated using a backward difference scheme, and the coupled system is solved using Newton's method; these problems can be transient or steady-state and linear or nonlinear. In contrast, in Abaqus explicit the heat transfer equations are integrated using an explicit forward-difference time integration rule, and the mechanical solution response is obtained using an explicit central difference integration rule. Fully coupled thermal stress analysis is always transient in Abaqus explicit. Cavity radiation effects cannot be included in a fully coupled thermal stress analysis.

Fully coupled thermal electrical structural analysis: A coupled thermal electrical structural procedure is used to solve the stress displacement, the electrical potential, and the temperature fields simultaneously. A coupled analysis is used when the thermal, electrical, and mechanical solutions affect each other strongly. These problems can be transient or steady-state and linear or nonlinear. Cavity radiation effects cannot be included in a fully coupled thermal electrical structural analysis. This procedure is only available in Abaqus standard.

Adiabatic analysis: An adiabatic mechanical analysis can be used in cases where mechanical deformation causes heating, but the event is so rapid that this heat has no time to diffuse through the material. Adiabatic analysis can be performed in Abaqus standard or Abaqus explicit.

Coupled thermal electrical analysis: A fully coupled thermal electrical analysis capability is provided in Abaqus standard for problems where heat is generated due to the flow of electrical current through a conductor.

Cavity radiation: In Abaqus standard, cavity radiation effects can be included (in addition to prescribed boundary radiation) in uncoupled heat transfer problems. The cavities can be open or closed, while symmetries and blocking within cavities can be modeled. view factors are calculated automatically and the motion of objects bounding a cavity can be prescribed during the analysis. Cavity radiation problems are nonlinear and can be either transient or steady-state.

In such coupled field analyses, most of the time the reason for exceeding numerical tolerance or for the existence of convergence issues is that the user has forgotten to define the physical constants of the model attributes. From the main menu bar **Model → Edit Attributes** shown in Fig. 4.10 and according to a consistent system of units as exemplified in Table 3.1.

The ***PHYSICAL CONSTANTS** option is used to define the physical constants necessary for an analysis; since Abaqus has no built-in units, no default values are provided. If a physical constant required for the analysis is not given, Abaqus will issue a fatal error message. The units used for the constants must be consistent with the remaining input data.

It is recommended to avoid performing thermal conduction analysis using the Stefan–Boltzmann with a temperature unit other than the degree Kelvin. If the temperature results need to be calculated in degrees Celsius, it is possible to enter this data in Fig. 4.10. In Table 4.5, for instance, in the SI (mm) units system, the Stefan–Boltzmann constant equals to 5.669e-14 $\mathrm{W} \cdot \mathrm{mm}^{-2} \cdot \mathrm{K}^{-4}$ and the absolute zero temperature equals to –273.15 °C to get the solution with a node temperature result computed in $^{\circ}C$ while retaining a consistent system of units defined.

Why will the results still be unit consistent? This is because Eq. 4.21 is a linear function and therefore Abaqus will recalculate the direct node temperature as function of this offset to convert $^{\circ}K$ to $^{\circ}C$.

$$T\,(^{\circ}K) = T\,(^{\circ}C) + 273.15 \tag{4.21}$$

4.12.1 Transient Analysis

The time increments can be selected automatically based on a user-prescribed maximum allowable nodal temperature change in an increment $\Delta\theta_{max}$. Abaqus standard will restrict the time increments to ensure that this value is not exceeded at any node (except nodes with boundary conditions) during any increment of the analysis.

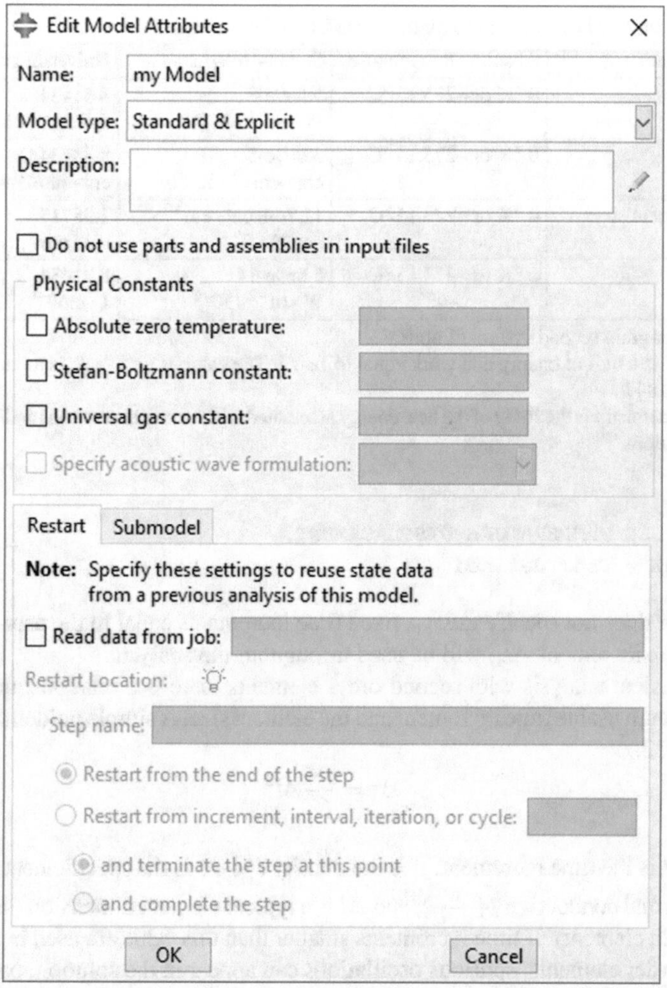

Fig. 4.10 Model attribute settings. In the Physical Constants portion of the dialog box, do the following: to specify surface emissivity and radiation conditions in heat transfer analyses, enter values for the **absolute zero temperature** by default equal to zero and the **Stefan–Boltzmann constant**; to specify the **universal gas constant**, enter a value in the Universal gas constant field; to identify the type of incident wave loading for an incident wave interaction in acoustic analyses, toggle on **Specify acoustic wave formulation**, click the arrow to the right of the text field, and select the formulation. There are two options for wave formulation: either, select Scattered wave to obtain the **scattered wave** field solution that will be produced by incident wave loading or select **Total wave** to obtain the total acoustic pressure wave solution

Table 4.5 Physical constants used for heat transfer models

Units system	Absolute temperature	Stefan–Boltzmann	Universal gas
SI	0 °K or –273.15 °C	5.669e-8 $W \cdot m^{-2} \cdot K^{-4}$	8.31434 $J \cdot mol^{-1} \cdot K^{-1}$
CGS[a]	0 °K or –273.15 °C	5.669e-5 $erg \cdot cm^{-2} \cdot K^{-4}$[b]	8.31434e7 $erg \cdot mol^{-1} \cdot K^{-1}$
Thermo chemistry[c]	0 °K or –273.15 °C	11.7e-8 $cal \cdot cm^{-2} \cdot day^{-1} \cdot K^{-4}$	1.98717 $cal \cdot mol^{-1} \cdot K^{-1}$
SI (mm)	0 °K or –273.15 °C	5.669e-14 $W \cdot m^{-2} \cdot K^{-4}$	8.31434 $J \cdot mol^{-1} \cdot K^{-1}$

[a]Centimeter gram second system of units

[b]The “erg” is a unit of energy and work equal to 1e-7 J. The calories unit “cal” is a unit of energy equal to 4.184 J

[c]Thermo chemistry is the study of the heat energy associated with chemical reactions and/or physical transformations

```
*COUPLED TEMPERATURE-DISPLACEMENT,
DELTMX = User defined
```

If the user does not specify $\Delta\theta_{max}$, fixed time increments equal to the user-specified initial time increment Δt_0, will be used throughout the analysis.

In transient analysis with second-order elements there is a relationship between the minimum usable time increment and the element size. A simple guideline is given by Eq. 4.22:

$$\Delta t = \frac{\rho c}{6k} \Delta l^2 \tag{4.22}$$

where Δt is the time increment, ρ is the density ($\frac{kg}{m^3}$), c is the specific heat ($\frac{J}{kg.^\circ C}$), k is the thermal conductivity ($\frac{W}{m.^\circ C}$), and Δl is a typical element dimension (such as the side length element). If time increments smaller than this value are used in a mesh of second-order elements, spurious oscillations can appear in the solution, particularly in the vicinity of boundaries with rapid temperature changes. These oscillations are nonphysical and may cause problems if the temperature is dependent on material properties. In transient analyses using first-order elements the heat capacity terms are lumped, which eliminates such oscillations but can also lead to locally inaccurate solutions for small-time increments. If smaller time increments are required, a finer mesh should be used in regions where the temperature changes rapidly. There is no upper limit on the time increment size (the integration procedure is unconditionally stable) unless nonlinearities cause convergence problems. Equation 4.22 can also be used in coupled thermal electrical analysis or in fully coupled thermal electrical structural analysis.

Some types of analyses may develop local instabilities, for instance, surface wrinkling, material instability, or local buckling. In such scenarios, it may not be possible to obtain a quasi-static solution even with the aid of automatic incrementation. Abaqus standard offers a method of stabilizing this class of problems by applying damping throughout the model in such a way that the viscous forces introduced are sufficiently large to prevent instantaneous buckling or collapse but small enough not to have a significant influence on the behavior while the problem is stable.[12]

In coupled problems where two different fields are active, it is important to take care when choosing the units of the problem. If the choice of units is such that the terms generated by the equations for each field are different by many orders of magnitude, the precision on some computers may be insufficient to resolve the numerical ill-conditioning of the coupled equations. Therefore, units that avoid ill-conditioned matrices are necessary. For example, units of MPa can be used as an alternative to Pascal for the stress equilibrium equations to reduce the disparity between the magnitudes of the stress equilibrium equations and the heat flux continuity equations.

4.12.1.1 Stability

The explicit procedure integrates through time by using multiple small-time increments. The central difference and forward-difference operators are conditionally stable. The stability limit for both operators (with no damping in the mechanical solution response) is obtained by choosing:

$$\Delta t \leq min\left(\frac{2}{\omega_{max}}; \frac{2}{\lambda_{max}}\right) \tag{4.23}$$

where ω_{max} is the highest frequency in the system of equations of the mechanical solution response and λ_{max} is the largest eigenvalue in the system of equations within the thermal solution response.

To estimate the time increment size there is an approximation of the stability limit for the forward-difference operator in the thermal solution response :

$$\Delta t \approx \frac{L_{min}^2}{2\alpha} \tag{4.24}$$

where L_{min} is the smallest element dimension in the mesh and $\alpha = \frac{k}{\rho c}$ is the thermal diffusivity of the material from Eq. 4.23. The parameters k, ρ, and c represent the material's thermal conductivity, density, and specific heat, respectively.

In most applications of explicit analysis, the mechanical response will govern the stability limit. The thermal response may govern the stability limit when material parameter values are nonphysical or when a very large amount of mass scaling is used.

[12]See the Automatic stabilization of unstable problems inside Abaqus Analysis User's Manual in Sect. 7.1.1 Solving nonlinear problems.

4.13 Fluid Dynamic Analysis

Coupled structure fluid analyses including or not including temperature can be performed, so it is important to get a general overview of the most common convergence issues involved in Abaqus CFD solver usage. Abaqus CFD can solve the following incompressible flow problems:

Laminar and turbulent, internal or external flows that are steady-state or transient, span a broad Reynolds number range, and involve complex geometry may be simulated with Abaqus/CFD. These include flow problems induced by spatially varying distributed body forces.

Thermal convective, problems that involve heat transfer and require an energy equation and that may involve buoyancy-driven flows (i.e., natural convection) can also be solved with Abaqus CFD. This type of problem includes turbulent heat transfer for a broad range of Prandtl numbers.

Deforming-mesh ALE Abaqus CFD includes the ability to perform deforming-mesh analyses using an arbitrary Lagrangian Eulerian (ALE) description of the equations of motion, heat transfer, and turbulent transport. Deforming-mesh problems may include prescribed boundary motion that induces fluid flow or Fluid–Structure Interaction FSI[13] problems where the boundary motion is relatively independent of the fluid flow.

There are the following limitations:

- While turbulence can be activated for a porous media flow problem, a rigorous volume-averaging procedure has not been implemented in Abaqus CFD to account for turbulence transport within the porous media. The equations governing the transport of the turbulence variables are solved by neglecting the effects of the presence of porous medium; in other words, the porous medium remains transparent (fully open) with regards to the transport of turbulence variables.
- When the arbitrary Lagrangian–Eulerian (ALE) and deforming-mesh algorithms are activated for a porous flow problem, changes in the porosity of the medium associated with large mesh/domain deformations are not taken into account. The model is strictly valid only for situations involving undeformable porous media.

4.13.1 Convergence Criteria and Diagnostics

Iterative solvers compute an approximate solution to a given set of equations; therefore, convergence criteria are required to determine whether the solution is acceptable. While default settings should be adequate for most problems, it is possible to

[13] Set TYPE = FSI to set the parameters that will be used in deforming-mesh problems involving moving boundaries or deforming geometries in Abaqus CFD as well as an Abaqus CFD to Abaqus standard or to Abaqus explicit co-simulation. This parameter setting can be used only in Abaqus CFD analyses.

modify the convergence criteria. Convergence history output is also available, which may be useful in terms of helping advanced users tune the solvers for performance or robustness. For the algebraic multigrid preconditioner, diagnostic information (such as the number of grids, grid sparsity, and largest eigenvalue and condition number estimates) is available upon request. The diagnostic information for the algebraic multigrid preconditioner is printed every time the preconditioner is computed.

Users can specify convergence criteria such as the linear convergence limit (also commonly referred to as the convergence tolerance), the frequency of convergence checking, and the maximum number of iterations that can be set in the system. The iterative solver will stop when the relative residual norm of the system of equations and the relative correction of the solution norm fall below the convergence limit.

```
*MOMENTUM EQUATION SOLVER
max iterations, frequency check, convergence limit
*TRANSPORT EQUATION SOLVER
max iterations, frequency check, convergence limit
*PRESSURE EQUATION SOLVER
max iterations, frequency check, convergence limit
```

Users also have to access the convergence output, and monitor the convergence of the iterative solver by accessing this factor. Once the convergence output is activated, the current relative residual norm and the relative solution correction norm are output each time the convergence is checked.

```
*MOMENTUM EQUATION SOLVER, CONVERGENCE=ON
*TRANSPORT EQUATION SOLVER, CONVERGENCE=ON
*PRESSURE EQUATION SOLVER, CONVERGENCE=ON
```

It is also possible to access diagnostic information. Diagnostic output is only useful for the algebraic multigrid preconditioner, while for other preconditioners a solver initialization message is printed for diagnostic output. For the algebraic multigrid preconditioner, the number of grids, grid sparsity, and largest eigenvalue and condition number estimates are output each time the preconditioner is computed.

```
*PRESSURE EQUATION SOLVER, TYPE=AMG, DIAGNOSTICS=ON
```

Three solver types are available for solving the pressure equation. The default AMG solver uses an algebraic multigrid preconditioner and offers a choice of three Krylov solvers: conjugate gradient, biconjugate gradient stabilized, and flexible generalized minimal residual. The SSORCG solver uses a symmetric successive over-relaxation preconditioner and conjugate gradient Krylov solver. The DSCG solver uses a diagonally scaled preconditioner and conjugate gradient Krylov solver. Finally, the AMG solver provides many additional options that are intended for advanced usage and in cases where convergence difficulties are encountered.

```
*PRESSURE EQUATION SOLVER, TYPE=AMG
*PRESSURE EQUATION SOLVER, TYPE=SSORCG
*PRESSURE EQUATION SOLVER, TYPE=DSCG
```

4.13.2 Time Increment Size Control

By default, Abaqus CFD uses an automatic time incrementation algorithm that continually adjusts the time increment size to satisfy the Courant–Friedrichs–Lewy (CFL) stability condition for advection. The default value, CFL $= 0.45$, guarantees the solution's stability. It is possible to further limit the automatically computed time increment size by specifying a maximum value, while an initial time increment size can also be specified. This value is automatically decreased as necessary to satisfy a maximum initial CFL value of 0.45 based on the starting conditions of the flow.

Alternatively, users can select fixed time incrementation and specify the time increment size. In this case, the time increment size remains constant throughout the step, but stability is not guaranteed.

In such situations, it is recommended to get a time- accurate analysis that the time integration parameters are all set by default to $\theta = 0.5$, which produces a second-order accurate semi-implicit method suitable for time-accurate transient analysis. When automatic time incrementation is used, $CFL \leq 2$ should be specified to maintain stability and time accuracy.

In a steady-state solution using the transient solver applied in analyses where the goal is to reach a steady-state solution, the fully implicit (backward Euler) method can be activated by setting all-time integration parameters to $\theta = 1$. This method is unconditionally stable, allowing you to specify large CFL values to significantly increase the time increment size. Strict guidelines for selecting the maximum allowable CFL number are not available, and this maximum value may vary for different flows and meshes. CFL values of 10 or more have been used successfully for some analyses where only the final result is of interest.

In a steady-state analysis using the steady-state solver set in Abaqus CFD, the steady-state solver is implemented using a second-order accurate SIMPLE based algorithm. The nonlinear transport equations are solved sequentially for a specified number of iterations. It is up to the user to manually terminate the steady-state iterations. In terms of the nonlinear convergence criteria for the SIMPLE algorithm employed in Abaqus CFD, the convergence behavior of the coupled nonlinear transport equations along with the pressure correction equation relies on the under- relaxation of the solution updates during successive iterations. Typically, this requires the specification of under-relaxation factors for the momentum, pressure correction, and other scalar transport equations such as temperature, turbulence, etc.

To specify the number of iterations and under-relaxation factors the following command lines can be used with Abaqus CFD. Abaqus CFD solves the nonlinear transport equations sequentially for a specified number of iterations. The default is 10000.

```
*CFD, INCOMPRESSIBLE NAVIER STOKES, STEADY STATE
number of nonlinear iterations
```

The under-relaxation factors are specified as the last data on the first data line of the corresponding linear equation solvers. Use the following options to specify the under-relaxation factors:

```
*MOMENTUM EQUATION SOLVER
data for all linear convergence criteria,
under-relaxation factor
*TRANSPORT EQUATION SOLVER
data for all linear convergence criteria,
under-relaxation factor
*PRESSURE EQUATION SOLVER
data for all linear convergence criteria,
under-relaxation factor
*ENERGY EQUATION SOLVER
data for all linear convergence criteria,
under-relaxation factor
```

4.14 Introduction to the User Subroutines

Abaqus provides to the users an extensive array of user subroutines regarding standard, explicit or CFD analyses, that allow them to adapt Abaqus to their particular analysis requirements.[14] The Fig. 4.11 shows the global flow in Abaqus standard which is interacting with the different user subroutines at the different steps of computed solution process.

The flowchart in Fig. 4.11 is idealized. In the first iteration of an increment, all of the user subroutines shown in the figure are called twice. During the first call the initial stiffness matrix is being formed using the configuration of the model at the start of the increment. Then during the second call a new stiffness, based on the updated configuration of the model, is created. In subsequent iterations, the subroutines are called only once, and in these subsequent iterations, the corrections to the models configuration are calculated using the stiffness from the end of the previous iteration.

To include user subroutines in an analysis, the analyst needs to specify the name of a file with the user parameter on the Abaqus execution command.

```
abaqus job=my_analysis user=my_subroutine
```

[14] See in documentation Abaqus User Subroutines Reference Guide.

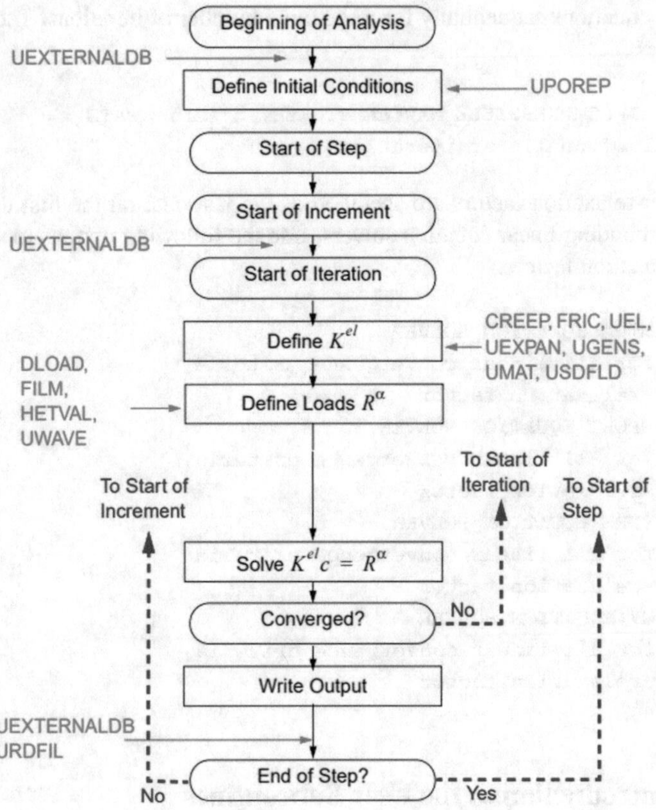

Fig. 4.11 Global flow in Abaqus standard

The file should be either source code **my_subroutine.f** or an object file **my_analysis.o**. The file extension can be included **user=my_analysis.f**; otherwise, Abaqus will determine automatically which type of file is specified.

When multiple user subroutines are needed in the analysis, the individual routines can be combined into a single file. A given user subroutine (such as **UMAT** or **FILM**) should appear only once in the specified user subroutine source or object code.

When an analysis that includes a user subroutine is restarted, the user subroutine must be specified again because the subroutine object or source code is not stored on the restart (.res) file.

In Abaqus standard user subroutines can write debug output to the message (.msg) file (unit 7) or to the printed output (.dat) file (unit 6). These units do not have to be opened within the user subroutine, they are opened by Abaqus. These unit numbers cannot be used by user subroutines in Abaqus explicit.

When a file is opened in a user subroutine, Abaqus assumes that it is located in the scratch directory created for the simulation; therefore, full path names must be used in the **OPEN** statements in the subroutine to specify the location of the files.

4.14.1 Installation of a Fortran Compiler

Abaqus and Fortran are linked to execute User Subroutines. The user can find various versions of Abaqus and Fortran available, here is the compatibility list:

- ABAQUS 2017—Intel Composer XE 2013 or above.
- Visual Studio 2010 or above.
- Intel Visual Fortran 12.0 or above.

Once the user found the compatible version of Visual Studio and Intel Visual Fortran to use with the installed Abaqus version, then the step by step procedure to link Abaqus with a Fortran compiler is explained below:

1. Configure Abaqus to Intel Fortran:

   a. Find the path for the file **ifortvars.bat**, which will be similar to:

   ```
   C:\Program Files (x86)\Intel\Compiler...
   ...\Fortran\9.1\em64t\bin\ifortvars.bat
   ```

   b. Click on the **Start Menu**, click on **Programs**, find Abaqus and right click on the ABAQUS Command window. Click on **Properties**. In the **Target** dialog after the

   ```
   /k
   ```

   symbol, add a blankspace, then the path to,

   ```
   ifortvars.bat
   ```

   between double quotation marks. The complete entry for **Target** will look like:

   ```
   C:\WINDOWS\SysWOW64\cmd.exe /k "C:\Program ..
   ...Files (x86)\Intel\Compiler\Fortran\9.1...
   ...\em64t\bin\ifortvars.bat"
   ```

   Click **OK**.

2. Set environment variables:

   a. Double click on the file **vcvarsall.bat**, which will be found on a path similar to

   ```
   C:\Program Files (x86)\Microsoft Visual Studio...
   ... 8\VC\vcvarsall.bat
   ```

b. Double click on the file **ifortvars.bat**, which will be found on a path similar to

```
C:\Program Files (x86)\Intel\Compiler\Fortran...
...\9.1\em64t\bin\ifortvars.bat
```

c. Click on the **Start Menu**, click on **Programs**, find Abaqus and double click on the **Abaqus Command** window to open one.
d. Type in the command window:

```
set >path.info
```

followed by a carriage return.

e. Open the file **path.info** which will be found in the Abaqus working directory, typically

```
C:\Temp
```

3. nameEdit environment variables:

a. Open a new document in Notepad. Find the paths after

```
path=
lib=
include=
```

and copy them into the Notepad document. Take care to copy all pathnames, since multiple lines will need to be copied.
b. Insert a carriage return after the closing semicolon for each pathname to separate them.
c. Delete any pathnames that are duplicated.
d. Delete the carriage returns to reproduce a continuous string of pathnames again. Take care not to delete any non-blank space characters.

4. Update environment variables:

a. Click on the **Start Menu** and open the **Control Panel**. Click on **System** and then the **Advanced** tab. Click on **Environment Variables**. Under the heading **System variables**, progressively select each of the environment variables **path**, **lib** and **include**. For each variable, click **Edit** and replace each variables with the entire string of pathnames from the Notepad document corresponding to the variable being edited. Click **OK** and then **OK**.
b. Log off and log back in again.

5. Verify user subroutine linkage:
   a. Open an Abaqus command window and type

   ```
   abaqus verify -user_std
   ```

   followed by a carriage return. Check that every test passes.
   b. Run a trial analysis using subroutines to check.

4.14.2 Run a Model Which Uses a User Subroutine

Using the following command:

```
abaqus job=job-id user= < subroutine-filename >
```

For the Linux Teaching System Computers give the extension .f and for PCs give the extension .for.

The user subroutine has to be in a separate file (say $my\_material.f$). For example:

```
abq614 job=my_model user=my_material
```

If there are more than one user subroutines then all the subroutines must be included in a single file.

If using CAE then in the **Job** Create a job and then click on the **General** tab and click on the browse button for user subroutine and choose the file containing the user subroutine and click on **OK** in the browser window and also in the dialog box. Then submit the job.

4.14.3 Debugging Techniques and Proper Programming Habits

After installing the Fortran compiler coupled with Abaqus, the user's main concern will be to know how to debug a subroutine. It is not something easy but here the user will be able to find some guidelines and good habits to develop an effective and efficient strategy to debug a Fortran code. The most simple and the best old fashion way to debug a program in Fortran is to use the command line **PRINT** and **PAUSE** at different line of code. In this way, the code will be executed through the subroutine and will return the results thanks to the **PRINT** function and will allow the user to follow step by step the execution of the code with the Enter button on the user's keyboard, thanks to the instruction **PAUSE**.

4.14.3.1 Fortran Statements

Every user subroutine in Abaqus standard must include the statement:

```
INCLUDE 'ABA_PARAM.INC'
```

as the first statement after the argument list.

The file *ABA_PARAM.INC* is installed on the computer system by the Abaqus installation procedure. The file specifies either **IMPLICIT REAL*8 (A-H, O-Z)** for double-precision machines or **IMPLICIT REAL (A-H, O-Z)** for single- precision machines. The Abaqus execution procedure, which compiles and links the user subroutine with the rest of Abaqus, will include the *ABA_PARAM.INC* file automatically. It is not necessary to find this file and copy it to any particular directory: Abaqus will know where to find it.

Every user subroutine in Abaqus explicit must include the statement:

```
include 'vaba_param.inc'
```

4.14.3.2 Naming Conventions

If user subroutines call other subroutines or use **COMMON** blocks to pass information, such subroutines or **COMMON** blocks should begin with the letter K since this letter is never used to start the name of any subroutine or **COMMON** block in Abaqus.

4.14.3.3 Subroutine Argument Lists

The variables passed into a user subroutine via the argument list are classified as either variable to be defined, variables that can be defined, or variables passed in for information.

The user must not alter the values of the variables passed in for information. Doing so will have unpredictable results.

4.14.3.4 Solution-Dependent State Variables

Solution-dependent state variables (SDVs) are values that can be defined to evolve with the solution of an analysis. An example of a solution-dependent state variable for the **UEL** subroutine is strain.

Several user subroutines allow the user to define SDVs. Within these user subroutines, the SDVs can be defined as functions of any variables passed into the user subroutine. It is the users responsibility to calculate the evolution of the SDVs within the subroutine; Abaqus just stores the variables for the user subroutine.

Space must be allocated to store each of the solution-dependent state variables defined in a user subroutine. For most subroutines the number of such variables required at the integration points or nodes is entered as the only value on the data line of the **DEPVAR** option.

```
*USER MATERIAL
*DEPVAR
8
```

For subroutines **UEL** and **UGENS** the **VARIABLES** parameter must be used on the **USER ELEMENT** and **SHELL GENERAL SECTION** options, respectively.

```
*USER ELEMENT, VARIABLES=8
```

For subroutine **FRIC** the number of variables is defined with the **DEPVAR** parameter on the **FRICTION** option.

```
*FRICTION, USER, DEPVAR=8
```

There are two methods available for defining the initial values of solution-dependent variables. The **INITIAL CONDITIONS, TYPE=SOLUTION** option can be used to define the variable field in a tabular format. For complicated cases user subroutine **SDVINI** can be used to define the initial values of the SDVs. Invoke this subroutine by adding the **USER** parameter to the **INITIAL CONDITIONS, TYPE=SOLUTION** option.

4.14.3.5 Testing Suggestions

Always develop and test user subroutines on the smallest possible model.

Do not include other complicated features, such as contact, unless they are absolutely necessary when testing the subroutine.

Test the most basic model of the user subroutine before adding additional complexity to the subroutine. Whenever a new feature is added to the model in a user subroutine, test it before adding an additional feature.

When appropriate, try to test the user subroutine in models where only values of the nodal degrees of freedom (displacement, rotations, temperature) are specified. Then test the subroutine in models where fluxes and nodal degrees of freedom are specified.

Ensure that arrays passed into a user subroutine with a given dimension are not used as if they had a larger dimension. For example, if a user subroutine is written such that the number of SDVs is 10 but only 8 SDVs are specified on the **DEPVAR** option, the user subroutine will overwrite data stored by Abaqus; the consequences of this accident will be unpredictable.

4.14.4 *Examples of User Subroutine with Abaqus Standard*

In order to give a short overview about user subroutine structure, some Fortran subroutine will be shortly presented here, and all user subroutine can be found in Abaqus User Subroutines Reference Guide v6.14 in Sect. 1.1 Abaqus/Standard subroutines.

4.14.4.1 User Subroutine DLOAD

User subroutine **DLOAD** is typically used when a load is a complex function of time and/or position. Loads that are simple functions of time can usually be modeled with the **AMPLITUDE** option. The subroutine can also be used to define a load that varies with element number and/or integration point number.

The subroutine is called when the **DLOAD** or ***DSLOAD** options contain a nonuniform load type label. For example,

```
*DLOAD
ELTOP,P1NU, 10.0
```

specifies that the elements in element set **ELTOP** will be subject to a force per area on the 1-face of solid (continuum) elements or a force per unit length in the beam local 1-direction when used with beam elements. The magnitude specified on the data line is passed into the subroutine as the value of variable F.

A list of the nonuniform distributed load types that are available for use with any particular element is given in the Abaqus standard Users Manual. The AMPLITUDE parameter cannot be used with the **DLOAD** or ***DSLOAD** options when the user subroutine is used to define the magnitude of the distributed load. The distributed load magnitude cannot be written as output with the **EL** FILE or **EL PRINT** options.

If the distributed load is dependent on the elements deformation rather than the position of the element, a stiffness is being defined and a user element subroutine **UEL**, not user subroutine **DLOAD**, is required.

Transient Internal Pressure Loading

The structure is a solid rocket motor, modeled as a long, hollow viscoelastic cylinder encased in a thin steel shell. The rocket's ignition is simulated by a transient internal pressure load acting at the inner diameter of the viscoelastic cylinder. The transient response of the structure is sought. The model is fully described in Abaqus Verification Guide v6.14 in Sect. 2.2.9 Transient internal pressure loading of a viscoelastic cylinder.

The transient pressure load is an exponential function of time according to Eq. 4.25

$$p = 10(1 - e^{-23.03t}) \tag{4.25}$$

The partial input data is shown below, with the parameter **P4NU** which is the parameter to apply a nonuniform **DLOAD** to face 4 of element 1.

```
*HEADING
:
*BOUNDARY
ALL,2
*STEP, INC=50
*VISCO, CETOL=7.E-3
0.01, 0.5
*DLOAD
1, P4NU
:
*END STEP
```

The Fortran code programmed inside the user subroutine **DLOAD.f** is shown below:

Listing 4.1 DLOAD.f

```
      SUBROUTINE DLOAD(F,KSTEP,KINC,TIME,NOEL,NPT,
     1                 LAYER,KSPT,COORDS,JLTYP,SNAME)
C
C          EXPONENTIAL PRESSURE LOAD
C
      INCLUDE ABA_PARAM.INC
C
      DIMENSION COORDS(3),TIME(2)
      CHARACTER*80 SNAME
      DATA TEN,ONE,CONST /10.,1.,-23.03/
      F=TEN*(ONE-(EXP(CONST*TIME(1))))
      IF (NPT.EQ.1)
        WRITE(6,*) 'LOAD APPLIED',F,'AT TIME=',TIME(1)
      RETURN
      END
```

The load in this model is defined as a function of step time, **time(1)**.

The load is applied only to the one element on the inner diameter of the rocket motor, element 1.

The load is monitored by writing output to the printed output (.dat) file, once per iteration, when the distributed load value is defined at the first integration point.

4.14.4.2 User Subroutine FILM

User subroutine FILM is typically used when either the film coefficient, h, or sink temperature, θ^s, is a complex function of time, position, and/or surface temperature. The parameters h and θ^s that are simple functions of time usually can be modeled with the **AMPLITUDE** option. The subroutine can also be used to define a load that varies with element number and/or integration point number.

The subroutine is called when the **FILM** or ***SFILM** options contain a nonuniform load type label or when the USER parameter is used with the ***CFILM** option. For example,

```
*FILM
ELLEFT, F6NU, 10.0, 1500
```

with **F6NU** the load type label, θ^s equals 10 and h equals 1500. It specifies that the elements in element set **ELLEFT** will be subject to a film load (convection boundary condition) on face six (6) of solid (continuum), heat transfer elements. The variable **H(1)** will be passed into the routine with the value of h specified on the data line of the FILM option. The variable **SINK** will be passed into the routine with the value of θ^s specified on the data line of the **FILM** option.

The interface to user subroutine FILM is given below:

Listing 4.2 FILM.f

```
      SUBROUTINE FILM(H,SINK,TEMP,KSTEP,KINC,TIME,
     1     NOEL,NPT,COORDS,JLTYP,FIELD,NFIELD,SNAME,
     2     NODE,AREA)
C
      INCLUDE 'ABA_PARAM.INC'
C
      DIMENSION H(2), TIME(2), COORDS(3), FIELD(NFIELD)
      CHARACTER*80 SNAME
      user coding to define H(1), H(2), and SINK
      RETURN
      END
```

An example of the usage of the FILM subroutine is given in Abaqus Verification Guide v6.14 Sect. 1.3.41 Temperature-dependent film condition.

4.14.5 Examples of User Subroutine with Abaqus Explicit

In order to give a short overview about user subroutine structure, some Fortran subroutine will be shortly presented here, and all user subroutine can be found in Abaqus User Subroutines Reference Guide v6.14 in Sect. 1.2 Abaqus/Explicit subroutines.

4.14.5.1 User Subroutine VDISP

It can be used to prescribe translational and rotational boundary conditions; for all degrees of freedom listed in the associated boundary condition; it allows the user to specify values for either the degree of freedom or its time derivatives such as velocity and acceleration. At the beginning of each step user subroutine VDISP is called once to establish the initial velocity; and then, it is called once on each configuration, including the initial configuration, to establish the nodal acceleration.

The variable to be defined is **rval(nDof, nblock)** which include all values of the prescribed variable for degrees of freedom 1–6 (translation and rotation) at the nodes. The variable can be displacement, velocity, or acceleration, depending on the type specified in the associated boundary condition. The variable type is indicated by jBCType. The variable **rval** has a default value that is computed as if the boundary condition is released. The user only need to reset the **rval** if the boundary condition is active. The user subroutine VDISP is presented below:

Listing 4.3 VDISP.f

```
      subroutine vdisp(
c__Read_only variables -
     1   nblock, nDof, nCoord, kstep, kinc,
     2   stepTime, totalTime, dtNext, dt,
     3   cbname, jBCType, jDof, jNodeUid, amp,
     4   coordNp, u, v, a, rf, rmass, rotaryI,
c__Write_only variable -
     5   rval )
c
      include 'vaba_param.inc'
c
      character*80 cbname
      dimension jDof(nDof), jNodeUid(nblock),
     1  amp(nblock), coordNp(nCoord,nblock),
     2  u(nDof,nblock),v(nDof,nblock),a(nDof,nblock),
     3  rf(nDof,nblock), rmass(nblock),
     4  rotaryI(3,3,nblock), rval(nDof,nblock)
c
      do 100 k = 1, nblock
         do 100 j = 1, nDof
            if( jDof(j) .gt. 0 ) then
c              user coding_to define rval(j, k)
            end if
 100  continue
c
      return
      end
```

4.14.5.2 User Subroutine VDLOAD

To contrast with the user subroutine DLOAD used with Abaqus standard and presented in Sect. 4.14.4.1, the equivalent user subroutine working with Abaqus elicit is presented here. It can be used to define the variation of the distributed load magnitude as a function of position, time, velocity and so on, for a group of points, each of which appears in an element-based or surface-based nonuniform load definition.

Variable to be defined is **value (nblock)** with a magnitude of the distributed load. Units are FL^{-2} for surface loads, FL^{-3} for body forces. The user subroutine is described below:

Listing 4.4 VDISP.f

```
      subroutine vdload (
C__Read_only (unmodifiable)variables -
     1 nblock, ndim, stepTime, totalTime,amplitude,
     2 curCoords, velocity, dirCos, jltyp, sname,
C__Write_only (modifiable) variable -
     3 value )
C
      include 'vaba_param.inc'
C
      dimension curCoords(nblock,ndim),
     1  velocity(nblock,ndim),
     2  dirCos(nblock,ndim,ndim), value(nblock)
      character*80 sname
C
      do 100 km = 1, nblock
C       user coding_to define_value
  100 continue

      return
      end
```

4.14.6 Examples of User Subroutine with Abaqus CFD

In Abaqus User Subroutines Reference Guide v6.14 in Sect. 1.3 Abaqus/CFD subroutines there are only two user subroutines, not coded in Fortran but in C++, the first user subroutine is titled SMACfdUserPressureBC coded to specify prescribed pressure boundary conditions and the second user subroutine is SMACfdUserVelocityBC to specify prescribed velocity boundary conditions.

References

1. Press WH, Vetterling WT, Teukolsky SA (2006) Numerical recipes in Fortran 77, the art of scientific computing, 2nd edn. Cambridge University Press, Cambridge, p 827
2. Hilber HM, Hughes TJR, Taylor RL (1977) Improved numerical dissipation for time integration algorithms in structural dynamics. Earthquake Eng Struct Dyn 5:283–292
3. Riks E (1979) An incremental approach to the solution of snapping and buckling problems. Int J Solids Struct 15:529–551

Part II
Stop Struggling with Specific Issues

In this second part, the user becomes more familiar with troubleshooting understandings in general. Therefore, it is time to dig more into some specific troubleshooting about classic convergence issues dealing with FEA nonlinearities like, for instance, with materials, mesh, and contact problems. Each chapter will cover the largest non-convergence events users may encounter, with procedures and ideas to get a proper solution directly related to error or warning messages which might cause the Abaqus solver stopped running correctly.

Chapter 5
Materials

5.1 Generalities

Most materials that exhibit ductile behavior (large inelastic strains) yield at stress levels that are orders of magnitude less than the elastic modulus of the material, which implies that the relevant stress and strain measures are true stress (Cauchy stress) and logarithmic strain as shown in Fig. 5.1. Therefore, material data for all of these models should be given in these measures otherwise Abaqus will not solve the correct material curve. To convert an isotropic material engineering data ($\sigma_{nom}, \varepsilon_{nom}$) which combines nominal stress–strain data with a uniaxial test, a simple conversion to true stress and logarithmic plastic strain is given by the equation below:

$$\begin{cases} \sigma_{true} = \sigma_{nom}\left(1+\varepsilon_{nom}\right) \\ \varepsilon_{true}^{p} = ln\left(1+\varepsilon_{nom}\right) - \frac{\sigma_{true}}{E} \end{cases} \tag{5.1}$$

Abaqus only calculates materials curve with a monotonic function[1] meaning with no variation of the first derivative as shown in Fig. 5.1, this is why the user needs to switch the material data from engineering curve to the true stress–strain curve because from region (EF) the first derivative ($\frac{d\sigma}{d\varepsilon}$) is positive and after the onset of necky the same first derivative becomes negative in region (FG). The monotonic function includes a first derivative value equal to zero, which represents a perfect plasticity zone almost identical to that shown in region (DE), because with real material data there are some small variations.

A classic material curve shape for steel material is given in Fig. 5.1 along with its different properties data such as

- Point A representing the limit of proportionality.
- Point B representing Elastic limit.

[1]In mathematics, a monotonic function (or monotone function) is a function between ordered sets that preserves or reverses the given order. This concept first arose in calculus, and was later generalized to the more abstract setting of order theory.

R. J. Boulbes, *Troubleshooting Finite-Element Modeling with Abaqus*,
https://doi.org/10.1007/978-3-030-26740-7_5

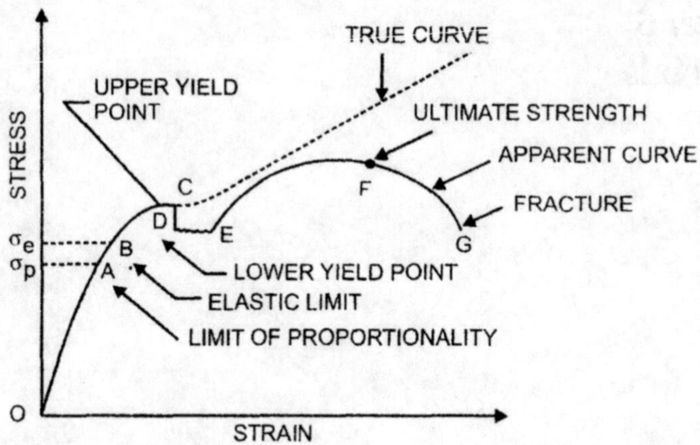

Fig. 5.1 Material engineering stress–strain data curve plotted with its calculated true curve

- Point C representing Upper yield point.
- Point D representing Lower yield point.
- Point E representing Strain hardening starts.
- Point F representing Ultimate point or maximum stress point.
- Point G representing Fracture point.

Each different region has a physics sense and is identified:

Region OA: The straight line it is the proportional region where Hook's law is valid in 1D, with the Young's modulus E.

Region OB: Elastic region until the yield stress.

Region BD: Elastoplastic region taking into account the yield strength which is the real value of the yield stress rescale at 0.2% offset.

Region DE: The transition region of the straight hardening. This is the zone in which the dislocations start moving inside the material microstructure. Anything that makes it more difficult for dislocations to move will increase strength and decrease ductility at the same time. A material is ductile if the fracture point is far away from this region and brittle otherwise.

Region EF: Power law hardening $\sigma = k\left(\varepsilon_p\right)^n$ with the strength coefficient k and the strain hardening exponent n material constants as a function of the material microstructure identified in the uniaxial tension test. This is the zone where the plastic deformation is uniform and homogeneous.

Region FG: Neck region, where following the onset of necky the plastic deformation stops being uniform and homogeneous to become non uniform with a progressive reduction of the cross section until it breaks at the fracture point.

Some norms and standards provide calculation procedures to get the true stress–strain curve data from the material's property data. It can also be useful to plot both material curves as shown in Fig. 5.1 to figure out whether or not the true curve represents a conservative material response compared to the test data. For example, in Fig. 5.1 the true material curve is higher than any material test responses and therefore, the true curve is not a conservative response. A conservative true curve response, in that case, would be a perfect plasticity (flat curve) from point B to fracture point G. It is possible to do the same from point A, however, in such cases, the conservatism of results will be ensured more effectively than point B results: from point B, there will be a conservative true material curve, while point A will produce a very conservative true material curve.

A simple way of quantifying the conservatism of a plotted curve as a function of another is to calculate the total work, meaning performing an integration of the full material curve ($\int \sigma\,(\varepsilon)\,d\varepsilon$). For instance, in Fig. 5.1, the work of the true curve will be higher than the work of the material engineering curve and therefore the true curve will not be a conservative response compared to the material data.

In the same way that the plastic curve is set in a table format with Abaqus, the last data sets $(\sigma_n, \varepsilon_n)$ can be extended from the last slope data curve up to 100% of the strain in order to bypass numerical difficulties to continue incrementing the solution. The final two extended data points $(\sigma_{n+1}, \varepsilon_{n+1})$ are calculated from the previous slope $(\sigma_{n-1}, \varepsilon_{n-1})$ as an extended linear function such as

$$\sigma_{n+1} = \frac{\sigma_{n-1} - \sigma_n}{\varepsilon_{n-1} - \varepsilon_n}\,(\varepsilon_{n+1} - \varepsilon_n) + \sigma_n \tag{5.2}$$

The plastic curve can be extended for a strain $\varepsilon_{n+1} = 1$ and thanks to Eq. 5.2 it is now possible to calculated the stress value σ_{n+1} to get the last data points in the plastic curve. Another solution is to set a perfect plastic curve $\sigma_n = \sigma_{n+1}$ up to 100% of the strain. In both cases, the analyst must know that the stress value will not represent the real fracture stress value as shown in Fig. 5.1. This is why some standards require material properties to be set with the ultimate tensile stress (the ultimate strength point F in Fig. 5.1) to conduct a more realistic and conservative calculation of stress analyses.

5.2 The Current Strain Increment Exceeds the Strain to First Yield

In the message file, there are warnings that the current strain increment exceeds the strain to first yield by more than 50 times: "the strain increment has exceeded fifty times the strain to cause first yield". This warning message is issued only if something in the model is causing relatively large deformations in a problem in which the material model is some form of plasticity. The causes may be erroneous input data, poor meshes, or instabilities developing in the model after some deformation.

- Does the message occur during the first increment? If so, the input data is probably incorrect. Check the following:

  - Is the loading excessive?
  - Is the magnitude of loading physically realistic?
  - Are the units correct with respect to the rest of the model?
  - If this is a dynamic analysis, are the units for density correct?
- If the message occurs after some plastic deformation has developed:
  - Check the stress–plastic strain data. Make sure the material data does not become perfectly plastic at too low a plastic strain. If it does, add a small amount of hardening up to very large plastic strains. If the user needs perfectly plastic material, it may be necessary to use the Riks procedure.
  - Is the mesh refinement adequate?
    Meshes that are adequate for linear elasticity may not be adequate for large-strain plasticity problems. If the elements for which this message is being given are in a region where the geometry changes rapidly, the mesh in this region may need to be refined.
  - If neither of the above problems is the cause, the message may be due to unstable deformation (possibly local). Look at the increments just prior to the occurrence of the problem and see if a physical instability seems reasonable. Including viscous stabilization in a static, coupled temperature displacement, soils or a quasi-static (Visco) procedure may stabilize the model sufficiently so that this problem does not occur.

5.3 Convergence Behavior of Models Using Hyperelastic Materials

Assuming that the UMAT subroutine coded in Fortran has been correctly written, the difference may be attributable to the way in which the initial stress stiffness portion of the system matrix is computed for the native ***HYPERELASTIC** material models.

For most material models in Abaqus, the initial stress stiffness during the first iteration of any increment is computed based on the stress at the end of the previous converged increment. From the second iteration onwards, the initial stress stiffness is based on the current $t + \Delta t$ estimate of the stress. Abaqus uses this approach based on experience with various classes of problems. The approach is also motivated by the fact that during the first iteration of any increment the displacement (and hence the stresses) is based, by default, on extrapolation from the previous increment, and hence may not be capturing the actual displacement history due to the applied loads. This approach is also followed by the UMAT implementation.

The exception is hyperelasticity. For ***HYPERELASTIC** materials, the current stress is always used to compute the initial stress stiffness. This difference can sometimes be sufficient to cause different convergence patterns to be observed between a native hyperelastic model and an equivalent UMAT.

This difference in the initial stress stiffness calculation can be eliminated by using **EXTRAPOLATION=NO** on the ***STEP** option. In this case, the initial guess for the solution of an increment is the converged state of the previous increment. Consequently, the stress at the beginning of the increment is used to evaluate the initial stress stiffness during the first iteration of the increment. This will be true regardless of material model used. This fact can be exploited to help debug a UMAT for hyperelasticity.

Because extrapolation is intended to improve convergence, discontinuing its use may introduce additional differences to the respective convergence behaviors that have no relation to the material behavior. For the comparison to be valid, both models (UMAT and native hyperelasticity) must use no extrapolation.

5.4 Models Using Incompressible or Nearly Incompressible Materials

This is the case when a model in Abaqus standard analysis failed to converge and materials with incompressible or nearly incompressible behavior are used. The analysis has probably failed in these situations because of a wrong element type. Indeed, if the wrong element types are used when the material is incompressible or nearly incompressible, the mesh may lock volumetrically; it will be overconstrained and will not allow arbitrary deformations. The response of the model will be too stiff.

To solve this problem, quilt plots of pressure stress should be created in Abaqus viewer. If the magnitude changes sign from one element to the next or if the magnitude has a regular pattern across a number of elements in a region, the mesh is probably locking.

For second-order elements in Abaqus viewer the averaging across elements should be set to zero while checking that the variation of pressure stress in an element is not large; in particular, check that the value is not changing from positive to negative across diagonals of the element. (If the user prints out pressure stress at the integration points for second-order elements, this behavior can be seen very easily when it occurs.)

Deformed mesh plots of volumetrically locking regions can look like hourglassing for second-order elements. If the incompressibility is primarily due to large inelastic straining, reduced-integration second-order elements or first-order elements can be used to reduce volumetric locking. Mesh refinement will also help delay the onset of locking for these elements. If these elements are still locking at large strains, one option is to try hybrid elements (those whose element name ends in H). Hybrid elements use a mixed formulation that treats the pressure field as an independent variable.

Is the material fully incompressible and over-confined, that is, the material cannot move due to the confinement? If so, the tolerance on the pressure terms of the hybrid elements will not be satisfied; the user can see this by looking in the message file, which will display will be a message about volumetric compatibility errors.

Add a small amount of compressibility to prevent locking of highly confined incompressible materials. An example of such a material is an incompressible rubber block that is bonded to a rigid surface and is then deformed. The problem area is frequently the region where the boundary condition changes from fully bonded to free. Frequently, only a small region around the elements having the problem needs to have the incompressibility constraint relaxed. The amount of compressibility required generally reduces the effective Poisson's ratio of the material from 0.5 or close to 0.5 to a value closer to 0.45.

5.5 Equivalence of Uniaxial Tension and Compression Hyperelastic Test Data

Here the analyst will be concerned with how to simulate the quasi-static compression behavior of a hyperelastic material only using the uniaxial tension test data available. In such situations, it is essential to know the maximum compression strain up to which the response of the user's model will be based on the available test data.

The discussion presented here is valid only for material models with strain energy potentials that do not depend on the second deviatoric strain invariant. In Abaqus standard or Abaqus explicit, this includes the Marlow, Neo-Hooke, Reduced Polynomial, Yeoh, and Arruda–Boyce representations. Moreover, the argument applies only to uniaxial loads (tension or compression) and assumes nearly incompressible material behavior.

For any of the aforementioned models, the first deviatoric strain invariant $\bar{I}_1$ can be written in terms of the deviatoric stretches and is defined as

$$\bar{I}_1 = \bar{\lambda}_1^2 + \bar{\lambda}_2^2 + \bar{\lambda}_3^2 \tag{5.3}$$

The deviatoric stretches are defined as

$$\bar{\lambda}_i = J^{-\frac{1}{3}}\lambda_i \tag{5.4}$$

where $J = \lambda_1\lambda_2\lambda_3$ is the total volume ratio and λ_i with $i = 1, 2, 3$ which are the principal stretches. For a nearly incompressible material $J \approx 1$.

For an uniaxial test, it becomes:

$$\lambda_1 = \lambda \qquad \lambda_2 = \lambda_3 = \frac{1}{\sqrt{\lambda}} \tag{5.5}$$

The uniaxial stretch λ is related to the nominal strain ε as

$$\lambda = 1 + \varepsilon \tag{5.6}$$

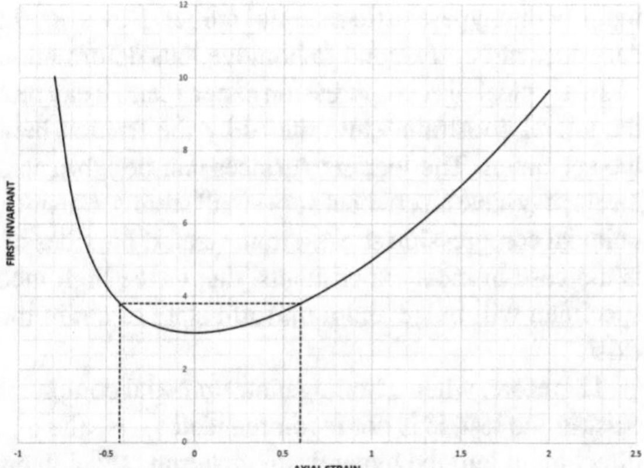

Fig. 5.2 First invariant versus axial strain

where the nominal strain ε is defined on the interval $-1 < \varepsilon < \infty$. Thus the first invariant becomes:

$$I_1 = \lambda^2 + \frac{2}{\lambda} = (1+\varepsilon)^2 + \frac{2}{(1+\varepsilon)} \tag{5.7}$$

It can be shown that there are two real roots for ε in Eq. 5.7. These are the values of nominal strain that give the same value of I_1, one corresponding to a state of uniaxial tension and the other corresponding to a state of uniaxial compression. Figure 5.2 presents a schematic plot of I_1 versus ε.

Under the assumption that only uniaxial tensile test data is available, it is possible to compute the value of the first invariant I_1 for the largest value of tensile nominal strain, and then compute the corresponding value of compressive nominal strain that produces the same value of the first invariant (since both the tension and compression responses of the material are functions of I_1). This value of the compressive nominal strain will be the maximum value to which the material response will be based on the experimental data. Beyond this level of strain, the material behavior is extrapolated from the test data.

In Fig. 5.2 the following example to obtain the equivalence between uniaxial tension and compression hyperelastic test data is given; if the maximum tensile strain is 0.6, the corresponding maximum compressive strain will be -0.425277 (I_1 is 3.81 in both cases).

5.5.1 *Uniaxial Compression Test Data for a Rubber Material*

Analysts in specific cases should understand how a uniaxial compression test data for a rubber material can be set as a direct input in the hyperelastic material model. For a homogenous material, homogenous deformation modes will suffice to characterize the material constants; hence, uniaxial test data for compression can be used as direct input in the hyperelastic material model. However, this is not the case when the test data does not correspond to homogenous deformation states.

In a typical disk compression experiment, it can be especially difficult to minimize the lateral constraints introduced by the friction between the loading surfaces and the specimen. The loading surfaces can be lubricated, but even very small friction levels may have an effect at very small strains, causing deviations from a homogenous uniaxial compression stress–strain state. If the sides of the specimen are glued, which is the case in most experiments, due to the high incompressibility the sides of the specimen will bulge and a uniform state of strain through the sample will still not exist.

Therefore, when a homogenous uniaxial compression stress–strain state does not exist in the test, it is not recommended to use the uniaxial compression test data as direct input into the hyperelastic material model. However, the test data can be used to validate a component FEA model that replicates the compression test conditions.

The deformation induced in an equibiaxial tension test will be the equivalent to that of a uniaxial compression test and this data can be used as an input in the hyperelastic model.

5.5.2 Specifying Tension or Compression Test Data for the Marlow Hyperelasticity Model

Defining the properties of a hyperelastic[2] material with the Marlow model can be difficult for analysts because they are not able to specify tension test data in combination with compression test data in 2D or 3D elements.

The mechanical response of a material is defined by choosing a strain energy potential to fit the particular material. The strain energy potential forms in Abaqus are written as separable functions of a deviatoric component and a volumetric component; i.e., $U = U_{dev}\left(\bar{I}_1, \bar{I}_2\right) + U_{vol}\left(J\right)$. Alternatively, in Abaqus standard, it is possible to define the strain energy potential with user subroutine **UHYPER**, in which case the strain energy potential need not be separable.

Generally, for the hyperelastic material models available in Abaqus, the user can either directly specify material coefficients or provide experimental test data and have Abaqus automatically determine appropriate values for the coefficients. An exception is the Marlow form. In this case the deviatoric part of the strain energy potential must be defined with test data. The different methods for defining the strain energy potential are described in detail below.

The properties of rubber like materials can vary significantly from one batch to another; therefore, if data is used from several experiments, all of the experiments should be performed on specimens taken from the same batch of material, regardless of whether the coefficients are computed by the user or Abaqus.

[2]A complement of information about the usage of hyperelastic materials can be found in the Abaqus Analysis User's Guide in Sect. 22.5 Hyperelasticity.

When specifying deviatoric test data for the Marlow model, only a single type of data (i.e., uniaxial, biaxial or planar) can be used. Further, for two- and three-dimensional elements, either data from tension tests or data from compression tests can be used, but not both. This is because for isotropic incompressible hyperelastic materials, certain types of experimental tests are equivalent. Specifically:

- Uniaxial tension ↔ Equibiaxial compression,
- Uniaxial compression ↔ Equibiaxial tension, and
- Planar tension ↔ Planar compression.

By equivalent, it means that the tests will induce equivalent deformations. For example, the deformation induced in a uniaxial compression test is equivalent to that of an equibiaxial tension test. Thus, if uniaxial tension is specified as well as uniaxial compression test data then implicitly one is simultaneously specifying uniaxial tensile behavior and biaxial tensile behavior, and this combination cannot be best-fitted by the Marlow model.

When only tensile or compressive data is specified, then a strain energy potential is constructed that will reproduce the test data exactly and have reasonable behavior for other deformation modes.

In one-dimensional elements (beams, trusses, and rebars), deformations in only one direction are considered, and therefore both tensile, as well as compressive uniaxial test data, can be best-fitted by the Marlow model.

Compressibility can be defined by specifying nonzero values for D_i (except for the Marlow model) either by setting the Poisson's ratio to a value lower than 0.5, or by providing test data that characterizes the compressibility. The test data method is described later in this section. If the Poisson's ratio is specified for hyperelasticity other than the Marlow model, Abaqus computes the initial bulk modulus from the initial shear modulus.

$$D_1 = \frac{2}{K_0} = \frac{3(1-2\nu)}{\mu_0(1+\nu)} \tag{5.8}$$

For the Marlow model, the specified Poisson's ratio represents a constant value, which determines the volumetric response throughout the deformation process. If D_1 is equal to zero, all of D_1 the must be equal to zero. In such cases, the material is assumed to be fully incompressible in Abaqus standard, while Abaqus explicit will assume compressible behavior with $\frac{K_0}{\mu_0} = 20$ (Poisson's ratio of 0.475).

5.5.3 Using Simple Shear Experimental Data for Hyperelastic Materials

To complete the tension and compression overview, the shear component in hyperelasticity material must be considered how should a user deal with an experimental stress–strain data from a simple shear test of a rubber specimen? Can this data be used in Abaqus?

The data can be used, but not directly. It must first be converted into equivalent planar test data. It must be noted that if the material that has been tested is an elastomeric foam, then the following does not apply to complete the data conversion we must assume that the tested material is incompressible.

The schematic description of a simple shear test is most often that of a block being sheared into a parallelogram. For an incompressible material, the principal stretch ratios in the simple shear deformation mode are the same as those of a pure shear deformation mode. The pure shear deformation mode does not involve any rotation of the principal axes of strain however; we may then say that a simple shear is equivalent to a pure shear together with a rotation [1].

In practice, a pure shear deformation mode is induced in a tensile planar test. The planar test involves the extension of a wide, thin strip that is clamped on its long edges. The conversion of simple shear data to equivalent planar test data can be completed as per Treloar [1].

Let T_s be the nominal stress in pure shear, λ_1 the first principal stretch ratio and τ_{xy} the shear stress per unit strained area in simple shear. The resultant equation is therefore:

$$\tau_{xy} = \frac{\lambda_1^2}{1+\lambda_1^2} T_s \tag{5.9}$$

The relationship between shear strain and the principal stretch ratios is discussed by Treloar [1] and Ogden [2] and is given as

$$\gamma = \lambda_1 - \frac{1}{\lambda_1} \tag{5.10}$$

The principal stretch and the nominal strain are related as

$$\lambda_1 = 1 + \varepsilon_1 \tag{5.11}$$

Combining equations from Eq. 5.9 and Eq. 5.11 can be solved to convert the known τ_{xy} and γ data to the T_s and ε_1 data necessary for Abaqus. Solving Eq. 5.10 for λ_1 and substituting into Eq. 5.11, it becomes:

$$\varepsilon_1 = \frac{\gamma}{2} - 1 + \sqrt{1 + \frac{\gamma^2}{4}} \tag{5.12}$$

Next, solving now Eq. 5.10 for λ_1 and substituting into Eq. 5.9 gives the nominal stress in pure shear component as follows:

$$T_s = \tau_{xy}\left[1 + \frac{1}{1 + \frac{\gamma^2}{2} + \gamma\sqrt{1 + \frac{\gamma^2}{4}}}\right] \tag{5.13}$$

5.6 Path Dependence of Nonlinear Results Using an Elastic Material

The analyst has conducted a nonlinear elastic analysis that has closed-loop loading. When the loading is returned to zero, a residual stress remains despite the fact that the user has not included plasticity in the model. Is this a bug?

The behavior observed in this case is not a bug, but a characteristic of the solution method. When a geometrically nonlinear analysis is performed (**NLGEOM** parameter included on ***STEP**), an incremental solution approach is taken. In each increment of the solution, an increment of strain is needed for the constitutive calculations. The increment of strain is calculated by integrating the strain rate over the time length of the increment [3] $\frac{\partial \varepsilon}{\partial t}$, which is essentially the first derivative in time of the strain Eq. 4.3.

When the ***ELASTIC** material model is used in nonlinear simulations, an updated incremental approach is used. Specifically, a Jaumann stress rate and velocity strain material description are used.[3] With this method, the reference configuration for the integration of the strain rate is the condition of the model at the end of the previous increment [4].

Under these conditions, when the strain rate is integrated in a closed-loop fashion it is known that the integral does not vanish. Thus, if a displacement cycle is applied to the model such that the cycle ends with the model in its initial configuration, a nonzero strain will result. Because there is a nonzero strain at the end of the cycle, there will be a corresponding nonzero stress. This will occur for both applied displacements and loads. This phenomenon will not occur with ***HYPERELASTIC** [5] material models because a total formulation, rather than an updated one, is used for these materials.

The ***ELASTIC** material model is intended to be used in small strain applications, typically to model the elastic response of a metal. In these situations, the amount of strain that remains at the completion of the loading loop is normally minimal and does not introduce a significant error. Further, reducing the size of the time increment used by the analysis will reduce the magnitude of the nonzero strains.

---

[3] In continuum mechanics, objective stress rates are time derivatives of stress that do not depend on the frame of reference [3]. The Jaumann rate of the Cauchy stress is a further specialization of the Lie derivative (Truesdell rate). For this rate, ω is the spin tensor (the skew part in Eq. 4.3 of the velocity gradient) as follows:

$$\tilde{\sigma} = \dot{\sigma} + \sigma.\omega - \omega.\sigma \tag{5.14}$$

Many constitutive equations are designed in the form of a relation between a stress rate and a strain rate (or the rate of deformation tensor). The mechanical response of a material should not depend on the frame of reference. In other words, material constitutive equations should be frame indifferent (objective). If the stress and strain measures are material quantities then objectivity is automatically satisfied. However, if the quantities are spatial, then the objectivity of the stress rate is not guaranteed even if the strain rate is objective.

For example, the stress in a bar element after it has been stretched is σ_{11}, and this stress does not change during the rigid body rotation. The X', Y' coordinate system that rotates as a result of the rigid body rotation is the corotational coordinate system. Consequently, the stress tensor and state variables are, therefore, computed directly and updated in user subroutine **VUMAT** using the strain tensor since all of these quantities are in the corotational system; these quantities do not have to be rotated by the user subroutine as is sometimes required in user subroutine **UMAT**.

The elastic response predicted by a rate form constitutive law depends on the objective stress rate employed. For example, the Green–Naghdi stress rate (looks with the same form of equation as used for the Jaumann stress rate) is used in **VUMAT**. However, the stress rate used for built-in material models may differ. For instance, most material models used with solid (continuum) elements in Abaqus explicit employ the Jaumann stress rate. This difference in the formulation will only cause significant differences in the results if finite rotation of a material point is accompanied by finite shear.

The objective stress rates used in Abaqus are summarized in Table 5.1. Objective rates are only relevant for rate form constitutive equations (e.g., elastoplasticity). For hyperelastic materials a total formulation is used; hence, the concept of an objective rate is not relevant for the constitutive law. However, when material orientations are defined, the objective rate governs the evolution of the orientations and the output will be affected.

Table 5.1 Objective stress rates

Solvers	Element types	Constitutive models	Objective rate
Standard	Solid (Continuum)	All built-in and user-defined materials	Jaumann[a]
	Structural (Shells, Membranes, Beams, Trusses)	All built-in and user-defined materials	Green–Naghdi[b]
Explicit	Solid (Continuum)	All except viscoelastic, brittle cracking, and VUMAT[c]	Jaumann[a]
		Viscoelastic, brittle cracking, and VUMAT[c]	Green–Naghdi[b]
	Structural (Shells, Membranes, Beams, Trusses)	All built-in and user-defined materials	Green–Naghdi[b]

[a]The Jaumann rate is used widely in computations primarily for two reasons, first it is relatively easy to implement, and second it leads to symmetric tangent moduli

[b]The GreenNaghdi rate of the Kirchhoff stress also has the form since the stretch is not taken into consideration

[c]It is a user subroutine to define material behavior only in Abaqus explicit. The use of this user subroutine generally requires considerable expertise. The analysts are cautioned that the implementation of any realistic constitutive model requires extensive development and testing. Initial testing on a single-element model with prescribed traction loading is strongly recommended

5.7 User Material Subroutine

Abaqus standard and Abaqus explicit have interfaces that allow the user to implement general constitutive equations. In Abaqus standard, the user-defined material model is implemented in user subroutine UMAT. In Abaqus explicit, the user-defined material model is implemented in user subroutine VUMAT. The usage of either UMAT or VUMAT when none of the existing material models included in the Abaqus material library accurately represents the behavior of the material to be modeled.

These interfaces make it possible to define any constitutive model of arbitrary complexity. User-defined material models can be used with any Abaqus structural element type. Multiple user materials can be implemented in a single UMAT or VUMAT routine and can be used together. In this section the implementation of material models in UMAT or VUMAT will be discussed and illustrated with a number of examples.

First of all, the motivation to establish a proper testing of advanced constitutive models to simulate experimental results often requires complex finite-element models, for instance, with advanced structural elements, complex loading conditions, thermo mechanical loading, contact and friction conditions, and static and dynamic analysis. Special analysis problems occur if the constitutive model simulates material instabilities and localization phenomena. For example, with special solution techniques are required for quasi-static analysis, robust element formulations should be available or explicit dynamic solution algorithms with robust, vectorized contact algorithms are desired. In addition, robust features are required to present and visualize the results with a contour and path plots of state variables, a X-Y plots or in a tabulated result. Thus the material model developer should be concerned only with the development of the material model and not the development and maintenance of the FE software. Such as the developments unrelated to material modeling, porting problems with new systems and the long-term program maintenance of user-developed code. Some references are given here as an example to illustrate this point regarding damage modeling [6], excavations mining [7], laminated composite [8], processing of metal powders [9] or an evolution of crystallographic texture [10].

The results are obtained with the modified stiffness matrix, because the errors in the tangent stiffness will only affect the convergence behavior, not the final result. When the elastic stiffness is used instead of the true stiffness, the number of iterations is increased. The results (when obtained) are still correct.

The convergence behavior of a model where a pressure loading is applied instead of a boundary condition is better controlled in displacement. Indeed, displacement control problems are more stable than load control problems, because when the elastic stiffness is used instead of the true stiffness, the displacement control analysis ran successfully but the load control analysis failed.

5.7.1 Guideline to Write a UMAT or a VMAT

Proper definition of the constitutive equation, which requires one of the following:

- Explicit definition of stress (Cauchy stress for large-strain applications).
- Definition of the stress rate only (in corotational framework).

Furthermore, it is likely to require:

- Definition of dependence on time, temperature, or field variables.
- Definition of internal state variables, either explicitly or in rate form.

The transformation of the constitutive rate equation into an incremental equation using a suitable integration procedure with forward Euler (explicit integration), or backward Euler (implicit integration) or the midpoint method.

The most difficult part is the forward Euler (explicit) integration methods because they are simple but have a stability limit according to Eq. 5.15,

$$|\Delta\varepsilon| < \Delta\varepsilon_{stab} \tag{5.15}$$

where is usually less than the elastic strain magnitude. For explicit integration, the time increment must be controlled. For implicit or midpoint integration the algorithm is more complicated and often requires local iteration. However, there is usually no stability limit. An incremental expression for the internal state variables must also be obtained.

The calculation of the (consistent) Jacobian (required for Abaqus standard UMAT only). For small-deformation problems (e.g., linear elasticity) or large-deformation problems with small volume changes (e.g., metal plasticity), the consistent Jacobian is shown in Eq. 5.16,

$$C = \frac{\partial \Delta\sigma}{\partial \Delta\varepsilon} \tag{5.16}$$

where is the increment in (Cauchy) stress and is the increment in strain. (In finite-strain problems, it is an approximation to the logarithmic strain.) This matrix may be nonsymmetric as a result of the constitutive equation or integration procedure. The Jacobian is often approximated such that a loss of quadratic convergence may occur.

It is easily calculated for forward integration methods (usually the elasticity matrix).

If large deformations with large volume changes are considered (e.g., pressure-dependent plasticity), the exact form of the consistent Jacobian should be used to ensure rapid convergence in accordance with Eq. 5.17. Here is the determinant of the deformation gradient.

$$C = \frac{1}{J}\frac{\partial \Delta(J\sigma)}{\partial \Delta\varepsilon} \tag{5.17}$$

The hyperelastic constitutive equations include a total-form of constitutive equations relating the Cauchy stress and the deformation gradient F are commonly used to model, for example, rubber elasticity. In this case, the consistent Jacobian is defined through Eq. 5.18,

$$\delta(J\sigma) = JC : \delta D \quad (5.18)$$

where $J = |F|$, and C is the material Jacobian, and δD is the virtual rate of deformation, defined as shown in Eq. 5.19.

$$\delta D = sym\{\delta F.F^{-1}\} \quad (5.19)$$

Then coding the UMAT or VUMAT need to follow the Fortran or C conventions. The user will have to make sure that the code can be vectorized (for VUMAT only, to be discussed later). Also to make sure that all variables are defined and initialized properly. That subroutine uses Abaqus utility routines as required, and assign enough storage space for state variables with the **DEPVAR** option.

As a good practice, it is also recommended to verify the UMAT or VUMAT with a small (one element) input file in order to cross-check the results with the theoretical calculations afterward.

1. Run tests with all displacements prescribed to verify the integration algorithm for stresses and state variables. Suggested tests include:
   - Uniaxial
   - Uniaxial in oblique direction
   - Uniaxial with finite rotation
   - Finite shear
2. Run similar tests with load prescribed to verify the accuracy of the Jacobian.
3. Compare test results with analytical solutions or standard Abaqus material models, if possible. If the above verification is successful, apply to more complicated problems.

5.8 UMAT Subroutine Examples

These input lines act as the interface to a UMAT in which isotropic hardening plasticity is defined.

```
*MATERIAL, NAME=ISOPLAS
*USER MATERIAL, CONSTANTS=8, (UNSYMM)
30.E6, 0.3, 30.E3, 0., 40.E3, 0.1, 50.E3, 0.5
*DEPVAR
13
*INITIAL CONDITIONS, TYPE=SOLUTION
Data line to specify initial solution-dependent
variables
*USER SUBROUTINES,(INPUT=file_name)
```

The **USER MATERIAL** option is used to input material constants for the UMAT. The unsymmetric equation solution technique will be used if the UNSYMM parameter is used.

The **DEPVAR** option is used to allocate space at each material point for solution-dependent state variables (SDVs). The **INITIAL CONDITIONS, TYPE= SOLUTION** option is used to initialize SDVs if they are starting at a nonzero value. Thus coding for the UMAT is supplied in a separate file. The UMAT is invoked with the ABAQUS execution procedure, as follows:

```
abaqus job=... user=...
```

The user subroutine must be invoked in a restarted analysis because user subroutines are not saved on the restart file.

If a constant material Jacobian is used and no other nonlinearity is present, reassembly can be avoided by invoking the quasi-Newton method with the input line:

```
*SOLUTION TECHNIQUE, REFORM KERNEL=n
```

with n the number of iterations done without reassembly. This does not offer advantages if other nonlinearities (such as contact changes) are present.

The solution-dependent state variables can be output with identifiers SDV1, SDV2, etc. Contour, path, and X-Y plots of SDVs can be plotted in Abaqus viewer.

The following quantities are available in UMAT:

- Stress, strain, and SDVs at the start of the increment
- Strain increment, rotation increment, and deformation gradient at the start and end of the increment
- Total and incremental values of time, temperature, and user-defined field variables
- Material constants, material point position, and a characteristic element length
- Element, integration point, and composite layer number (for shells and layered solids)
- Current step and increment numbers.

The following quantities must be defined the stress, SDVs, and material Jacobian.

The following variables may be defined the strain energy, plastic dissipation, and creep dissipation, and the suggested new (reduced) time increment.

The header is usually followed by dimensioning of local arrays. It is good practice to define constants via parameters and to include comments. The **PARAMETER** assignments yield accurate floating-point constant definitions on any platform as shown below:

```
PARAMETER(ZERO=0.D0, ONE=1.D0, TWO=2.D0, THREE=3.D0,
1 SIX=6.D0, ENUMAX=.4999D0, NEWTON=10, TOLER=1.0D-6)
```

Utility routines like **SINV**, **SPRINC**, **SPRIND**, and **ROTSIG** can be called to assist in coding UMAT.

- **SINV** will return the first and second invariants of a tensor.
- **SPRINC** will return the principal values of a tensor.
- **SPRIND** will return the principal values and directions of a tensor.
- **ROTSIG** will rotate a tensor with an orientation matrix.
- **XIT** will terminate an analysis and close all files associated with the analysis properly.

The UMAT conventions in indexing variable are the following:

- Stresses and strains are stored as vectors.
  - For plane stress elements: $\sigma_{11}, \sigma_{22}, \sigma_{12}$
  - For (generalized) plane strain and axisymmetric elements: $\sigma_{11}, \sigma_{22}, \sigma_{33}, \sigma_{12}$
  - For three-dimensional elements: $\sigma_{11}, \sigma_{22}, \sigma_{33}, \sigma_{12}, \sigma_{13}, \sigma_{23}$
- The shear strain is stored as engineering shear strain, $\gamma_{12} = 2\varepsilon_{12}$
- The deformation gradient, F_{ij}, is always stored as a three-dimensional matrix.

The different UMAT formulation aspects must be considered for the geometrically nonlinear analysis the strain increment and the incremental rotation passed into the routine are based on the Hughes–Winget formulae. Including the linearized strain and rotation increments are calculated in the mid-increment configuration, and the approximations made, particularly if rotation increments are large: more accurate measures can be obtained from the deformation gradient if desired. Moreover, the user must define the Cauchy stress: when this stress reappears during the next increment, it will have been rotated with the incremental rotation, **DROT**, passed into the subroutine. The stress tensor can be rotated back using the utility routine **ROTSIG** if this is not desired. If the **ORIENTATION** option is used in conjunction with UMAT, stress and strain components will be in the local system (again, this basis system rotates with the material in finite-strain analysis). Thus the tensor state variables must be rotated in the subroutine (use **ROTSIG**). If UMAT is used with reduced-integration elements or shear flexible shell or beam elements, the hourglass stiffness, and the transverse shear stiffness must be specified with the **HOURGLASS STIFFNESS** and **TRANSVERSE SHEAR STIFFNESS** options, respectively.

At the start of a new increment, the strain increment is extrapolated from the previous increment. This extrapolation, which may sometimes cause trouble, is turned off with **STEP, EXTRAPOLATION=NO**. If the strain increment is too large, the variable **PNEWDT** can be used to suggest a reduced time increment. The code will abandon the current time increment in favor of a time increment given by **PNEWDT** multiplied by **DTIME**. The characteristic element length can be used to define softening behavior based on fracture energy concepts.

5.8.1 UMAT Subroutine for Isotropic Isothermal Elasticity

The governing equations of the problem are listed below:

1. Isothermal elasticity in Eq. 5.20 (with Lamé's constants):

$$\sigma_{ij} = \lambda\delta_{ij}\varepsilon_{kk} + 2\mu\varepsilon_{ij} \tag{5.20}$$

   or in a Jaumann (corotational) rate form in Eq. 5.21:

$$\dot{\sigma}^{J}_{ij} = \lambda\delta_{ij}\dot{\varepsilon}_{kk} + 2\mu\dot{\varepsilon}_{ij} \tag{5.21}$$

2. The Jaumann rate Eq. 5.22 is integrated in a corotational framework:

$$\Delta\sigma_{ij} = \lambda\delta_{ij}\Delta\varepsilon_{kk} + 2\mu\Delta\varepsilon_{ij} \tag{5.22}$$

The appropriate coding is shown below:

Listing 5.1 IIE.f

```
      SUBROUTINE UMAT(STRESS,STATEV,DDSDDE,SSE,SPD,
     1 SCD,RPL,DDSDDT,DRPLDE,DRPLDT,STRAN,DSTRAN,TIME,
     2 DTIME,TEMP,DTEMP,PREDEF,DPRED,CMNAME,NDI,NSHR,
     3 NTENS,NSTATV,PROPS,NPROPS,COORDS,DROT,PNEWDT,
     4 CELENT,DFGRD0,DFGRD1,NOEL,NPT,LAYER,KSPT,KSTEP,
     5 KINC)
C
      INCLUDE 'ABA_PARAM.INC'
C
      CHARACTER*8 CMNAME
C
      DIMENSION STRESS(NTENS), STATEV(NSTATV),
     1 DDSDDE(NTENS, NTENS), DDSDDT(NTENS),
     2 DRPLDE(NTENS), STRAN(NTENS), DSTRAN(NTENS),
     3 PREDEF(1), DPRED(1), PROPS(NPROPS), COORDS(3),
     4 DROT(3, 3), DFGRD0(3, 3), DFGRD1(3, 3)
C ------------------------------------------------------------
C UMAT FOR ISOTROPIC ELASTICITY
C CANNOT BE USED FOR PLANE STRESS
C ------------------------------------------------------------
C PROPS(1) - E
C PROPS(2) - NU
C ------------------------------------------------------------
C
      IF (NDI.NE.3) THEN
        WRITE(7, *) "THIS UMAT MAY ONLY BE USED FOR
     1 ELEMENTS WITH THREE DIRECT STRESS COMPONENTS"
        CALL XIT
      ENDIF
C
C ELASTIC PROPERTIES
      EMOD=PROPS(1)
```

```
      ENU=PROPS(2)
      EBULK3=EMOD/(ONE-TWO*ENU)
      EG2=EMOD/(ONE+ENU)
      EG=EG2/TWO
      EG3=THREE*EG
      ELAM=(EBULK3-EG2)/THREE
C
C ELASTIC STIFFNESS
C
      DO K1=1,NDI
         DO K2=1,NDI
            DDSDDE(K2,K1)=ELAM
         END DO
         DDSDDE(K1,K1)=EG2+ELAM
      END DO
      DO K1=NDI+1,NTENS
         DDSDDE(K1,K1)=EG
      END DO
C
C CALCULATE STRESS
C
      DO K1=1,NTENS
         DO K2=1,NTENS
            STRESS(K2)=STRESS(K2)+DDSDDE(K2,K1)*
     1      DSTRAN(K1)
         END DO
      END DO
C
      RETURN
      END
```

This very simple UMAT yields has exactly the same results as the Abaqus **ELASTIC** option. This is true even for large-strain calculations: all necessary large-strain contributions are generated by Abaqus. The routine can be used with and without the **ORIENTATION** option. It is usually straightforward to write a single routine that handles (generalized) plane strain, axisymmetric, and three-dimensional geometries.

Generally, plane stress must be treated as a separate case because the stiffness coefficients are different. The routine is written in incremental form as a preparation for subsequent elastic–plastic examples. Even for linear analysis, UMAT is called twice for the first iteration of each increment: once for assembly and once for recovery. Subsequently, it is called once per iteration: assembly and recovery are combined.

A check is performed on the number of direct stress components, and the analysis is terminated by calling the subroutine, **XIT**. A message is written to the message file (unit = 7).

5.8.2 UMAT Subroutine for Non-isothermal Elasticity

The governing equations of the problem are listed below:

1. Non-isothermal elasticity in Eq. 5.23:

$$\sigma_{ij} = \lambda(T)\delta_{ij}\varepsilon_{kk}^{el} + 2\mu(T)\varepsilon_{ij}^{el} \tag{5.23}$$

with the elastic strain defined in Eq. 5.24:

$$\varepsilon_{ij}^{el} = \varepsilon_{ij} - \alpha T\delta_{ij} \tag{5.24}$$

or in a Jaumann (corotational) rate form Eq. 5.25:

$$\dot{\sigma}_{ij}^{J} = \lambda\delta_{ij}\dot{\varepsilon}_{kk}^{el} + 2\mu\dot{\varepsilon}_{ij}^{el} + \dot{\lambda}\delta_{ij}\varepsilon_{kk}^{el} + 2\dot{\mu}\varepsilon_{ij}^{el} \tag{5.25}$$

with the elastic strain rate given by Eq. 5.26:

$$\dot{\varepsilon}_{ij}^{el} = \dot{\varepsilon}_{ij} - \alpha\dot{T}\delta_{ij} \tag{5.26}$$

2. The Jaumann rate Eq. 5.27 is integrated in a corotational framework:

$$\Delta\sigma_{ij} = \lambda\delta_{ij}\Delta\varepsilon_{kk} + 2\mu\Delta\varepsilon_{ij} + \Delta\lambda\delta_{ij}\varepsilon_{kk} + 2\Delta\mu\varepsilon_{ij} \tag{5.27}$$

The appropriate coding is shown below:

Listing 5.2 NIE.f

```
      SUBROUTINE UMAT(STRESS,STATEV,DDSDDE,SSE,SPD,
     1 SCD,RPL,DDSDDT,DRPLDE,DRPLDT,STRAN,DSTRAN,TIME,
     2 DTIME,TEMP,DTEMP,PREDEF,DPRED,CMNAME,NDI,NSHR,
     3 NTENS,NSTATV,PROPS,NPROPS,COORDS,DROT,PNEWDT,
     4 CELENT,DFGRD0,DFGRD1,NOEL,NPT,LAYER,KSPT,KSTEP,
     5 KINC)
C
      INCLUDE 'ABA_PARAM.INC'
C
      CHARACTER*8 CMNAME
C
      DIMENSION STRESS(NTENS), STATEV(NSTATV),
     1 DDSDDE(NTENS, NTENS), DDSDDT(NTENS),
     2 DRPLDE(NTENS), STRAN(NTENS), DSTRAN(NTENS),
     3 PREDEF(1), DPRED(1), PROPS(NPROPS), COORDS(3),
     4 DROT(3, 3), DFGRD0(3, 3), DFGRD1(3, 3)
C LOCAL ARRAYS
C ------------------------------------------------------------
C EELAS - ELASTIC STRAINS
C ETHERM - THERMAL STRAINS
C DTHERM - INCREMENTAL THERMAL STRAINS
C DELDSE - CHANGE_IN STIFFNESS DUE_TO TEMPERATURE
```

```
C ---------------------------------------------------------------
      DIMENSION EELAS(6),ETHERM(6),DTHERM(6),
     1          DELDSE(6,6)
C
      PARAMETER(ZERO=0.D0, ONE=1.D0, TWO=2.D0,
     1          THREE=3.D0, SIX=6.D0)
C ---------------------------------------------------------------
C ISOTROPIC THERMO-ELASTICITY WITH LINEARLY VARYING
C MODULI - CANNOT BE USED FOR PLANE STRESS
C ---------------------------------------------------------------
C PROPS(1) - E(T0)
C PROPS(2) - NU(T0)
C PROPS(3) - T0
C PROPS(4) - E(T1)
C PROPS(5) - NU(T1)
C PROPS(6) - T1
C PROPS(7) - ALPHA
C PROPS(8) - T_INITIAL
C ELASTIC PROPERTIES AT START OF INCREMENT
C
      FAC1=(TEMP-PROPS(3))/(PROPS(6)-PROPS(3))
      IF (FAC1 .LT. ZERO) FAC1=ZERO
      IF (FAC1 .GT. ONE) FAC1=ONE
      FAC0=ONE-FAC1
      EMOD=FAC0*PROPS(1)+FAC1*PROPS(4)
      ENU=FAC0*PROPS(2)+FAC1*PROPS(5)
      EBULK3=EMOD/(ONE-TWO*ENU)
      EG20=EMOD/(ONE+ENU)
      EG0=EG20/TWO
      ELAM0=(EBULK3-EG20)/THREE
C
C ELASTIC PROPERTIES AT_END OF INCREMENT
C
      FAC1=(TEMP+DTEMP-PROPS(3))/(PROPS(6)-PROPS(3))
      IF (FAC1 .LT. ZERO) FAC1=ZERO
      IF (FAC1 .GT. ONE) FAC1=ONE
      FAC0=ONE-FAC1
      EMOD=FAC0*PROPS(1)+FAC1*PROPS(4)
      ENU=FAC0*PROPS(2)+FAC1*PROPS(5)
      EBULK3=EMOD/(ONE-TWO*ENU)
      EG2=EMOD/(ONE+ENU)
      EG=EG2/TWO
      ELAM=(EBULK3-EG2)/THREE
C
C ELASTIC STIFFNESS AT_END OF INCREMENT_AND
c STIFFNESS CHANGE
C
      DO K1=1,NDI
          DO K2=1,NDI
              DDSDDE(K2,K1)=ELAM
              DELDSE(K2,K1)=ELAM-ELAM0
          END DO
          DDSDDE(K1,K1)=EG2+ELAM
```

```
            DELDSE(K1,K1)=EG2+ELAM-EG20-ELAM0
      END DO
      DO K1=NDI+1,NTENS
            DDSDDE(K1,K1)=EG
            DELDSE(K1,K1)=EG-EG0
      END DO
C
C CALCULATE THERMAL EXPANSION
C
      DO K1=1,NDI
            ETHERM(K1)=PROPS(7)*(TEMP-PROPS(8))
            DTHERM(K1)=PROPS(7)*DTEMP
      END DO
      DO K1=NDI+1,NTENS
            ETHERM(K1)=ZERO
            DTHERM(K1)=ZERO
      END DO
C
C CALCULATE STRESS, ELASTIC STRAIN_AND THERMAL
c STRAIN
C
      DO K1=1, NTENS
            DO K2=1, NTENS
                  STRESS(K2)=STRESS(K2)+DDSDDE(K2,K1)*
            1           (DSTRAN(K1)-DTHERM(K1))+DELDSE(K2,K1)*
            2           (STRAN(K1)-ETHERM(K1))
            END DO
            ETHERM(K1)=ETHERM(K1)+DTHERM(K1)
            EELAS(K1)=STRAN(K1)+DSTRAN(K1)-ETHERM(K1)
      END DO
C
c STORE ELASTIC AND THERMAL STRAINS_IN STATE
c VARIABLE ARRAY
C
      DO K1=1, NTENS
            STATEV(K1)=EELAS(K1)
            STATEV(K1+NTENS)=ETHERM(K1)
      END DO
      RETURN
      END
```

This UMAT yields has exactly the same results as the **ELASTIC** option with temperature dependence. The routine is written in incremental form, which allows generalization to more complex temperature dependence.

5.8.3 UMAT Subroutine for Neo-Hookean Hyperelasticity

As discussed in Sect. 5.5.2 the user will also find additional information in Abaqus documentation manuals.[4] The **ELASTIC** option does not work well for finite elastic

[4]See Abaqus Theory Guide v6.14 in Sect. 4.6.2 Fitting of hyperelastic and hyperfoam constants.

strains because a proper finite-strain energy function is not defined. Hence, it is possible to define a proper strain energy density function as shown in Eq. 5.28.

$$U = U(I_1, I_2, J) = C_{10}(I_1 - 3) + \frac{1}{D_1}(J - 1)^2 \tag{5.28}$$

Here I_1 in Eq. 5.29, I_2 in Eq. 5.30, and J in Eq. 5.31 are the three strain invariants, expressed in terms of the left Cauchy–Green tensor in Eq. 5.32, $\underline{B}$:

$$I_1 = trace(\underline{B}) \tag{5.29}$$

$$I_2 = \frac{1}{2}\left(I_1^2 - trace\left(\underline{B}.\underline{B}\right)\right) \tag{5.30}$$

$$J = det\left(\underline{F}\right) \tag{5.31}$$

$$\underline{B} = \underline{F}\,\underline{F}^T \tag{5.32}$$

The deviatoric invariants as defined in Abaqus documentation[5] the constitutive equation can be written directly in terms of the deformation gradient in Eq. 5.33,

$$\sigma_{ij} = \frac{2}{J}C_{10}\left(\bar{B}_{ij} - \frac{1}{3}\delta_{ij}\bar{B}_{kk}\right) + \frac{2}{D_1}(J - 1)\,\delta_{ij} \tag{5.33}$$

with $\bar{B}$ equals to

$$\bar{B}_{ij} = \frac{B_{ij}}{J^{\frac{2}{3}}} \tag{5.34}$$

The Eq. 5.35 defines the virtual rate of deformation as

$$\delta D_{ij} = \frac{1}{2}\left(\delta F_{im} F_{mj}^{-1} + F_{mi}^{-1}\delta F_{jm}\right) \tag{5.35}$$

The Eq. 5.36 defines the Kirchhoff stress as

$$\tau_{ij} = J\sigma_{ij} \tag{5.36}$$

The material Jacobian derives from the variation in Kirchhoff stress according to Eq. 5.37:

$$\delta\tau_{ij} = J C_{ijkl}\delta D_{kl} \tag{5.37}$$

where C_{ijkl} in Eq. 5.38 are the components of the Jacobian. Using the Neo-Hookean model,

[5] See Abaqus Theory Guide v6.14 in Sect. 4.6.3 Anisotropic hyperelastic material behavior.

$$C_{ijkl} = \frac{2}{D_1}(2J-1)\,\delta_{ij}\delta_{kl} +$$
$$\frac{2}{J}C_{10}\left[\frac{1}{2}\left(\delta_{ik}\bar{B}_{jl} + \bar{B}_{ik}\delta_{jl} + \delta_{il}\bar{B}_{jk} + \bar{B}_{il}\delta_{jk}\right) - \frac{2}{3}\delta_{ij}\bar{B}_{kl} - \frac{2}{3}\bar{B}_{ij}\delta_{kl} + \frac{2}{9}\delta_{ij}\delta_{kl}\bar{B}_{mm}\right] \tag{5.38}$$

The expression shown in Eq. 5.38 is fairly complex, but it is straightforward to implement. So the appropriate coding is shown below:

Listing 5.3 NHH.f

```
      SUBROUTINE UMAT(STRESS,STATEV,DDSDDE,SSE,SPD,
     1 SCD,RPL,DDSDDT,DRPLDE,DRPLDT,STRAN,DSTRAN,TIME,
     2 DTIME,TEMP,DTEMP,PREDEF,DPRED,CMNAME,NDI,NSHR,
     3 NTENS,NSTATV,PROPS,NPROPS,COORDS,DROT,PNEWDT,
     4 CELENT,DFGRD0,DFGRD1,NOEL,NPT,LAYER,KSPT,KSTEP,
     5 KINC)
C
      INCLUDE 'ABA_PARAM.INC'
C
      CHARACTER*8 CMNAME
C LOCAL ARRAYS
C ----------------------------------------------------------------
C EELAS - LOGARITHMIC ELASTIC STRAINS
C EELASP - PRINCIPAL ELASTIC STRAINS
C BBAR - DEVIATORIC RIGHT CAUCHY-GREEN TENSOR
C BBARP - PRINCIPAL VALUES OF BBAR
C BBARN - PRINCIPAL DIRECTION OF BBAR (AND_EELAS)
C DISTGR - DEVIATORIC DEFORMATION GRADIENT
C (DISTORTION TENSOR)
C ----------------------------------------------------------------
C
      DIMENSION EELAS(6), EELASP(3), BBAR(6), BBARP(3),
     1          BBARN(3, 3), DISTGR(3,3)
C
      PARAMETER(ZERO=0.D0, ONE=1.D0, TWO=2.D0,
     1          THREE=3.D0, FOUR=4.D0, SIX=6.D0)
C
C ----------------------------------------------------------------
C UMAT FOR COMPRESSIBLE NEO-HOOKEAN HYPERELASTICITY
C CANNOT BE USED FOR PLANE STRESS
C ----------------------------------------------------------------
C PROPS(1) - E
C PROPS(2) - NU
C ----------------------------------------------------------------
C
C ELASTIC PROPERTIES
C
      EMOD=PROPS(1)
      ENU=PROPS(2)
      C10=EMOD/(FOUR*(ONE+ENU))
      D1=SIX*(ONE-TWO*ENU)/EMOD
C
```

```
C JACOBIAN AND DISTORTION TENSOR
C
      DET=DFGRD1(1,1)*DFGRD1(2,2)*DFGRD1(3,3)
      1 -DFGRD1(1,2)*DFGRD1(2,1)*DFGRD1(3,3)
      IF(NSHR.EQ.3) THEN
         DET=DET+DFGRD1(1,2)*DFGRD1(2,3)*DFGRD1(3,1)
      1          +DFGRD1(1,3)*DFGRD1(3,2)*DFGRD1(2,1)
      2          -DFGRD1(1,3)*DFGRD1(3,1)*DFGRD1(2,2)
      3          -DFGRD1(2,3)*DFGRD1(3,2)*DFGRD1(1,1)
      END IF
      SCALE=DET**(-ONE/THREE)
      DO K1=1, 3
         DO K2=1, 3
            DISTGR(K2,K1)=SCALE*DFGRD1(K2,K1)
         END DO
      END DO
C CALCULATE DEVIATORIC LEFT CAUCHY-GREEN DEFORMATION
C TENSOR
C
      BBAR(1)=DISTGR(1,1)**2+DISTGR(1,2)**2
      1      +DISTGR(1,3)**2
      BBAR(2)=DISTGR(2,1)**2+DISTGR(2,2)**2
      1      +DISTGR(2,3)**2
      BBAR(3)=DISTGR(3,3)**2+DISTGR(3,1)**2
      1      +DISTGR(3,2)**2
      BBAR(4)=DISTGR(1,1)*DISTGR(2,1)+DISTGR(1,2)
      1      *DISTGR(2,2)  +DISTGR(1,3)*DISTGR(2,3)
      IF(NSHR.EQ.3) THEN
         BBAR(5)=DISTGR(1,1)*DISTGR(3,1)+DISTGR(1,2)
      1          *DISTGR(3,2)+DISTGR(1,3)*DISTGR(3,3)
         BBAR(6)=DISTGR(2,1)*DISTGR(3,1)+DISTGR(2,2)
      1          *DISTGR(3,2)+DISTGR(2,3)*DISTGR(3,3)
      END IF
C
C CALCULATE THE STRESS
C
      TRBBAR=(BBAR(1)+BBAR(2)+BBAR(3))/THREE
      EG=TWO*C10/DET
      EK=TWO/D1*(TWO*DET-ONE)
      PR=TWO/D1*(DET-ONE)
      DO K1=1,NDI
         STRESS(K1)=EG*(BBAR(K1)-TRBBAR)+PR
      END DO
      DO K1=NDI+1,NDI+NSHR
         STRESS(K1)=EG*BBAR(K1)
      END DO
C CALCULATE THE STIFFNESS
C
      EG23=EG*TWO/THREE
      DDSDDE(1,1)=  EG23*(BBAR(1)+TRBBAR)+EK
      DDSDDE(2,2)=  EG23*(BBAR(2)+TRBBAR)+EK
      DDSDDE(3,3)=  EG23*(BBAR(3)+TRBBAR)+EK
      DDSDDE(1,2)= -EG23*(BBAR(1)+BBAR(2)-TRBBAR)+EK
```

```
      DDSDDE(1,3)=-EG23*(BBAR(1)+BBAR(3)-TRBBAR)+EK
      DDSDDE(2,3)=-EG23*(BBAR(2)+BBAR(3)-TRBBAR)+EK
      DDSDDE(1,4)= EG23*BBAR(4)/TWO
      DDSDDE(2,4)= EG23*BBAR(4)/TWO
      DDSDDE(3,4)=-EG23*BBAR(4)
      DDSDDE(4,4)= EG*(BBAR(1)+BBAR(2))/TWO
      IF(NSHR.EQ.3) THEN
          DDSDDE(1,5)= EG23*BBAR(5)/TWO
          DDSDDE(2,5)=-EG23*BBAR(5)
          DDSDDE(3,5)= EG23*BBAR(5)/TWO
          DDSDDE(1,6)=-EG23*BBAR(6)
          DDSDDE(2,6)= EG23*BBAR(6)/TWO
          DDSDDE(3,6)= EG23*BBAR(6)/TWO
          DDSDDE(5,5)= EG*(BBAR(1)+BBAR(3))/TWO
          DDSDDE(6,6)= EG*(BBAR(2)+BBAR(3))/TWO
          DDSDDE(4,5)= EG*BBAR(6)/TWO
          DDSDDE(4,6)= EG*BBAR(5)/TWO
          DDSDDE(5,6)= EG*BBAR(4)/TWO
      END IF
      DO K1=1, NTENS
          DO K2=1, K1-1
              DDSDDE(K1,K2)=DDSDDE(K2,K1)
          END DO
      END DO
C
C CALCULATE LOGARITHMIC ELASTIC STRAINS (OPTIONAL)
C
      CALL SPRIND(BBAR, BBARP, BBARN, 1, NDI, NSHR)
      EELASP(1)=LOG(SQRT(BBARP(1))/SCALE)
      EELASP(2)=LOG(SQRT(BBARP(2))/SCALE)
      EELASP(3)=LOG(SQRT(BBARP(3))/SCALE)
      EELAS(1)=EELASP(1)*BBARN(1,1)**2+EELASP(2)
     1        *BBARN(2,1)**2 +EELASP(3)*BBARN(3,1)**2
      EELAS(2)=EELASP(1)*BBARN(1,2)**2+EELASP(2)
     1        *BBARN(2,2)**2+EELASP(3)*BBARN(3, 2)**2
      EELAS(3)=EELASP(1)*BBARN(1,3)**2+EELASP(2)
     1        *BBARN(2,3)**2+EELASP(3)*BBARN(3,3)**2
      EELAS(4)=TWO*(EELASP(1)*BBARN(1,1)*BBARN(1,2)
     1        +EELASP(2)*BBARN(2,1)*BBARN(2,2)
     2        +EELASP(3)*BBARN(3,1)*BBARN(3,2))
      IF(NSHR.EQ.3) THEN
          EELAS(5)=TWO*(EELASP(1)*BBARN(1,1)*BBARN(1,3)
     1            +EELASP(2)*BBARN(2,1)*BBARN(2,3)
     2            +EELASP(3)*BBARN(3,1)*BBARN(3,3))
      EELAS(6)=TWO*(EELASP(1)*BBARN(1,2)*BBARN(1,3)
     1        +EELASP(2)*BBARN(2,2)*BBARN(2,3)
     2        +EELASP(3)*BBARN(3,2)*BBARN(3,3))
      END IF
C
C STORE ELASTIC STRAINS IN STATE VARIABLE ARRAY
C
      DO K1=1, NTENS
          STATEV(K1)=EELAS(K1)
      END DO
```

```
C
      RETURN
      END
```

This UMAT yields has exactly the same results as the **HYPERELASTIC** option with $N = 1$ and $C_{01} = 0$. Please note the use of the utility **SPRIND** with the code:

```
CALL SPRIND(BBAR, BBARP, BBARN, 1, NDI, NSHR)
```

Here, the tensor BBAR consists of NDI direct components and NSHR shear components. **SPRIND** returns the principal values and direction cosines of the principal directions of BBAR in BBARP and BBARN, respectively. A value equal to one is used as the fourth argument to indicate that BBAR contains stresses. (A value of two is used for strains.) The hyperelastic materials are often implemented more easily in user subroutine **UHYPER**.[6]

5.8.4 UMAT Subroutine for Kinematic Hardening Plasticity

The governing equations of the problem are listed below:

1. The elasticity equation in Eq. 5.39:

$$\sigma_{ij} = \lambda \delta_{ij} \varepsilon_{kk}^{el} + 2\mu \varepsilon_{ij}^{el} \tag{5.39}$$

   or in a Jaumann (corotational) rate form in Eq. 5.40:

$$\dot{\sigma}_{ij}^{J} = \lambda \delta_{ij} \dot{\varepsilon}_{kk}^{el} + 2\mu \dot{\varepsilon}_{ij}^{el} \tag{5.40}$$

2. The Jaumann rate equation in Eq. 5.41 is integrated in a corotational framework:

$$\Delta\sigma_{ij}^{J} = \lambda \delta_{ij} \Delta\varepsilon_{kk}^{el} + 2\mu \Delta\varepsilon_{ij}^{el} \tag{5.41}$$

3. The plasticity equations are given in Eqs. 5.42–5.45,
   - The yield function:

$$\sqrt{\frac{3}{2}\left(S_{ij} - \alpha_{ij}\right)\left(S_{ij} - \alpha_{ij}\right)} - \sigma_y = 0 \tag{5.42}$$

   - The equivalent plastic strain rate:

$$\dot{\bar{\varepsilon}}^{pl} = \sqrt{\frac{2}{3} \dot{\varepsilon}_{ij}^{pl} \dot{\varepsilon}_{ij}^{pl}} \tag{5.43}$$

[6]See Abaqus User Subroutines Reference Guide v6.14 Sect. 1.1.38 UHYPER User subroutine to define a hyperelastic material.

- The plastic flow law:

$$\dot{\varepsilon}_{ij}^{pl} = \frac{3}{2}\dot{\bar{\varepsilon}}^{pl}\frac{(S_{ij} - \alpha_{ij})}{\sigma_y} \tag{5.44}$$

- The Prager–Ziegler (linear) kinematic hardening:

$$\dot{\alpha}_{ij} = \frac{2}{3}h\dot{\varepsilon}_{ij}^{pl} \tag{5.45}$$

The calculation procedure is integrated following the different steps formulated below from Eqs. 5.46 to 5.56:

1. First, the subroutine calculates the equivalent stress based on purely elastic behavior (elastic predictor):

$$\bar{\sigma}^{pr} = \sqrt{\frac{3}{2}\left(S_{ij}^{pr} - \alpha_{ij}^{0}\right)\left(S_{ij}^{pr} - \alpha_{ij}^{0}\right)} \tag{5.46}$$

$$S_{ij}^{pr} = S_{ij}^{0} + 2\mu\Delta e_{ij} \tag{5.47}$$

2. The plastic flow occurs if the elastic predictor is larger than the yield stress. The backward Euler method is used to integrate the equations:

$$\Delta\varepsilon_{ij}^{pl} = \frac{3}{2}\Delta\bar{\varepsilon}^{pl}\frac{S_{ij}^{pr} - \alpha_{ij}^{0}}{\bar{\sigma}^{pr}} \tag{5.48}$$

3. After some manipulation, we obtain a closed-form expression for the equivalent plastic strain increment:

$$\Delta\bar{\varepsilon}^{pl} = \frac{\bar{\sigma}^{pr} - \sigma_y}{h + 3\mu} \tag{5.49}$$

4. This leads to the following update equations for the shift tensor, the stress, and the plastic strain:

$$\eta_{ij} = \frac{S_{ij}^{pr} - \alpha_{ij}^{0}}{\bar{\sigma}^{pr}} \tag{5.50}$$

$$\Delta\varepsilon_{ij}^{pl} = \frac{3}{2}\eta_{ij}\Delta\bar{\varepsilon}^{pl} \tag{5.51}$$

$$\Delta\alpha_{ij} = \eta_{ij}h\Delta\bar{\varepsilon}^{pl} \tag{5.52}$$

$$\sigma_{ij} = \alpha_{ij}^{0} + \Delta\alpha_{ij} + \eta_{ij}\sigma_y + \frac{1}{3}\delta_{ij}\sigma_{kk}^{pr} \tag{5.53}$$

5. In addition, the user can readily obtain the consistent Jacobian:

$$\mu^* = \mu \frac{\sigma_y + h\Delta\bar{\varepsilon}^{pl}}{\bar{\sigma}^{pr}} \tag{5.54}$$

$$\lambda^* = k - \frac{2}{3}\mu^* \tag{5.55}$$

$$\Delta\dot{\sigma}_{ij} = \lambda^*\delta_{ij}\Delta\dot{\varepsilon}_{kk} + 2\mu^*\Delta\dot{\varepsilon}_{ij} + \left(\frac{h}{1+\frac{h}{3\mu}} - 3\mu^*\right)\eta_{ij}\eta_{kl}\Delta\dot{\varepsilon}_{kl} \tag{5.56}$$

The appropriate coding is shown below:

Listing 5.4 KHP.f

```
      SUBROUTINE UMAT(STRESS,STATEV,DDSDDE,SSE,SPD,
     1 SCD,RPL,DDSDDT,DRPLDE,DRPLDT,STRAN,DSTRAN,TIME,
     2 DTIME,TEMP,DTEMP,PREDEF,DPRED,CMNAME,NDI,NSHR,
     3 NTENS,NSTATV,PROPS,NPROPS,COORDS,DROT,PNEWDT,
     4 CELENT,DFGRD0,DFGRD1,NOEL,NPT,LAYER,KSPT,KSTEP,
     5 KINC)
C
      INCLUDE 'ABA_PARAM.INC'
C
      CHARACTER*8 CMNAME
C LOCAL ARRAYS
C ----------------------------------------------------------------
C EELAS - ELASTIC STRAINS
C EPLAS - PLASTIC STRAINS
C ALPHA - SHIFT TENSOR
C FLOW - PLASTIC FLOW DIRECTIONS
C OLDS - STRESS AT START OF INCREMENT
C OLDPL - PLASTIC STRAINS AT START OF INCREMENT
C
      DIMENSION EELAS(6), EPLAS(6), ALPHA(6), FLOW(6),
     1          OLDS(6), OLDPL(6)
C
      PARAMETER(ZERO=0.D0, ONE=1.D0, TWO=2.D0,
     1          THREE=3.D0, SIX=6.D0, ENUMAX=.4999D0,
     2          TOLER=1.0D-6)
C
C ----------------------------------------------------------------
C UMAT FOR ISOTROPIC ELASTICITY_AND MISES PLASTICITY
C WITH KINEMATIC HARDENING
C CANNOT BE USED FOR PLANE STRESS
C ----------------------------------------------------------------
C PROPS(1) - E
C PROPS(2) - NU
C PROPS(3) - SYIELD
C PROPS(4) - HARD
C ----------------------------------------------------------------
C
```

```
C ELASTIC PROPERTIES
C
      EMOD=PROPS (1)
      ENU=MIN(PROPS (2), ENUMAX)
      EBULK3=EMOD/(ONE-TWO*ENU)
      EG2=EMOD/(ONE+ENU)
      EG=EG2/TWO
      EG3=THREE*EG
      ELAM=(EBULK3-EG2)/THREE
C
C ELASTIC STIFFNESS
C
      DO K1=1, NDI
        DO K2=1, NDI
          DDSDDE(K2, K1)=ELAM
        END DO
        DDSDDE(K1, K1)=EG2+ELAM
      END DO
      DO K1=NDI+1, NTENS
        DDSDDE(K1, K1)=EG
      END DO
C
C RECOVER ELASTIC STRAIN, PLASTIC STRAIN_AND SHIFT
C TENSOR_AND ROTATE
C NOTE: USE_CODE_1 FOR (TENSOR) STRESS, CODE_2 FOR
C(ENGINEERING) STRAIN
C
      CALL ROTSIG(STATEV(1),DROT,EELAS,2,NDI,NSHR)
      CALL ROTSIG(STATEV(NTENS+1),DROT,EPLAS,2,NDI,NSHR)
      CALL ROTSIG(STATEV(2*NTENS+1),DROT,ALPHA,1,NDI,NSHR)
C
C SAVE_STRESS_AND PLASTIC STRAINS_AND
C CALCULATE PREDICTOR STRESS_AND ELASTIC STRAIN
C
      DO K1=1, NTENS
        OLDS(K1)=STRESS(K1)
        OLDPL(K1)=EPLAS(K1)
        EELAS(K1)=EELAS(K1)+DSTRAN(K1)
        DO K2=1, NTENS
          STRESS(K2)=STRESS(K2)+DDSDDE(K2, K1)
     1                 *DSTRAN(K1)
        END DO
      END DO
C
C CALCULATE EQUIVALENT VON MISES STRESS
C
      SMISES=(STRESS(1)-ALPHA(1)-STRESS(2)+ALPHA(2))**2
     1      +(STRESS(2)-ALPHA(2)-STRESS(3)+ALPHA(3))**2
     2      +(STRESS(3)-ALPHA(3)-STRESS(1)+ALPHA(1))**2
      DO K1=NDI+1,NTENS
        SMISES=SMISES+SIX*(STRESS(K1)-ALPHA(K1))**2
      END DO
      SMISES=SQRT(SMISES/TWO)
```

```
C
C GET YIELD STRESS_AND HARDENING MODULUS
C
      SYIELD=PROPS(3)
      HARD=PROPS(4)
C
C DETERMINE_IF ACTIVELY YIELDING
C
      IF(SMISES.GT.(ONE+TOLER)*SYIELD) THEN
C
C ACTIVELY YIELDING
C SEPARATE THE HYDROSTATIC FROM THE DEVIATORIC STRESS
C CALCULATE THE FLOW DIRECTION
C
        SHYDRO=(STRESS(1)+STRESS(2)+STRESS(3))/THREE
        DO K1=1,NDI
          FLOW(K1)=(STRESS(K1)-ALPHA(K1)-SHYDRO)
     1             /SMISES
        END DO
        DO K1=NDI+1,NTENS
          FLOW(K1)=(STRESS(K1)-ALPHA(K1))/SMISES
        END DO
C
C SOLVE FOR EQUIVALENT PLASTIC STRAIN INCREMENT
C
        DEQPL=(SMISES-SYIELD)/(EG3+HARD)
C
C UPDATE SHIFT TENSOR, ELASTIC_AND PLASTIC STRAINS
C_AND STRESS
C
        DO K1=1,NDI
          ALPHA(K1)=ALPHA(K1)+HARD*FLOW(K1)*DEQPL
          EPLAS(K1)=EPLAS(K1)+
     1              THREE/TWO*FLOW(K1)*DEQPL
          EELAS(K1)=EELAS(K1)-
     1              THREE/TWO*FLOW(K1)*DEQPL
          STRESS(K1)=ALPHA(K1)+
     1              FLOW(K1)*SYIELD+SHYDRO
        END DO
        DO K1=NDI+1,NTENS
          ALPHA(K1)=ALPHA(K1)+HARD*FLOW(K1)*DEQPL
          EPLAS(K1)=EPLAS(K1)+THREE*FLOW(K1)*DEQPL
          EELAS(K1)=EELAS(K1)-THREE*FLOW(K1)*DEQPL
          STRESS(K1)=ALPHA(K1)+FLOW(K1)*SYIELD
        END DO
C CALCULATE PLASTIC DISSIPATION
C
        SPD=ZERO
        DO K1=1,NTENS
          SPD=SPD+(STRESS(K1)+OLDS(K1))*(EPLAS(K1)
     1        -OLDPL(K1))/TWO
        END DO
C
C FORMULATE THE JACOBIAN (MATERIAL TANGENT)
```

```
C FIRST CALCULATE EFFECTIVE MODULI
C
        EFFG=EG*(SYIELD+HARD*DEQPL)/SMISES
        EFFG2=TWO*EFFG
        EFFG3=THREE*EFFG
        EFFLAM=(EBULK3-EFFG2)/THREE
        EFFHRD=EG3*HARD/(EG3+HARD)-EFFG3
        DO K1=1, NDI
            DO K2=1, NDI
                DDSDDE(K2, K1)=EFFLAM
            END DO
            DDSDDE(K1, K1)=EFFG2+EFFLAM
        END DO
        DO K1=NDI+1, NTENS
            DDSDDE(K1, K1)=EFFG
        END DO
        DO K1=1, NTENS
            DO K2=1, NTENS
                DDSDDE(K2,K1)=DDSDDE(K2,K1)+EFFHRD*
     1                        FLOW(K2)*FLOW(K1)
            END DO
        END DO
      ENDIF
C
C STORE ELASTIC STRAINS, PLASTIC STRAINS_AND SHIFT
C TENSOR IN_STATE VARIABLE ARRAY
C
      DO K1=1,NTENS
        STATEV(K1)=EELAS(K1)
        STATEV(K1+NTENS)=EPLAS(K1)
        STATEV(K1+2*NTENS)=ALPHA(K1)
      END DO
C
      RETURN
      END
```

This UMAT yields has exactly the same results as the **PLASTIC** option with KINEMATIC hardening. This is also true for large-strain calculations. The necessary rotations of stress and strain are taken care of by Abaqus. Rotation of the shift tensor and the elastic and plastic strains is accomplished by the calls to **ROTSIG**. The call function,

```
CALL ROTSIG(STATEV(1), DROT, EELAS, 2, NDI, NSHR)
```

applies the incremental rotation, DROT, to STATEV and stores the result in ELAS. STATEV consists of NDI direct components and NSHR shear components. A value of one is used as the fourth argument to indicate that the transformed array contains tensor shear components such as α_{ij}. A value of two indicates that the array contains engineering shear components, such as ε_{kl}^{pl}. The rotation should be applied prior to the integration procedure. The routine is written for linear hardening because the classical Prager–Ziegler theory is limited to this case. More complex nonlinear kinematic hardening models are much more difficult to integrate. However, once a suitable integration procedure is obtained, the implementation in UMAT is straightforward and follows the examples discussed here.

5.8.5 UMAT Subroutine for Isotropic Hardening Plasticity

The governing equations of elasticity for this problem are the same as described in Eq. 5.39 or 5.40, and Eq. 5.41.

The plasticity equations are given from Eqs. 5.57 to 5.61.

- The yield function:

$$\sqrt{\frac{3}{2}S_{ij}S_{ij}} - \sigma_y\left(\bar{\varepsilon}^{pl}\right) = 0 \tag{5.57}$$

$$S_{ij} = \sigma_{ij} - \frac{1}{3}\delta_{ij}\sigma_{kk} \tag{5.58}$$

- The equivalent plastic strain rate:

$$\dot{\bar{\varepsilon}}^{pl} = \sqrt{\frac{2}{3}\dot{\varepsilon}_{ij}^{pl}\dot{\varepsilon}_{ij}^{pl}} \tag{5.59}$$

$$\bar{\varepsilon}^{pl} = \int_0^t \dot{\bar{\varepsilon}}^{pl} dt \tag{5.60}$$

- The plastic flow law:

$$\dot{\varepsilon}_{ij}^{pl} = \frac{3}{2}\dot{\bar{\varepsilon}}^{pl}\frac{S_{ij}}{\sigma_y} \tag{5.61}$$

The calculation procedure is integrated following the different steps formulated below from Eqs. 5.44 to 5.53:

1. First, it calculates the von Mises stress based on purely elastic behavior (elastic predictor):

$$\bar{\sigma}^{pr} = \sqrt{\frac{3}{2}S_{ij}S_{ij}} \tag{5.62}$$

$$S_{ij}^{pr} = S_{ij}^{0} + 2\mu\Delta e_{ij} \tag{5.63}$$

2. If the elastic predictor is larger than the current yield stress, plastic flow occurs. The backward Euler method is used to integrate the equations. After some manipulation, the user can reduce the problem to a single equation in terms of the incremental equivalent plastic strain:

$$\bar{\sigma}^{pr} - 3\mu\Delta\bar{\varepsilon}^{pl} = \sigma_y\left(\bar{\varepsilon}^{pl}\right) \tag{5.64}$$

   This equation is solved with Newtons method.
3. After the equation is solved, the following update equations for the stress and the plastic strain can be used:

$$\eta_{ij} = \frac{S_{ij}^{pr}}{\bar{\sigma}^{pr}} \tag{5.65}$$

$$\Delta\varepsilon_{ij}^{pr} = \frac{3}{2}\eta_{ij}\Delta\bar{\varepsilon}^{pl} \tag{5.66}$$

$$\sigma_{ij} = \eta_{ij}\sigma_y + \frac{1}{3}\delta_{ij}\sigma_{kk}^{pr} \tag{5.67}$$

4. In addition, the user can readily obtain the consistent Jacobian:

$$\mu^* = \mu\frac{\sigma_y}{\bar{\sigma}^{pr}} \tag{5.68}$$

$$\lambda^* = k - \frac{2}{3}\mu^* \tag{5.69}$$

$$h = \frac{d\sigma_y}{d\bar{\varepsilon}^{pl}} \tag{5.70}$$

$$\Delta\dot{\sigma}_{ij} = \lambda^*\delta_{ij}\Delta\dot{\varepsilon}_{kk} + 2\mu^*\Delta\dot{\varepsilon}_{ij} + \left(\frac{h}{1+\frac{h}{3\mu}} - 3\mu^*\right)\eta_{ij}\eta_{kl}\Delta\dot{\varepsilon}_{kl} \tag{5.71}$$

The appropriate coding is shown below:

Listing 5.5 IHP.f

```
      SUBROUTINE UMAT(STRESS,STATEV,DDSDDE,SSE,SPD,
     1 SCD,RPL,DDSDDT,DRPLDE,DRPLDT,STRAN,DSTRAN,TIME,
     2 DTIME,TEMP,DTEMP,PREDEF,DPRED,CMNAME,NDI,NSHR,
     3 NTENS,NSTATV,PROPS,NPROPS,COORDS,DROT,PNEWDT,
     4 CELENT,DFGRD0,DFGRD1,NOEL,NPT,LAYER,KSPT,KSTEP,
     5 KINC)
C
      INCLUDE 'ABA_PARAM.INC'
C
      CHARACTER*8 CMNAME
C LOCAL ARRAYS
C ----------------------------------------------------------------
C EELAS - ELASTIC STRAINS
C EPLAS - PLASTIC STRAINS
C FLOW - DIRECTION OF PLASTIC FLOW
C ----------------------------------------------------------------
C
      DIMENSION EELAS(6),EPLAS(6),FLOW(6), HARD(3)
C
      PARAMETER(ZERO=0.D0, ONE=1.D0, TWO=2.D0,
     1          THREE=3.D0, SIX=6.D0, ENUMAX=.4999D0,
     2          NEWTON=10, TOLER=1.0D-6)
C
C ----------------------------------------------------------------
```

```
C UMAT FOR ISOTROPIC ELASTICITY_AND ISOTROPIC MISES
C PLASTICITY CANNOT BE USED FOR PLANE STRESS
C ----------------------------------------------------------
C PROPS(1) - E
C PROPS(2) - NU
C PROPS(3..) - SYIELD AN HARDENING_DATA
C CALLS UHARD FOR CURVE OF YIELD STRESS VS.
C PLASTIC STRAIN
C ----------------------------------------------------------
C
C ELASTIC PROPERTIES
C
      EMOD=PROPS(1)
      ENU=MIN(PROPS(2), ENUMAX)
      EBULK3=EMOD/(ONE-TWO*ENU)
      EG2=EMOD/(ONE+ENU)
      EG=EG2/TWO
      EG3=THREE*EG
      ELAM=(EBULK3-EG2)/THREE
C
C ELASTIC STIFFNESS
C
      DO K1=1, NDI
         DO K2=1, NDI
            DDSDDE(K2, K1)=ELAM
         END DO
         DDSDDE(K1, K1)=EG2+ELAM
      END DO
      DO K1=NDI+1, NTENS
         DDSDDE(K1, K1)=EG
      END DO
C RECOVER ELASTIC AND PLASTIC STRAINS AND ROTATE
C FORWARD ALSO RECOVER EQUIVALENT PLASTIC STRAIN
C
      CALL ROTSIG(STATEV(1),DROT,EELAS,2,NDI,NSHR)
      CALL ROTSIG(STATEV(NTENS+1),DROT,EPLAS,2,NDI,NSHR)
      EQPLAS=STATEV(1+2*NTENS)
C
C CALCULATE PREDICTOR STRESS AND ELASTIC STRAIN
C
      DO K1=1, NTENS
         DO K2=1, NTENS
            STRESS(K2)=STRESS(K2)+DDSDDE(K2, K1)
     1                 *DSTRAN(K1)
         END DO
         EELAS(K1)=EELAS(K1)+DSTRAN(K1)
      END DO
C
C CALCULATE EQUIVALENT VON MISES STRESS
C
      SMISES=(STRESS(1)-STRESS(2))**2+(STRESS(2)
     1      -STRESS(3))**2+(STRESS(3)-STRESS(1))**2
      DO K1=NDI+1,NTENS
```

```
          SMISES=SMISES+SIX*STRESS(K1)**2
        END DO
        SMISES=SQRT(SMISES/TWO)
C
C GET YIELD STRESS FROM THE SPECIFIED HARDENING CURVE
C
        NVALUE=NPROPS/2-1
        CALL UHARD(SYIEL0,HARD,EQPLAS,EQPLASRT,TIME,DTIME,
     1            TEMP,DTEMP,NOEL,NPT,LAYER,KSPT,KSTEP,
     2            KINC,CMNAME,NSTATV,STATEV,NUMFIELDV,
     3            PREDEF,DPRED,NVALUE,PROPS(3))
C
C DETERMINE_IF ACTIVELY YIELDING
C
        IF (SMISES.GT.(ONE+TOLER)*SYIEL0) THEN
C
C ACTIVELY YIELDING
C SEPARATE THE HYDROSTATIC FROM THE DEVIATORIC STRESS
C CALCULATE THE FLOW DIRECTION
C
          SHYDRO=(STRESS(1)+STRESS(2)+STRESS(3))/THREE
          DO K1=1,NDI
            FLOW(K1)=(STRESS(K1)-SHYDRO)/SMISES
          END DO
          DO K1=NDI+1, NTENS
            FLOW(K1)=STRESS(K1)/SMISES
          END DO
C
C SOLVE FOR EQUIVALENT VON MISES STRESS
C_AND EQUIVALENT PLASTIC STRAIN INCREMENT USING
C NEWTON ITERATION
C
          SYIELD=SYIEL0
          DEQPL=ZERO
          DO KEWTON=1, NEWTON
            RHS=SMISES-EG3*DEQPL-SYIELD
            DEQPL=DEQPL+RHS/(EG3+HARD(1))
            CALL UHARD(SYIELD,HARD,EQPLAS+DEQPL,
     1                 EQPLASRT,TIME,DTIME,TEMP,DTEMP,
     2                 NOEL,NPT,LAYER,KSPT,KSTEP,KINC,
     3                 CMNAME,NSTATV,STATEV,NUMFIELDV,
     4                 PREDEF,DPRED,NVALUE,PROPS(3))
            IF(ABS(RHS).LT.TOLER*SYIEL0) GOTO 10
          END DO
C
C_WRITE WARNING MESSAGE_TO THE MSG_FILE C
          WRITE(7,2) NEWTON
     1      FORMAT(//,30X,'***WARNING - PLASTICITY'
     2             'ALGORITHM DID NOT CONVERGE AFTER '
     3             ,I3,' ITERATIONS')
   10     CONTINUE
C C UPDATE STRESS, ELASTIC_AND PLASTIC STRAINS_AND C EQUIVALENT
PLASTIC STRAIN C
```

```
        DO K1=1,NDI
            STRESS(K1)=FLOW(K1)*SYIELD+SHYDRO
            EPLAS(K1)=EPLAS(K1)+THREE/TWO*FLOW(K1)*DEQPL
            EELAS(K1)=EELAS(K1)-THREE/TWO*FLOW(K1)*DEQPL
        END DO
        DO K1=NDI+1,NTENS
            STRESS(K1)=FLOW(K1)*SYIELD
            EPLAS(K1)=EPLAS(K1)+THREE*FLOW(K1)*DEQPL
            EELAS(K1)=EELAS(K1)-THREE*FLOW(K1)*DEQPL
        END DO
        EQPLAS=EQPLAS+DEQPL
C
C CALCULATE PLASTIC DISSIPATION
C
        SPD=DEQPL*(SYIEL0+SYIELD)/TWO
C
C FORMULATE THE JACOBIAN (MATERIAL TANGENT)
C FIRST CALCULATE
EFFECTIVE MODULI
C
        EFFG=EG*SYIELD/SMISES
        EFFG2=TWO*EFFG
        EFFG3=THREE/TWO*EFFG2
        EFFLAM=(EBULK3-EFFG2)/THREE
        EFFHRD=EG3*HARD(1)/(EG3+HARD(1)) - EFFG3
        DO K1=1, NDI
            DO K2=1, NDI
                DDSDDE(K2, K1)=EFFLAM
            END DO
            DDSDDE(K1, K1)=EFFG2+EFFLAM
        END DO
        DO K1=NDI+1, NTENS
            DDSDDE(K1, K1)=EFFG
        END DO
        DO K1=1, NTENS
            DO K2=1, NTENS
                DDSDDE(K2, K1)=DDSDDE(K2, K1)+EFFHRD*
            1                     FLOW(K2)*FLOW(K1)
            END DO
        END DO
    ENDIF
C
C STORE ELASTIC_AND (EQUIVALENT) PLASTIC STRAINS
C_IN STATE
VARIABLE ARRAY
C
    DO K1=1, NTENS
        STATEV(K1)=EELAS(K1)
        STATEV(K1+NTENS)=EPLAS(K1)
    END DO
    STATEV(1+2*NTENS)=EQPLAS
C
    RETURN
```

```
      END

      SUBROUTINE UHARD(SYIELD,HARD,EQPLAS,EQPLASRT,TIME,
     1 DTIME,TEMP,DTEMP,NOEL,NPT,LAYER,KSPT,KSTEP,KINC,
     2 CMNAME,NSTATV,STATEV,NUMFIELDV,PREDEF,DPRED,
     3 NVALUE,TABLE)
      INCLUDE 'ABA_PARAM.INC'
      CHARACTER*80 CMNAME
      DIMENSION HARD(3),STATEV(NSTATV),TIME(*),
     1          PREDEF(NUMFIELDV),DPRED(*)
C
      DIMENSION TABLE(2, NVALUE)
C
      PARAMETER(ZERO=0.D0)
C
C SET YIELD STRESS TO LAST VALUE OF TABLE,
C HARDENING TO ZERO
C
      SYIELD=TABLE(1, NVALUE)
      HARD(1)=ZERO
C IF MORE THAN ONE ENTRY, SEARCH TABLE
C
      IF(NVALUE.GT.1) THEN
          DO K1=1, NVALUE-1
              EQPL1=TABLE(2,K1+1)
              IF(EQPLAS.LT.EQPL1) THEN
                  EQPL0=TABLE(2, K1)
                  IF(EQPL1.LE.EQPL0) THEN
                      WRITE(7, 1)
     1                FORMAT(//, 30X, '***ERROR - PLASTIC ',
     1                       'STRAIN MUST BE ENTERED IN ',
     1                       'ASCENDING ORDER')
                  CALL XIT
                  ENDIF
C
C CURRENT YIELD STRESS AND HARDENING
C
                  DEQPL=EQPL1-EQPL0
                  SYIEL0=TABLE(1, K1)
                  SYIEL1=TABLE(1, K1+1)
                  DSYIEL=SYIEL1-SYIEL0
                  HARD(1)=DSYIEL/DEQPL
                  SYIELD=SYIEL0+(EQPLAS-EQPL0)*HARD(1)
                  GOTO 10
              ENDIF
          END DO
 10       CONTINUE
      ENDIF
      RETURN
      END
```

This UMAT yields has exactly the same results as the **PLASTIC** option with ISOTROPIC hardening. This result is also true for large-strain calculations. The

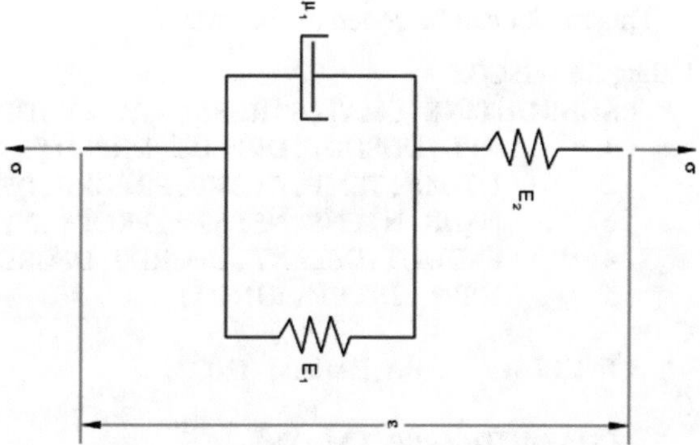

Fig. 5.3 Simple linear viscoelastic model

necessary rotations of stress and strain are taken care of by Abaqus. The rotation of elastic and plastic strain, prior to integration, is accomplished by the calls to ROTSIG. The routine calls user subroutine UHARD to recover a piecewise linear hardening curve. It is straightforward to replace the piecewise linear curve by an analytic description. A local Newton iteration is used to determine the current yield stress and hardening modulus. If the data are not given in ascending order of strain, the routine XIT is called, which closes all files and terminates execution.

5.8.6 UMAT Subroutine for Simple Linear Viscoelastic Material

As a simple example of the coding of user subroutine UMAT, consider the linear, viscoelastic model shown in Fig. 5.3. Although this is not a very useful model for real materials, it serves to illustrate how to code the routine.

The user must refer to the Abaqus documentation to have a full description of this example case.[7] The behavior of the one-dimensional model shown in the Fig. 5.3 is given by the Eq. 5.72.

$$\sigma + \frac{\mu_1}{E_1 + E_2}\dot{\sigma} = \frac{E_2\mu_1}{E_1 + E_2}\dot{\varepsilon} + \frac{E_1 E_2}{E_1 + E_2}\varepsilon \tag{5.72}$$

For this simple case, a user material definition can be used to read in the five constants in the order PROPS(1)=λ, PROPS(2)=μ, PROPS(3)=$\tilde{\lambda}$, PROPS(4)=$\tilde{\mu}$, and PROPS(4)=$\tilde{\nu}$.

[7]See Abaqus User Subroutines Reference Guide v6.14 in Sect. 1.1.41 UMAT User subroutine to define a material's mechanical behavior.

This model can be coded as follows:

Listing 5.6 VISCO.f

```
      SUBROUTINE UMAT( STRESS , STATEV , DDSDDE , SSE , SPD , SCD ,
     1           RPL , DDSDDT , DRPLDE , DRPLDT , STRAN , DSTRAN , TIME ,
     2           DTIME , TEMP , DTEMP , PREDEF , DPRED , CMNAME, NDI ,
     3           NSHR , NTENS , NSTATV , PROPS , NPROPS , COORDS , DROT ,
     4           PNEWDT , CELENT , DFGRD0 , DFGRD1 , NOEL , NPT , LAYER ,
     5           KSPT , JSTEP , KINC )
C
      INCLUDE  'ABA_PARAM. INC '
C
      CHARACTER*80  CMNAME
      DIMENSION  STRESS ( NTENS ) , STATEV ( NSTATV ) ,
     1           DDSDDE ( NTENS , NTENS ) , DDSDDT ( NTENS ) ,
     2           DRPLDE ( NTENS ) , STRAN ( NTENS ) , DSTRAN ( NTENS ) ,
     3           TIME ( 2 ) , PREDEF ( 1 ) , DPRED ( 1 ) , PROPS ( NPROPS ) ,
     4           COORDS ( 3 ) , DROT ( 3 , 3 ) , DFGRD0 ( 3 , 3 ) ,
     5           DFGRD1 ( 3 , 3 ) , JSTEP ( 4 )
      DIMENSION  DSTRES ( 6 ) , D( 3 , 3 )
C
C    EVALUATE NEW  STRESS  TENSOR
C
      EV  =  0.
      DEV  =  0.
      DO  K1 = 1 , NDI
          EV  =  EV  +  STRAN ( K1 )
          DEV  =  DEV  +  DSTRAN ( K1 )
      END  DO
C
      TERM1  =  .5 * DTIME  +  PROPS ( 5 )
      TERM1I  =  1. / TERM1
      TERM2  =  ( .5 * DTIME*PROPS ( 1 )+PROPS ( 3 ) ) * TERM1I*DEV
      TERM3  =  ( DTIME*PROPS ( 2 )+2. * PROPS ( 4 ) ) * TERM1I
C
      DO  K1 = 1 , NDI
         DSTRES ( K1 )  =  TERM2+TERM3*DSTRAN ( K1 )
     1                     +DTIME*TERM1I * ( PROPS ( 1 ) * EV
     2                     +2. * PROPS ( 2 ) * STRAN ( K1) − STRESS ( K1 ) )
          STRESS ( K1 )  =  STRESS ( K1 )  +  DSTRES ( K1 )
      END  DO
C
      TERM2  =  ( .5 * DTIME*PROPS ( 2 )  +  PROPS ( 4 ) ) * TERM1I
      I1  =  NDI
      DO  K1 = 1 , NSHR
          I1  =  I1 +1
          DSTRES ( I1 )  =  TERM2*DSTRAN ( I1 )+
     1                      DTIME*TERM1I * ( PROPS ( 2 ) * STRAN ( I1 )
     2                     −STRESS ( I1 ) )
```

```
            STRESS(I1) = STRESS(I1)+DSTRES(I1)
        END DO
C
C     CREATE NEW JACOBIAN
C
        TERM2 = (DTIME*(.5*PROPS(1)+PROPS(2))+PROPS(3)+
     1        2.*PROPS(4))*TERM1I
        TERM3 = (.5*DTIME*PROPS(1)+PROPS(3))*TERM1I
        DO K1=1,NTENS
            DO K2=1,NTENS
                DDSDDE(K2,K1) = 0.
            END DO
        END DO
C
        DO K1=1,NDI
            DDSDDE(K1,K1) = TERM2
        END DO
C
        DO K1=2,NDI
            N2 = K1   1
            DO K2=1,N2
                DDSDDE(K2,K1) = TERM3
                DDSDDE(K1,K2) = TERM3
            END DO
        END DO
        TERM2 = (.5*DTIME*PROPS(2)+PROPS(4))*TERM1I
        I1 = NDI
        DO K1=1,NSHR
            I1 = I1+1
            DDSDDE(I1 ,I1) = TERM2
        END DO
C
C     TOTAL CHANGE IN SPECIFIC ENERGY
C
        TDE = 0.
        DO K1=1,NTENS
            TDE=TDE+(STRESS(K1)-.5*DSTRES(K1))*DSTRAN(K1)
        END DO
C
C     CHANGE_IN SPECIFIC ELASTIC STRAIN ENERGY
C
        TERM1 = PROPS(1) + 2.*PROPS(2)
        DO  K1=1,NDI
            D(K1,K1) = TERM1
        END DO
        DO K1=2,NDI
            N2 = K1-1
            DO K2=1,N2
                D(K1,K2) = PROPS(1)
```

```
            D(K2,K1) = PROPS(1)
        END DO
    END DO
    DEE = 0.
    DO K1=1,NDI
        TERM1 = 0.
        TERM2 = 0.
        DO K2=1,NDI
            TERM1 = TERM1 + D(K1,K2)*STRAN(K2)
            TERM2 = TERM2 + D(K1,K2)*DSTRAN(K2)
        END DO
        DEE = DEE + (TERM1+.5*TERM2)*DSTRAN(K1)
    END DO
    I1 = NDI
    DO K1=1,NSHR
        I1 = I1+1
        DEE = DEE + PROPS(2)*(STRAN(I1)+
    1                .5*DSTRAN(I1))*DSTRAN(I1)
    END DO
    SSE = SSE + DEE
    SCD = SCD + TDE - DEE
    RETURN
    END
```

5.9 VUMAT Subroutine Examples

These input lines act as the interface to a VUMAT in which kinematic hardening plasticity is defined.

```
*MATERIAL, NAME=KINPLAS
*USER MATERIAL, CONSTANTS=4
30.E6, 0.3, 30.E3, 40.E3
*DEPVAR
5
*INITIAL CONDITIONS, TYPE=SOLUTION
Data line to specify initial solution-dependent
variables
```

The input lines are identical to those for the UMAT interface. The user subroutine must be kept in a separate file, and is invoked with the Abaqus execution procedure, as follows:

```
abaqus job=... user=....
```

The user subroutine must be invoked in a restarted analysis because user subroutines are not saved in the restart file. The solution-dependent state variables can be output with identifiers SDV1, SDV2, and so on. The contour, path, and X-Y plots of SDVs can be plotted in Abaqus viewer. There is only a single VUMAT subroutine running per analysis. If more than one material must be defined, test on the material name in the VUMAT routine and branch.

The following quantities are available in VUMAT, but they cannot be redefined such as the stress, stretch, and SDVs at the start of the increment, the relative rotation vector and deformation gradient at the start and end of an increment and strain increment, the total and incremental values of time, temperature, and the user-defined field variables at the start and end of an increment, the material constants, density, material point position, and a characteristic element length, the internal and dissipated energies at the beginning of the increment, the number of material points to be processed in a call to the routine (NBLOCK), and a flag indicating whether the routine is being called during an annealing process.

The following quantities must be defined, the stress and SDVs at the end of an increment.

The following variables may be defined, the internal and dissipated energies at the end of the increment.

Many of these variables are equivalent or similar to those in UMAT and using the same convention as described in Sect. 5.8.

There are a number of significant differences between the UMAT and VUMAT interfaces. The VUMAT uses a two-state architecture: the initial values are in the OLD arrays, the new values must be put in the NEW arrays. The VUMAT interface is written to take advantage of vector processing. The material Jacobian does not need to be defined. There is no information provided about element numbers. The time increment cannot be redefined, and the utility routines are not available because they would prevent vectorization.

There are two VUMAT formulation aspects. First, the vectorized interface and second the corotational formulation. The vectorized interface in VUMAT means that the data are passed in and out in large blocks (dimension NBLOCK). NBLOCK typically is equal to 64 or 128. Each entry in an array of length NBLOCK corresponds to a single material point. All material points in the same block have the same material name and belong to the same element type. This structure allows vectorization of the routine.

A vectorized VUMAT should make sure that all operations are done in vector mode with NBLOCK the vector length. In vectorized code, branching inside loops should be avoided. The element type-based branching should be outside the NBLOCK loop.

The corotational formulation includes the constitutive equation which is formulated in a corotational framework, based on the Jaumann stress rate. The strain increment is obtained with Hughes–Winget[8] [11]. Other measures can be obtained from the deformation gradient. The user must define the Cauchy stress: this stress reappears during the next increment as the old stress. There is no need to rotate tensor state variables.

[8] See in Abaqus Theory Guide v6.14 in Sect. 3.2.2 Solid element formulation.

5.9.1 VUMAT Subroutine for Kinematic Hardening Plasticity

The governing equations and integration procedure are the same as in Sect. 5.8.4, the Jacobian is not required.

This model can be coded as follows:

Listing 5.7 VKHP.f

```
      SUBROUTINE VUMAT(
C_Read_only -
     1   NBLOCK, NDIR, NSHR, NSTATEV, NFIELDV, NPROPS,
     2   LANNEAL, STEPTIME, TOTALTIME, DT, CMNAME, COORDMP,
     3   CHARLENGTH, PROPS, DENSITY, STRAININC, FIELDNEW,
     4   RELSPININC, TEMPOLD, STRETCHOLD, DEFGRADOLD,
     5   FIELDOLD, STRESSOLD, STATEOLD, ENERINTERNOLD,
     6   ENERINELASOLD, TEMPNEW, STRETCHNEW, DEFGRADNEW,
C_Write_only -
     7   STRESSNEW, STATENEW, ENERINTERNNEW, ENERINELASNEW)
C
      INCLUDE 'VABA_PARAM.INC'
C
      DIMENSION PROPS(NPROPS), DENSITY(NBLOCK),
     1   COORDMP(NBLOCK), CHARLENGTH(NBLOCK),
     2   STRAININC(NBLOCK, NDIR+NSHR),
     3   RELSPININC(NBLOCK, NSHR),
     4   TEMPOLD(NBLOCK), FIELDOLD(NBLOCK, NFIELDV),
     5   STRETCHOLD(NBLOCK, NDIR+NSHR),
     6   DEFGRADOLD(NBLOCK, NDIR+NSHR+NSHR),
     7   STRESSOLD(NBLOCK, NDIR+NSHR),
     8   STATEOLD(NBLOCK, NSTATEV),
     9   ENERINTERNOLD(NBLOCK),
     1   ENERINELASOLD(NBLOCK), TEMPNEW(NBLOCK),
     2   STRETCHNEW(NBLOCK, NDIR+NSHR),
     3   STATENEW(NBLOCK, NSTATEV),
     4   DEFGRADNEW(NBLOCK, NDIR+NSHR+NSHR),
     5   FIELDNEW(NBLOCK, NFIELDV),
     6   STRESSNEW(NBLOCK, NDIR+NSHR),
     7   ENERINTERNNEW(NBLOCK),
     8   ENERINELASNEW(NBLOCK)
C
      CHARACTER*8 CMNAME
C
      E     =PROPS(1)
      XNU   =PROPS(2)
      YIELD =PROPS(3)
      HARD  =PROPS(4)
C
C ELASTIC CONSTANTS
C
      TWOMU  =E/(ONE+XNU)
      THREMU =THREE_HALFS*TWOMU
```

```
      SIXMU   =THREE*TWOMU
      ALAMDA  =TWOMU*(E-TWOMU)/(SIXMU-TWO*E)
      TERM    =ONE/(TWOMU*(ONE+HARD/THREMU))
      CON1    =SQRT(TWO_THIRDS)
C
C_If stepTime equals_to zero, assume the material
C pure elastic and_use initial elastic modulus
C
      IF( STEPTIME .EQ. ZERO ) THEN
         DO I = 1,NBLOCK
C Trial Stress
            TRACE=STRAININC(I,1)+STRAININC(I,2)
     1            +STRAININC(I,3)
            STRESSNEW(I,1)=STRESSOLD(I,1)+
     1            ALAMDA*TRACE+TWOMU*STRAININC(I,1)
            STRESSNEW(I,2)=STRESSOLD(I,2) +
     1            ALAMDA*TRACE+TWOMU*STRAININC(I,2)
            STRESSNEW(I,3)=STRESSOLD(I,3) +
     1            ALAMDA*TRACE+TWOMU*STRAININC(I,3)
            STRESSNEW(I,4)=STRESSOLD(I,4)
     1                  +TWOMU*STRAININC(I,4)
         END DO
      ELSE
C
C PLASTICITY CALCULATIONS_IN BLOCK_FORM
C
         DO I = 1, NBLOCK
C Elastic predictor stress
            TRACE=STRAININC(I,1)+STRAININC(I,2)
     1            +STRAININC(I,3)
            SIG1=STRESSOLD(I,1)+ALAMDA*TRACE
     1           +TWOMU*STRAININC(I,1)
            SIG2=STRESSOLD(I,2)+ALAMDA*TRACE
     1           +TWOMU*STRAININC(I,2)
            SIG3=STRESSOLD(I,3)+ALAMDA*TRACE
     1           +TWOMU*STRAININC(I,3)
            SIG4=STRESSOLD(I,4)+TWOMU*STRAININC(I,4)
C Elastic predictor stress measured from the back stress
            S1=SIG1-STATEOLD(I,1)
            S2=SIG2-STATEOLD(I,2)
            S3=SIG3-STATEOLD(I,3)
            S4=SIG4-STATEOLD(I,4)
C Deviatoric part of predictor stress measured from the
C back stress
            SMEAN=THIRD*(S1+S2+S3)
            DS1=S1-SMEAN
            DS2=S2-SMEAN
            DS3=S3-SMEAN
C Magnitude of the deviatoric predictor stress difference
            DSMAG=SQRT(DS1**2+DS2**2+DS3**2+TWO*S4**2)
C Check for yield by determining the factor for
```

```
C plasticity, zero for elastic, one for yield
            RADIUS=CON1*YIELD
            FACYLD=ZERO
            IF(DSMAG-RADIUS .GE. ZERO) FACYLD=ONE
C Add a protective addition factor to_prevent a
C divide by zero when DSMAG is zero. If_DSMAG is zero,
C we will_not have exceeded the yield stress
C_and FACYLD will be zero.
            DSMAG=DSMAG+(ONE-FACYLD)
C Calculated increment in_gamma,
C this explicitly includes the time step.
            DIFF=DSMAG-RADIUS
            DGAMMA=FACYLD*TERM*DIFF
C Update equivalent plastic strain
            DEQPS=CON1*DGAMMA
            STATENEW(I,5)=STATEOLD(I,5)+DEQPS
C Divide DGAMMA by DSMAG so that the deviatoric
C stresses are explicitly converted to_tensors of_unit
C magnitude in_the following calculations
            DGAMMA=DGAMMA/DSMAG
C Update back stress
            FACTOR=HARD*DGAMMA*TWO_THIRDS
            STATENEW(I,1)=STATEOLD(I,1)+FACTOR*DS1
            STATENEW(I,2)=STATEOLD(I,2)+FACTOR*DS2
            STATENEW(I,3)=STATEOLD(I,3)+FACTOR*DS3
            STATENEW(I,4)=STATEOLD(I,4)+FACTOR*S4
C Update stress
            FACTOR=TWOMU*DGAMMA
            STRESSNEW(I,1)=SIG1-FACTOR*DS1
            STRESSNEW(I,2)=SIG2-FACTOR*DS2
            STRESSNEW(I,3)=SIG3-FACTOR*DS3
            STRESSNEW(I,4)=SIG4-FACTOR*S4
C Update the specific internal energy -
            STRESS_POWER=HALF*(
     1  (STRESSOLD(I,1)+STRESSNEW(I,1))*STRAININC(I,1)
     2  +(STRESSOLD(I,2)+STRESSNEW(I,2))*STRAININC(I,2)
     3  +(STRESSOLD(I,3)+STRESSNEW(I,3))*STRAININC(I,3)
     4  +TWO*(STRESSOLD(I,4)+STRESSNEW(I,4))
     5  *STRAININC(I,4))
            ENERINTERNNEW(I)=ENERINTERNOLD(I)
     1                  +STRESS_POWER/DENSITY(I)
C Update the dissipated inelastic specific energy -
            SMEAN=THIRD*(STRESSNEW(I,1)+STRESSNEW(I,2)
     1           +STRESSNEW(I,3))
            EQUIV_STRESS=SQRT(THREE_HALFS
     1                  *((STRESSNEW(I,1)-SMEAN)**2
     2                  +(STRESSNEW(I,2)-SMEAN)**2
     3                  +(STRESSNEW(I,3)-SMEAN)**2
     4                  +TWO*STRESSNEW(I,4)**2))
            PLASTIC_WORK_INC=EQUIV_STRESS*DEQPS
            ENERINELASNEW(I)=ENERINELASOLD(I)
```

```
     1                    +PLASTIC_WORK_INC / DENSITY ( I )
C
        END DO
     END IF
     RETURN
     END
```

In the datacheck phase, VUMAT is called with a set of fictitious strains and a TOTALTIME and STEPTIME both equal to zero. A check is done on the user's constitutive relation, and an initial stable time increment is determined based on calculated equivalent initial material properties. Ensure that elastic properties are used in this call to VUMAT; otherwise, too large an initial time increment may be used, leading to instability. A warning message is printed to the status (.sta) file informing the user that this check is being performed.

Special coding techniques are used to obtain vectorized coding. All small loops inside the material routine are unrolled. The same code is executed regardless of whether the behavior is purely elastic or elastic–plastic.

Special care must be taken to avoid divides by zero. No external subroutines are called inside the loop. The use of local scalar variables inside the loop is allowed. The compiler will automatically expand these local scalar variables to local vectors. Iterations should be avoided.

If iterations cannot be avoided, use a fixed number of iterations and do not test on convergence.

5.9.2 VUMAT Subroutine for Isotropic Hardening Plasticity

The governing equations and integration procedure are the same as in Sect. 5.8.5. The increment of equivalent plastic strain is obtained explicitly through Eq. 5.73,

$$\Delta\bar{\varepsilon}^{pl} = \frac{\bar{\sigma}^{pr} - \sigma_y}{3\mu + h} \tag{5.73}$$

where σ_y is the yield stress and $h = \frac{d\bar{\sigma}^{pr}}{d\bar{\varepsilon}^{pl}}$ is the plastic hardening at the beginning of the increment. The Jacobian is not required.

This model can be coded as follows:

Listing 5.8 VIHP.f

```
     SUBROUTINE VUMAT(
C_Read_only -
     1    NBLOCK, NDIR, NSHR, NSTATEV, NFIELDV, NPROPS,
     2    LANNEAL, STEPTIME, TOTALTIME, DT, CMNAME, COORDMP,
     3    CHARLENGTH, PROPS, DENSITY, STRAININC, FIELDNEW,
     4    RELSPININC, TEMPOLD, STRETCHOLD, DEFGRADOLD,
     5    FIELDOLD, STRESSOLD, STATEOLD, ENERINTERNOLD,
     6    ENERINELASOLD, TEMPNEW, STRETCHNEW, DEFGRADNEW,
C_Write_only -
```

```
     7       STRESSNEW, STATENEW, ENERINTERNNEW, ENERINELASNEW)
C
      INCLUDE 'VABA_PARAM.INC'
C
      DIMENSION PROPS(NPROPS), DENSITY(NBLOCK),
     1          COORDMP(NBLOCK), CHARLENGTH(NBLOCK),
     2          STRAININC(NBLOCK, NDIR+NSHR),
     3          RELSPININC(NBLOCK, NSHR),
     4          TEMPOLD(NBLOCK), FIELDOLD(NBLOCK, NFIELDV),
     5          STRETCHOLD(NBLOCK, NDIR+NSHR),
     6          DEFGRADOLD(NBLOCK, NDIR+NSHR+NSHR),
     7          STRESSOLD(NBLOCK, NDIR+NSHR),
     8          STATEOLD(NBLOCK, NSTATEV),
     9          ENERINTERNOLD(NBLOCK),
     1          ENERINELASOLD(NBLOCK), TEMPNEW(NBLOCK),
     2          STRETCHNEW(NBLOCK, NDIR+NSHR),
     3          STATENEW(NBLOCK, NSTATEV),
     4          DEFGRADNEW(NBLOCK, NDIR+NSHR+NSHR),
     5          FIELDNEW(NBLOCK, NFIELDV),
     6          STRESSNEW(NBLOCK, NDIR+NSHR),
     7          ENERINTERNNEW(NBLOCK),
     8          ENERINELASNEW(NBLOCK)
C
      CHARACTER*8 CMNAME
C
      parameter (zero=0.d0, one=1.d0, two=2.d0,
     1           third=1.d0/3.d0, half=0.5d0, op5=1.5d0)
C
C For plane strain, axisymmetric, and_3D cases using
C the J2 Mises Plasticity with piecewise-linear
C isotropic hardening.
C
C The state variable is stored as:
C
C STATE(*,1) = equivalent plastic strain
C
C User needs to_input
C props(1) Young s modulus
C props(2) Poisson s ratio
C props(3..) syield and_hardening_data
C calls vuhard for curve of yield stress vs. plastic strain
      e      =props(1)
      xnu    =props(2)
      twomu  =e/(one+xnu)
      alamda =xnu*twomu/(one-two*xnu)
      thremu =op5*twomu
      nvalue =nprops/2-1
C
      if ( stepTime .eq. zero ) then
        do k = 1, nblock
          trace=strainInc(k,1)+ strainInc(k,2)
```

```
     1          +strainInc(k,3)
           stressNew(k,1)= stressOld(k,1)
     1           +twomu*strainInc(k,1)+ alamda*trace
           stressNew(k,2)= stressOld(k,2)
     1           +twomu*strainInc(k,2)+ alamda*trace
           stressNew(k,3)= stressOld(k,3)
     1           +twomu*strainInc(k,3)+ alamda*trace
           stressNew(k,4)= stressOld(k,4)
     1           +twomu*strainInc(k,4)
           if ( nshr .gt. 1 ) then
                stressNew(k,5)= stressOld(k,5)
           1                    +twomu*strainInc(k,5)
                stressNew(k,6)= stressOld(k,6)
           1                    +twomu*strainInc(k,6)
           end if
        end do
     else
        do k = 1, nblock
           peeqOld=stateOld(k,1)
           call vuhard(yieldOld, hard, peeqOld,
     1                 props(3), nvalue)
           trace=strainInc(k,1)+ strainInc(k,2)
     1          +strainInc(k,3)
           s11=stressOld(k,1)+twomu*strainInc(k,1)
     1       +alamda*trace
           s22=stressOld(k,2)+twomu*strainInc(k,2)
     1       +alamda*trace
           s33=stressOld(k,3)+twomu*strainInc(k,3)
     1       +alamda*trace
           s12=stressOld(k,4)+twomu*strainInc(k,4)
           if ( nshr .gt. 1 ) then
                s13=stressOld(k,5)+twomu*strainInc(k,5)
                s23=stressOld(k,6)+twomu*strainInc(k,6)
           end if
C
           smean=third*(s11+s22+s33)
           s11=s11-smean
           s22=s22-smean
           s33=s33-smean
           if ( nshr .eq. 1 ) then
                vmises=sqrt(op5*(s11*s11+s22*s22+s33*
           1                 s33+two*s12*s12) )
           else
                vmises=sqrt(op5*(s11*s11+s22*s22+s33*
           1                   s33+two*s12*s12+two*s13*
           2                   s13+two*s23*s23))
           end if
C
           sigdif=vmises-yieldOld
           facyld=zero
           if ( sigdif .gt. zero ) facyld=one
```

```
            deqps=facyld*sigdif/(thremu+hard)
C
C Update the stress
C
            yieldNew=yieldOld+hard*deqps
            factor=yieldNew/(yieldNew+thremu*deqps)
            stressNew(k,1)=s11*factor+smean
            stressNew(k,2)=s22*factor+smean
            stressNew(k,3)=s33*factor+smean
            stressNew(k,4)=s12*factor
            if ( nshr .gt. 1 ) then
               stressNew(k,5)=s13*factor
               stressNew(k,6)=s23*factor
            end if
C
C Update the state variables
C
            stateNew(k,1)=stateOld(k,1)+deqps
C
C Update the specific internal energy -
C
            if ( nshr .eq. 1 ) then
               stressPower=half*(
     1 (stressOld(k,1)+stressNew(k,1))*strainInc(k,1)+
     2 (stressOld(k,2)+stressNew(k,2))*strainInc(k,2)+
     3 (stressOld(k,3)+stressNew(k,3))*strainInc(k,3))+
     4 (stressOld(k,4)+stressNew(k,4))*strainInc(k,4)
            else
               stressPower=half*(
     1 (stressOld(k,1)+stressNew(k,1))*strainInc(k,1)+
     2 (stressOld(k,2)+stressNew(k,2))*strainInc(k,2)+
     3 (stressOld(k,3)+stressNew(k,3))*strainInc(k,3))+
     4 (stressOld(k,4)+stressNew(k,4))*strainInc(k,4)+
     5 (stressOld(k,5)+stressNew(k,5))*strainInc(k,5)+
     6 (stressOld(k,6)+stressNew(k,6))*strainInc(k,6)
            end if
            enerInternNew(k)=enerInternOld(k)
     1                      +stressPower/density(k)
C
C Update the dissipated inelastic specific energy -
C
            plasticWorkInc=half*(yieldOld+yieldNew)
     1                   *deqps
            enerInelasNew(k)=enerInelasOld(k)
     1                      +plasticWorkInc/density(k)
         end do
      end if
C
      return
      end
```

```
      subroutine vuhard(syield,hard,eqplas,table,nvalue)
      include 'vaba_param.inc'
c
      dimension table(2, nvalue)
c
      parameter(zero=0.d0)
c
c set yield stress to_last_value of table, hardening
C_to zero
c
      syield=table(1, nvalue)
      hard=zero
c
c if_more than one_entry, search table
c
      if(nvalue.gt.1) then
          do k1=1, nvalue-1
              eqpl1=table(2,k1+1)
              if(eqplas.lt.eqpl1) then
                  eqpl0=table(2,k1)
c
c yield stress and_hardening
c
                  deqpl  =eqpl1-eqpl0
                  syiel0 =table(1,k1)
                  syiel1 =table(1,k1+1)
                  dsyiel =syiel1-syiel0
                  hard   =dsyiel/deqpl
                  syield =syiel0+(eqplas-eqpl0)*hard
                  goto 10
              endif
          end do
      10  continue
      endif
      return
      end
```

This VUMAT yields the same results as the PLASTIC option with ISOTROPIC hardening. This result is also true for large-strain calculations. The necessary rotations of stress and strain are taken care of by Abaqus. The routine calls user subroutine VUHARD to recover a piecewise linear hardening curve. It is straightforward to replace the piecewise linear curve by an analytic description.

References

1. Treloar LRG (1975) The physics of rubber elasticity. Clarendon Press, Oxford, p 85
2. Ogden RW (1997) Non-linear elastic deformations. Dover Publications, New York, pp 101–103
3. Gurtin ME, Fried E, Anand L (2010) The mechanics and thermodynamics of continua. Cambridge University Press, Cambridge, p 151, 242

4. Bathe KJ (1996) Finite element procedures. Prentice-Hall, Upper Saddle River, p 612
5. Weber G, Anand L (1990) Finite deformation constitutive equations and a time integration procedure for isotropic, hyperelastic-viscoplastic solids. Comput Methods Appl Mech Eng 79:173–202
6. McKenna RF, Jordaan IJ, Xiao J (1990) Finite element modelling of the damage process in ice. In: ABAQUS users conference proceedings
7. Snyman MF, Mitchell GP, Martin JB (1991) The numerical simulation of excavations in deep level mining. In: ABAQUS users conference proceedings
8. HajAli RM, Pecknold DA, Ahmad MF (1993) Combined micromechanical and structural finite element analysis of laminated composites. In: ABAQUS users conference proceedings
9. Govindarajan RM, Aravas N (1993) Deformation processing of metal powders: cold and hot isostatic pressing. Private communication
10. Kalidindi SR, Anand L (1993) Macroscopic shape change and evolution of crystallographic texture in pre-textured FCC metals. Acta Metall
11. Hughes TJR, Winget J (1980) Finite rotation effects in numerical integration of rate constitutive equations arising in large deformation analysis. Int J Numer Methods Eng 15:1862–1867

Chapter 6
Mesher and Meshing

6.1 Generalities

Meshing is rarely based on specific geometric criterion so the quality of meshing is primary based on visual inspection. In observation for a good practice of the structure meshed, some constraints in the model must be focused on

- Make the necessary assumptions with the designer to modify the initial design without going against the function of each component making the assembly. The design simplifications have been made to represent the approximation of the geometry of your structure according to the boundary and loading conditions.
- If the structure assembly has been imported from a CAD file, then track the small discontinuity lines or misalignment with the designer should be tracked to fix it. The second option to fix a bad construction strategy to construct the CAD model is to play with the virtual topology features with Abaqus mesher. If the structure mesh is not sufficiently well mapped according to the user's assumptions, then it will be necessary make the CAD model with Abaqus in order to control the CAD construction design and the control mesh afterward.
- Use the same element shape pattern to mesh the structure. The best-recommended option to control the structure mesh is the quadratic structured as this will minimize the mesh transition.
- Evaluate a geometric aspect ratio in accordance with your design to control the approach of the mesh transition zones.
- The partitioning strategy should be adapted progressively to control the meshed structure.
- A uniformed mesh of the structure around the critical analysis areas.

The mesh control module is, of course, excellently explained inside the Abaqus CAE user's manual Chap. 17.

R. J. Boulbes, *Troubleshooting Finite-Element Modeling with Abaqus*,
https://doi.org/10.1007/978-3-030-26740-7_6

The control mesh is split into three main categories:

1. The Element shape to select the elementary geometry (triangle, square, hexahedral, tetrahedral, or wedge) of one element for the approximation of the design.
2. The technique is the function of the partitioning strategy that the user will define to keep control of some strategic analysis areas. Indeed, a good partition strategy will be capable of meshing the design in a proper computed form. The partitioning is highly dependent on the preliminary knowledge of the response of the structure under boundary and loading conditions or a theoretical estimation of the response if applicable.
3. The algorithm is the control and adjustment under specific criterion programmed inside Abaqus according to mesh theory related to Delaunay[1] triangulation [1] plus the internal geometric limit for element distortion.

6.1.1 Mesh Control Options

The control panel used to select an element shape will only define the strategy employed for dealing with the defined area or volume and combining the elementary shapes to approximate the real geometry by triangular, quadrilateral in 2D, or tetrahedral and hexagonal shapes in 3D.

6.1.2 Mesh Controls for a 2D Structure

Figures 6.2, 6.3, and 6.4 show the element shape options in Fig. 6.1 for a planar structure to see what type of mesh pattern can be established for an element shape selected to map a structure with **Quad**, **Quad-dominated**, and **Tri**, respectively.

6.1.3 Mesh Controls for a 3D Structure

The same type of mesh control exists in 3D as shown in Figs. 6.5, 6.6, 6.7, and 6.8 show the different types of mapped structure the user can get from the element shape options in three-dimensional structures with **Hex**, **Hex-dominated**, **Tet**, and **Wedge** element types as shown in Fig. 6.9.

---

[1]Suppose the software generates a random set of nodes within the problem region. It is possible to connect as many pairs of nodes without ever crossing a previous line. The result is a (maximal) triangulation of the nodes. The process seems pretty arbitrary, and in fact, there are many possible triangulations of a set of points. The user may wonder how to automate this process; a natural way is to start by creating a giant triangle that encloses all the points the user needs to mesh. Then add the first node. Connect it to each vertex of the enclosing triangle, and finally the user will get a maximal triangulation. Add the second node. It falls into one of the triangles that the user already created, so the user subdivides that triangle. Keep going. In the end, remove the enclosing triangle, and any edges that connect to it, and the user will obtain a maximal triangulation of the nodes.

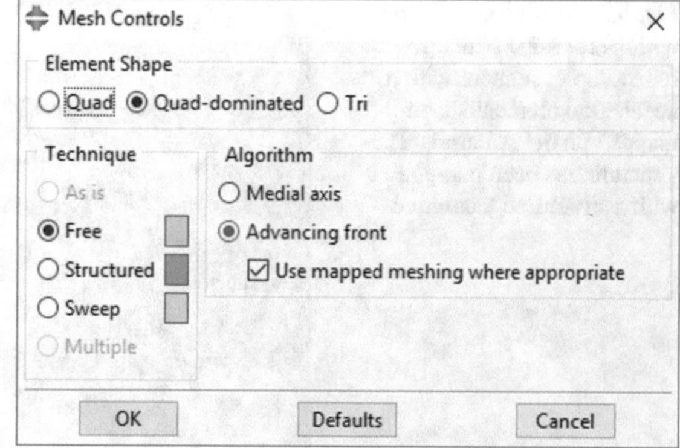

Fig. 6.1 Mesh control panel options for a 2D structure, for example, set here with a quadrilateral-dominated element shape mapped on the structure. The structure has been mapped with a free technique using an advancing front algorithm where meshing is the appropriate enforcement option

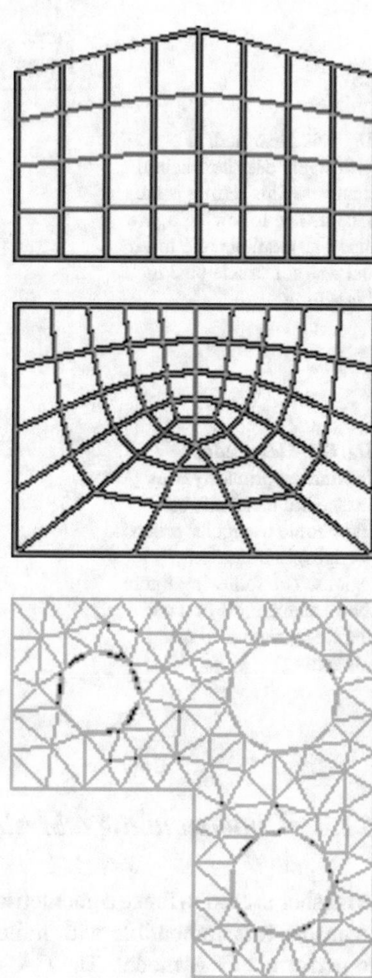

Fig. 6.2 **Quadrilateral** uses quadrilateral elements exclusively. The following figure shows an example of a mesh that was constructed using this setting

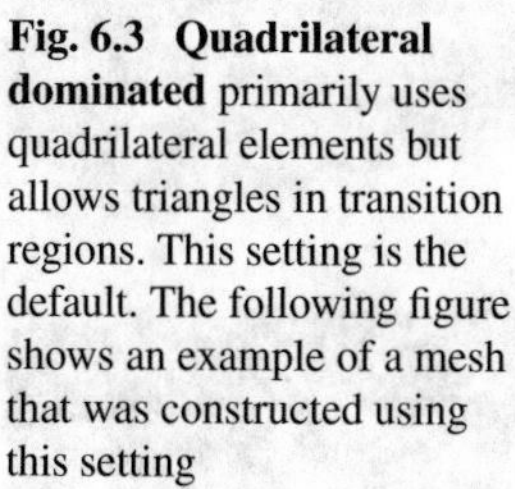

Fig. 6.3 **Quadrilateral dominated** primarily uses quadrilateral elements but allows triangles in transition regions. This setting is the default. The following figure shows an example of a mesh that was constructed using this setting

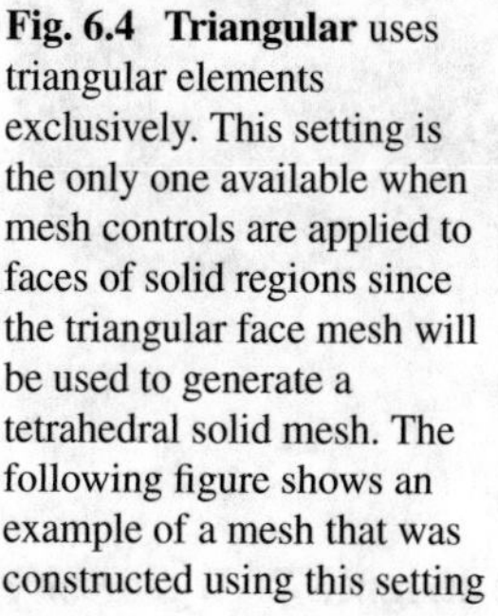

Fig. 6.4 **Triangular** uses triangular elements exclusively. This setting is the only one available when mesh controls are applied to faces of solid regions since the triangular face mesh will be used to generate a tetrahedral solid mesh. The following figure shows an example of a mesh that was constructed using this setting

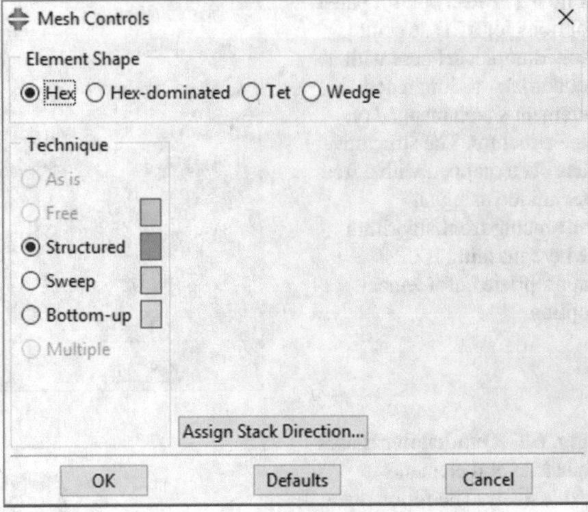

Fig. 6.5 Mesh control panel options for a 2D structure, for example, set here with a hexahedral element shape mapped on the structure. The structure has been mapped with a structured technique

Fig. 6.6 **Hexahedral** exclusively uses hexahedral elements. This setting is the default. The following figure shows an example of a mesh that was constructed using this setting

Fig. 6.7 **Hexahedral dominated** primarily uses hexahedral elements, but allow some triangular prisms (wedges) in transition regions. The following figure shows an example of a mesh that was constructed using this setting

6.1.4 Understanding a Mesher

A mesher is used to make a geometric approximation of the designed CAD geometry to ensure it is compatible with finite elements—with simple geometries—in order to perform a FEA model. The CAD model is a function of standard methods and protocols, and the FEA mesher will only obey a single unique math theory out of any

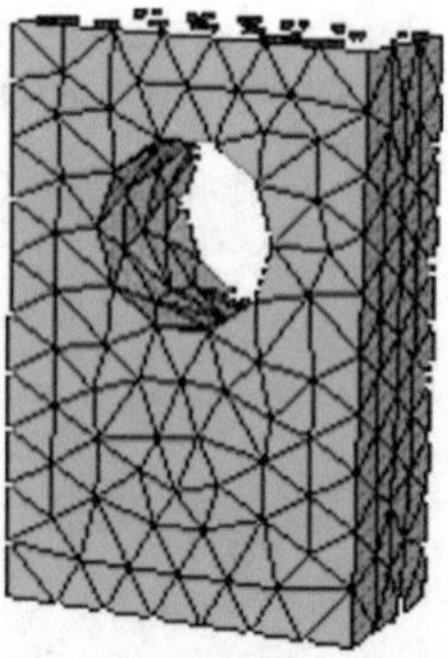

Fig. 6.8 **Tetrahedral** uses tetrahedral elements exclusively. The following figure shows an example of a mesh that was constructed using this setting

Fig. 6.9 **Wedge** uses wedge elements exclusively. The following figure shows an example of a single-element mesh that was constructed using this setting

standard methods or protocols used to make the CAD model. Therefore, the methods used to make and analyze the designed geometry are **NOT** compatible.

Indeed, finite-element analysis (FEA) is based on a **unique** mathematical method which requires a solution to be computed from elementary element shapes—simple 1D, 2D, or 3D geometries. The mesher is, therefore, the tool to make possible an approximation of the CAD model because the realistic geometries are **TOO** complicated to be generated from simple shapes and compute a numerical solution for a designed assembly.

To make the life of the mesher easier and therefore, help computation accuracy plus save time in analysis, it is crucial to think about geometry and topology to transform the **Real Geometry**—which is defined as the entities characterized by a direct definition of their geometry—into a **Faceted Geometry**—which is defined as the entities characterized ONLY by an indirect definition with respect to an underlying grid.

Geometrical types and topology features are based on

- Vertex (a point with coordinates),
- Edge (two or more vertices),
- Face (three or more edges), and
- Volume (four or more faces).

A **Bottom-up** approach is generated with low-dimensional entities and builds higher dimensional entities on top of them.

A **Top-bottom** approach is generated with upper dimensional entities and uses a Boolean operation to define the other entities.

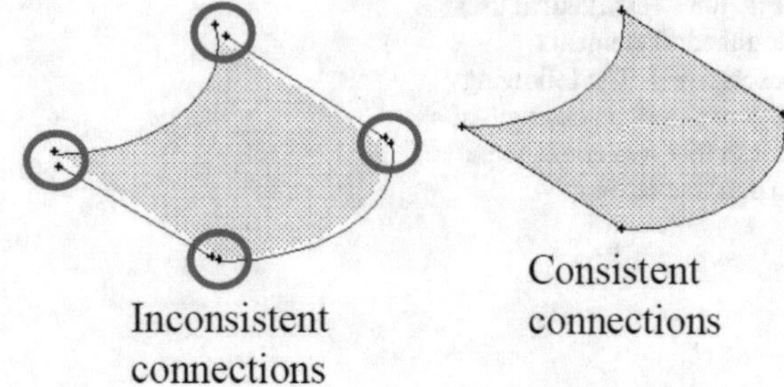

Fig. 6.10 An example of discontinuity in the design geometry

6.1.4.1 Connect Geometry

Discontinuity in the geometry is **NOT** permitted to compute a FEA solution. Building upper topological entities from the lower ones requires that they are properly connected. Sometimes, as shown in Fig. 6.10, the level of details within the CAD model is so small that it is difficult to detect discontinuity in geometry or inconsistent connections.

To avoid bad mesh continuity or inconsistent connections in the model region, the analyst must closely track and detect discontinuities with an imported model. Discontinuities in the model can be due to the import file itself or the method used to design the assembly. Methods used by the designer to make parts in accordance with some standards would be unfitted with the unique finite-element method used to compute the structural solution. As a result, both the analyst and designer should work together to make corrections if needed or to make a design model to fulfill some gaps. An FEA software is not a CAD software, and a CAD software is not a FEA software, even though both share certain functions and features to perform rough operations in each fields of expertise.

To track such discontinuity in geometry, there is no automatic pre-programmed way of performing such operations for a large model. The most efficient way of doing this is to make a visual inspection first then mesh the geometry and make a test load with a pure elastic material only with small deformations in order to detect any error or warning messages related to the meshed structure. However, Abaqus has some powerful tools in the Mesh module and especially with **Virtual topology**[2] features, which, for instance, can combine faces and edges.

6.1.4.2 Import Geometries

Many file formats exist to convert a CAD model file readable by an FEA software, the type of file to work with is also mainly a function of the type of the software license

[2]To know more about all virtual topology features please refer to the Abaqus CAE User's Guide in Sect. 75. The Virtual Topology toolset.

associated with the user's FEA software, for instance, with Abaqus it is preferable to use the STEP file format.

- STEP (STandard for Exchange of Product model data; ISO standard)
- IGES (Initial Graphics Exchange Specification; ANSI standard)
- STL (STereo Lithography; Rapid Prototyping Standard).

6.1.4.3 Clean-Up a CAD Model

A clean model is easier to mesh, which means fewer problems related to the geometry and fewer numerical difficulties. The main issues to check in order to clean a CAD model are listed below:

- Eliminate components not exposed to the flow.
- Eliminate duplicated entities.
- Eliminate small details.
- Water-proofing the surfaces.
- Rebuild geometrical connectivity between parts.

An example is given in Fig. 6.11 regarding edge connections.

6.1.4.4 Clean-Up a CAD Model Cracks

As demonstrated in Fig. 6.12, a crack is defined as a geometry consisting of an edge pair which meets the following criteria.

- Each edge in the pair serves as a boundary edge for a separate face.
- The edges share common endpoint vertices at one or both ends.
- The edges are separated along their lengths by a small gap.

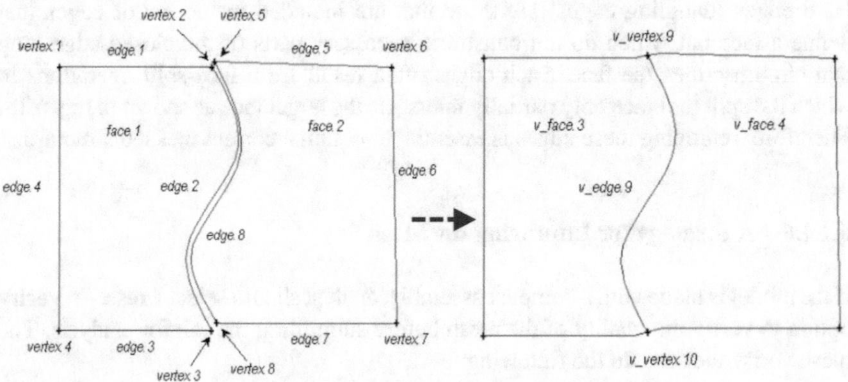

Fig. 6.11 An example of an edge connecting operation

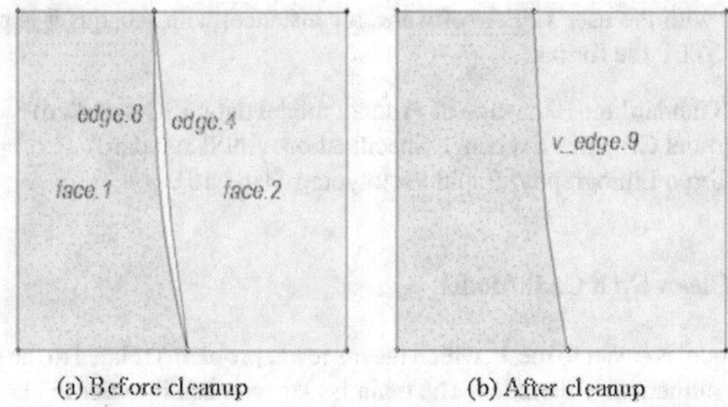

Fig. 6.12 An example of cleaning up the cracks

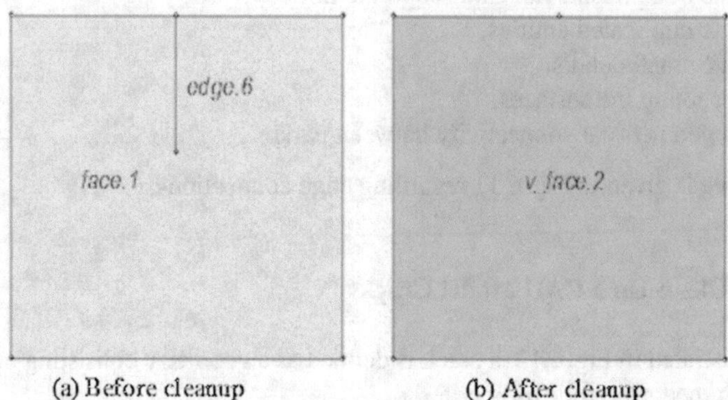

Fig. 6.13 An example of cleaning up the hard edges

6.1.4.5 Clean-Up Hard Edges

Hard edges (dangling edges) are those that are included in the list of edges that define a face but which do not constitute necessary parts of the closed edge loop that circumscribes the face. Such edges often result from face-split operations in which the split tool face only partially intersects the target face as shown in Fig. 6.13. Therefore, removing these edges is essential to obtain a correct meshed structure.

6.1.4.6 A Strategy for Improving the Mesh

If the model is made with a complex assembly, analyst should select **Mesh** → **Verify** option to verify the quality of the mesh before submitting the job for analysis. The mesh verify tool can do the following:

- Highlight elements of a selected shape that do not meet specified criteria such as aspect ratio.
- Print mesh statistics such as the total number of elements of the chosen shape, the number of highlighted elements, and the average and worst values of the selection criterion.
- Highlight elements that do not pass the mesh quality tests that are included with the input file processor in Abaqus standard and Abaqus explicit.

If the mesh verify tool indicates that the user should try to improve the quality of the mesh, first try the following before turning to the Edit Mesh toolset:

- Change the seed distribution.
- Add or modify partitions.
- Change the mesh technique.

In addition, the analyst could try modifying the parts in the Part module, or the user might try using the Virtual Topology toolset and regenerating the mesh. The analyst should treat the Edit Mesh toolset as the final step in the meshing process and use it only to make minor adjustments to nodes and elements. Abaqus CAE tries to preserve attributes, for instance, loads and boundary conditions, if changes are made to the mesh. If user modifies a part, Abaqus CAE deletes the mesh when it returns to the Mesh module; as a result, the analyst will lose any edits that users have made to the mesh.

6.1.5 Mesh as Grid Generation

As shown in Fig. 6.1, a grid generation (a mesh) is mainly based on structured or unstructured grid, defined as below:

Structured grids: Ordered set of (locally orthogonal) lines

- Several techniques can be used to map a computational domain into a physical domain: Transfinite Interpolation, Morphing, PDE Based, etc.
- The grid lines are curved to fit the shape of the boundaries.

Unstructured grids: Unorganized collection of polygons (polyhedron)

- Three main techniques are available to generate triangles automatically (tetrahedral): Delaunay triangulation, Advancing front, and so on.
- Paving for automatic generation of quads in 2D.

The unstructured grids with the triangulations technique called Free in Fig. 6.1 refer to an unstructured in opposition to the structured technique shown in the same Figure.

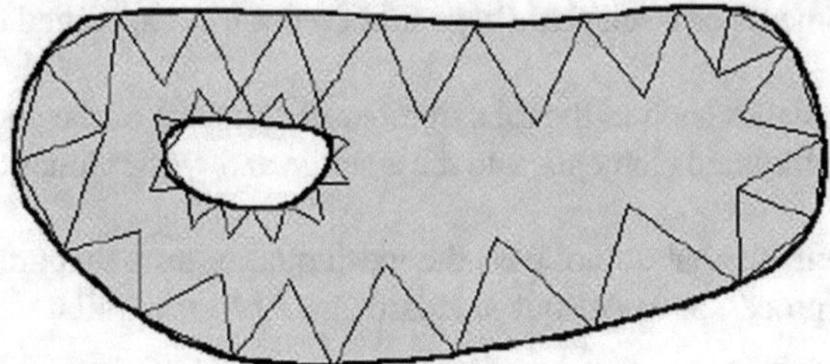

Fig. 6.14 Advancing front principles. The meshed structure can be considered as good to run when a surface grid is preserved with specialized layers near the surfaces. A bad to run mesh will result in computational complex difficulties and low quality in results

6.1.5.1 Advancing Front Algorithm

The principles of the algorithm option shown in Fig. 6.1 are summarized below:

- Triangles are built inward from the boundary surfaces.
- The last layer of elements constitutes the active front.
- An optimal location for a new node is generated for each segment on the front; the new node is generated by checking all existing nodes and this new optimal location.
- Intersection checks are required to avoid front overlap.

The advanced front algorithm shown in Fig. 6.14 works as shown below with five major steps to mesh the structure. First, it begins with an initial front (shown in Fig. 6.15) with a boundary mesh defined as the initial front For each edge (face) on the front, locate ideal node C based on front AB. The second step involves creating a first advanced front as shown in Fig. 6.16 to determine whether any other nodes on the current front are within the search radius (r) of the ideal location C (Choose D instead of C). The next advanced front shown in Fig. 6.17 is the so-called Book-Keeping, where the new front edges are added and deleted from the front as triangles which are formed and continue until no front edges remain on front. The iterating advanced front shown in Fig. 6.18 still continues to follow the same logic as used for the next advanced front to continue the mesh mapping by triangulation iterations. As shown in Fig. 6.19, all transition shapes during the iteration technique can have multiple choices available to ensure the best mesh quality (closest shape to equilateral). The criterion for rejecting a triangle iteration meshed is any triangles that would intersect the existing front or any inverted triangles such as the surface value calculated by using the determinant formula that do not satisfy the non-inverted condition $det\left(\overrightarrow{AB} \times \overrightarrow{AC}\right) > 0$. The non-inverted condition, where the determinant of the Jacobian of the second Bézier polynomial form is positive everywhere, is useful for finite-element analysis as this condition implies that the mesh is globally invertible—for every region in the domain there exists a unique element meshed in the mesh and a unique meshed triangle. If the mesh does not overlap and the mesh

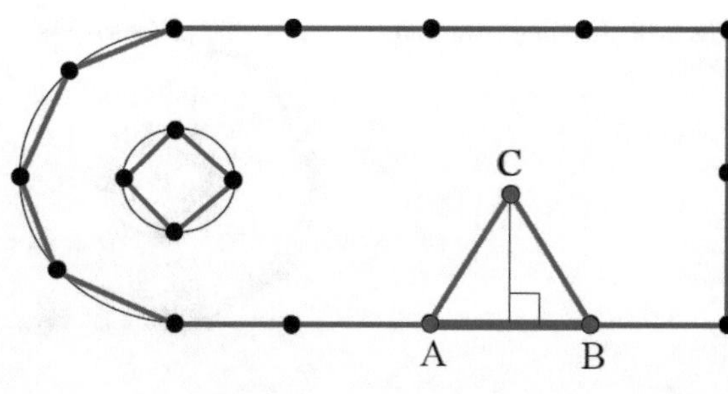

Fig. 6.15 Initial front

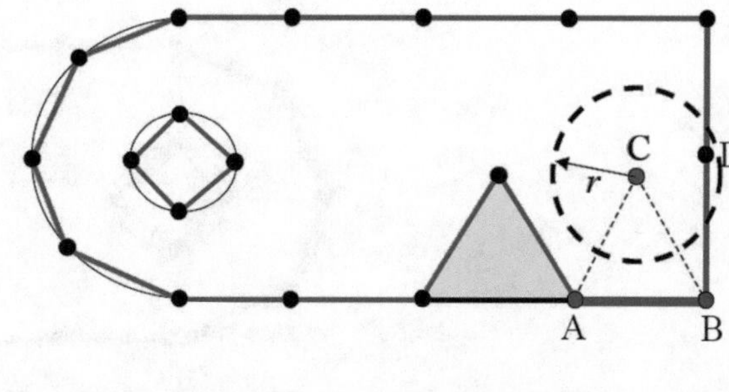

Fig. 6.16 First advanced front

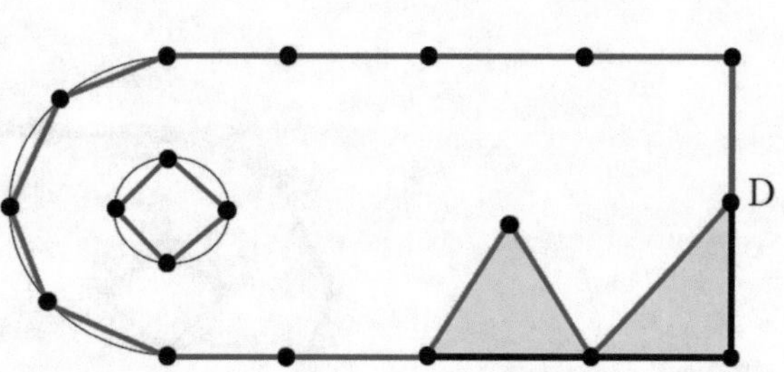

Fig. 6.17 Next advanced front

is globally invertible, then it is also the case that each element is non-inverted. If a second-degree Bézier triangle is non-inverted by the next iteration, then it is globally invertible.

6.1.5.2 Quad-dominated

The mesh is controlled by an element shape defined as Quad-dominated for quadrilateral dominated. As shown in Fig. 6.1, this construct has the following principles:

- Squares containing the boundaries are recursively subdivided until the desired resolution is obtained.
- Irregular cells (or triangulation) are generated near the surface where the squares intersect the boundary.

Figure 6.20 presents a representation of a mesh grid on a structure using a quadrilateral-dominated technique, a mesh criterion used to evaluate how effective

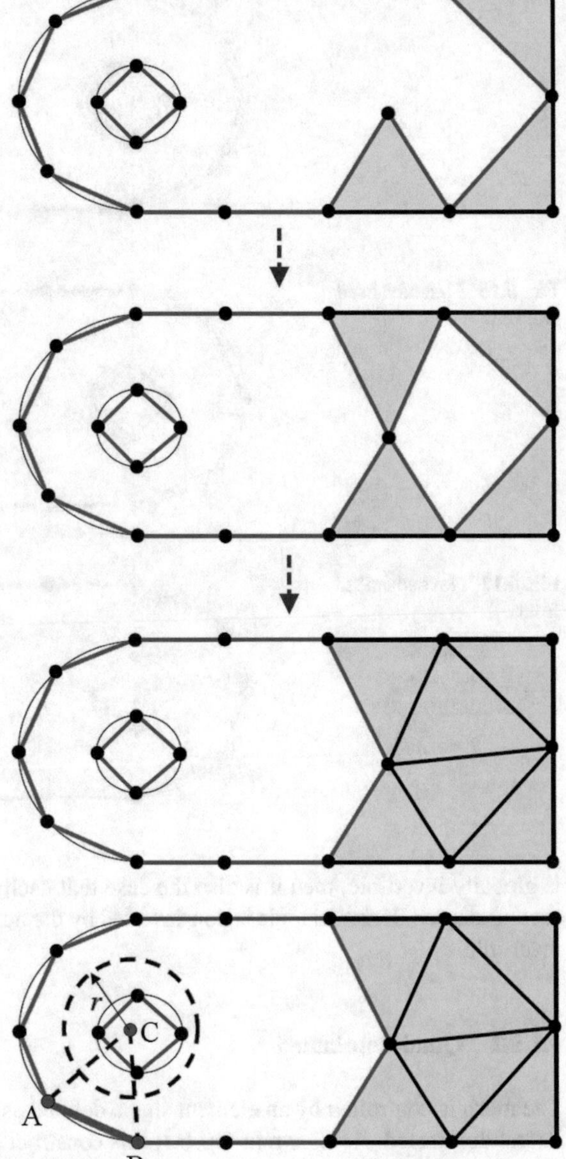

Fig. 6.18 Iterating advanced front

Fig. 6.19 Transition shape detected by the last advanced front increment

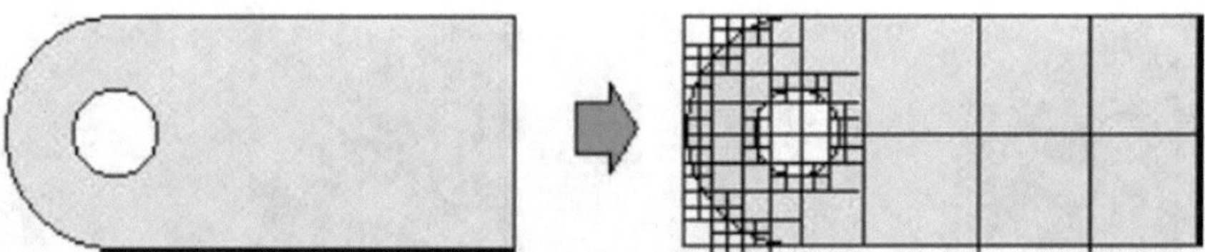

Fig. 6.20 Representation of a quad-dominated grid generation

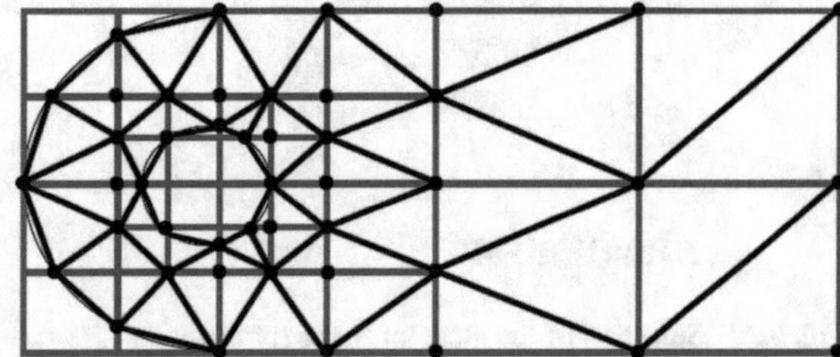

Fig. 6.21 Representation of a quad-dominated grid generation meshed with triangles

the mesh technique is. It requires the least surface representation and is highly automated through the structure. Otherwise, it cannot match the surface grid properly and therefore, leads to a low-quality meshed structure close to the design surface's limit.

Figure 6.21 gives an example of the surface designed in Fig. 6.20 with the quadrilateral-dominated mesh technique used with triangle elements. It is evident that the meshed structure obeys to the following step procedure:

1. Define the initial bounding box (root of quad tree);
2. Recursively break it into four leaves per root to resolve geometry;
3. Find intersections of leaves with geometry boundary;
4. Mesh each leaf using corners, side nodes, and intersections with geometry;
5. Delete Outside.

6.1.5.3 Unstructured Grids: Paving

When the mesh is a function of the element shape alone, from triangular to quadrilateral or vice-et-versa as shown in Fig. 6.22

- Advancing front technique based on quads (instead of triangles)
- Only in 2D.

6.1.5.4 Unstructured-Quad

The unstructured quadrilateral is set with the mesher as shown in Fig. 6.1 but this time using a Quad element shape instead of Quad-dominated, with a Free technique and

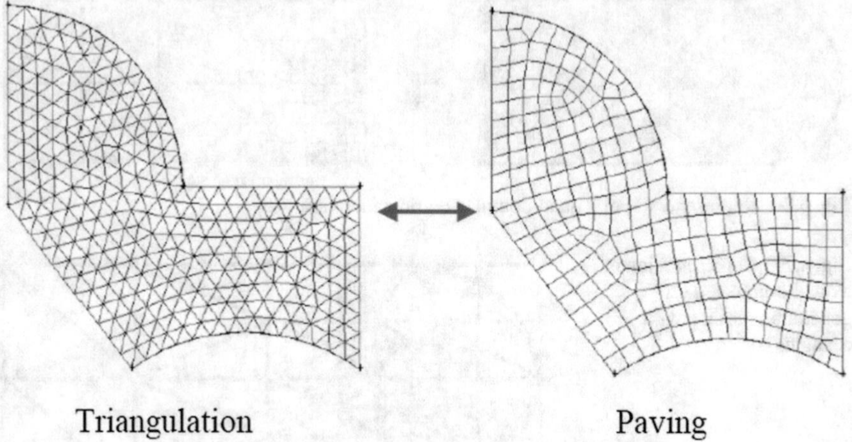

Fig. 6.22 Selection of the element shape for meshing

the Advancing front algorithm. Figure 6.23 demonstrates the different step-by-step meshing procedure used with the settings selected in Fig. 6.23. Here, the advancing front begins with the front at the boundary to form rows of elements based on front angles. Subsequently, the meshed structure must have an even number of intervals for all quadrilateral meshes in order to approach the designed surface with the appropriate quadrilateral element geometry.

6.1.5.5 Unstructured Grids: Sweeping

This option is available in the technique called sweep in Fig. 6.1. The geometry requirements for a proper sweep technique as shown in Fig. 6.24 are

- source and target surfaces which are topologically similar,
- linking mappable or sub-mappable surfaces.

6.1.5.6 Grid Generation Using Medial Axis

Figure 6.25 shows how the Medial axis algorithm option set in Fig. 6.1 works to mesh a part.

When a Medial axis algorithm is set via the mesher control options, a medial object is determined on the surface to mesh and roll a maximal circle in 2D or a sphere in 3D through the model. The center traces the medial object, which is tracked to cover the whole structure to mesh. Once this medial object trajectory is known, the mesher will define the model's anatomy spine to establish the according element mesh pattern. The medial object is used as a tool to automatically decompose the model into simpler mappable or sweepable parts.

Fig. 6.23 An example of an unstructured-Quad mesh

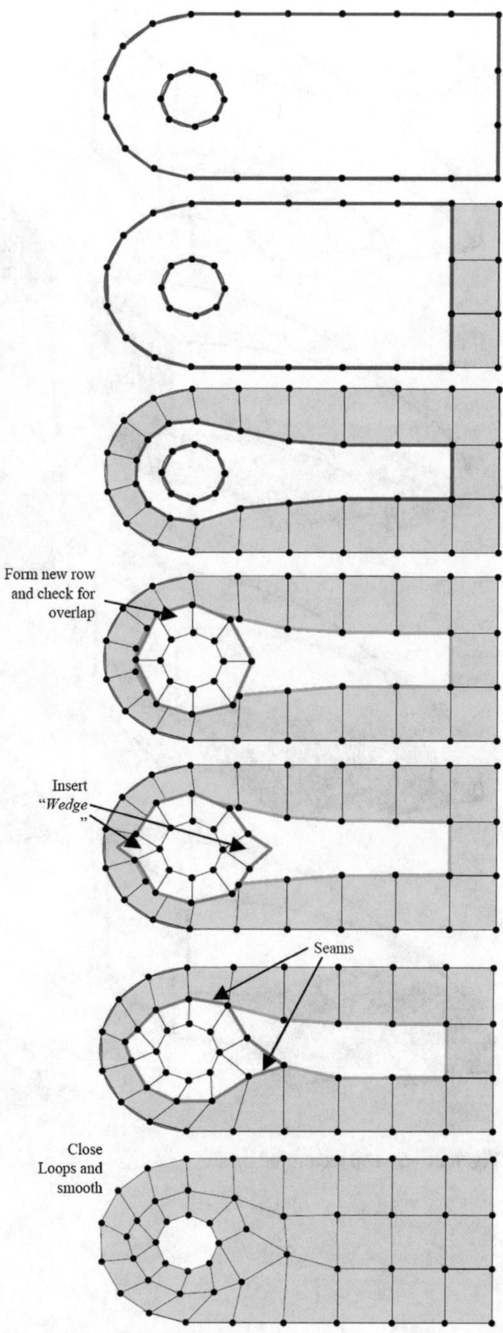

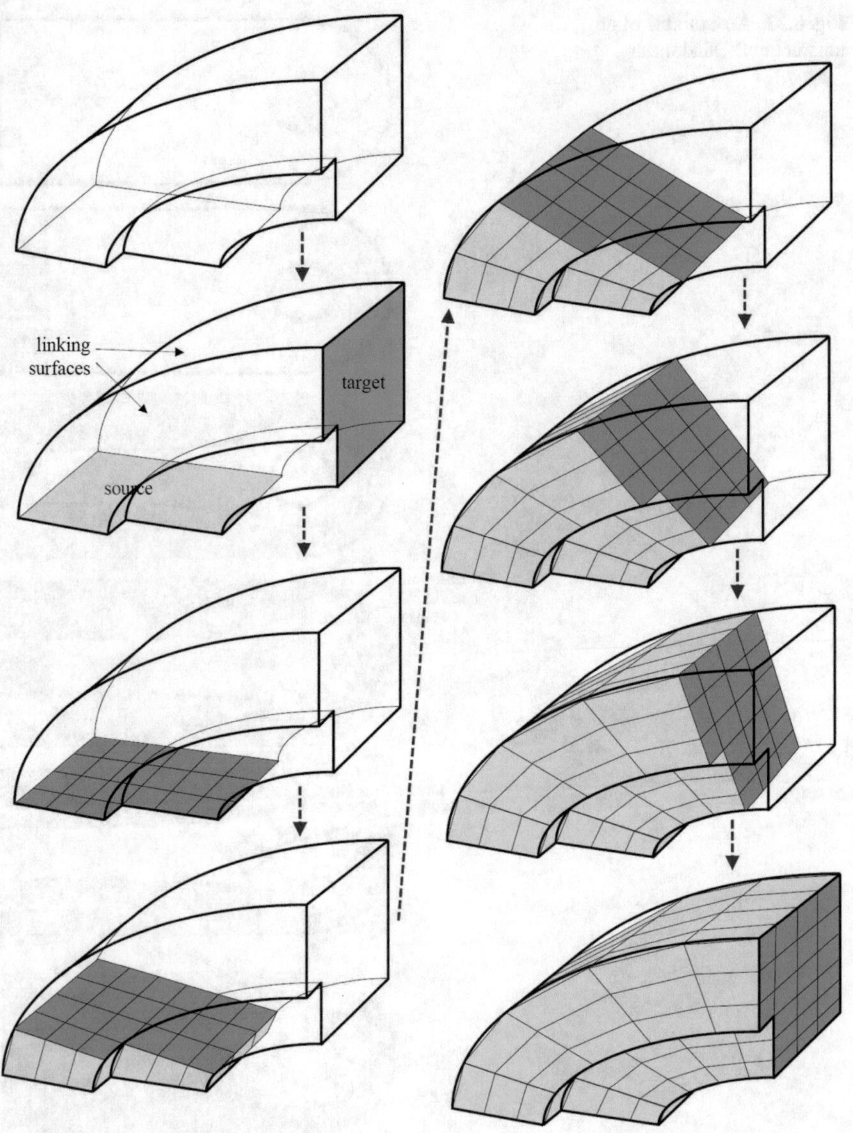

Fig. 6.24 Sweep technique mesh

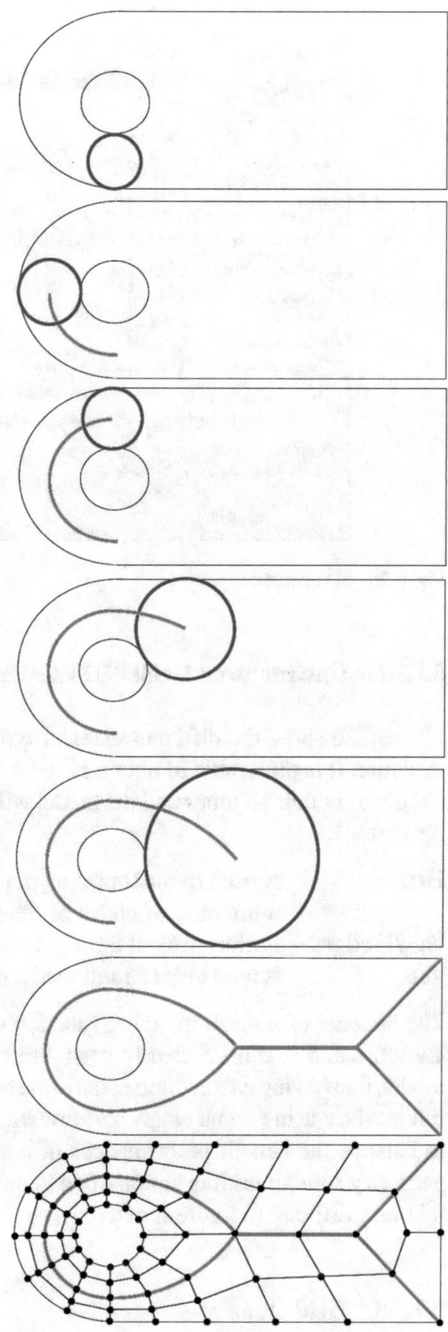

Fig. 6.25 Medial axis algorithm mesh

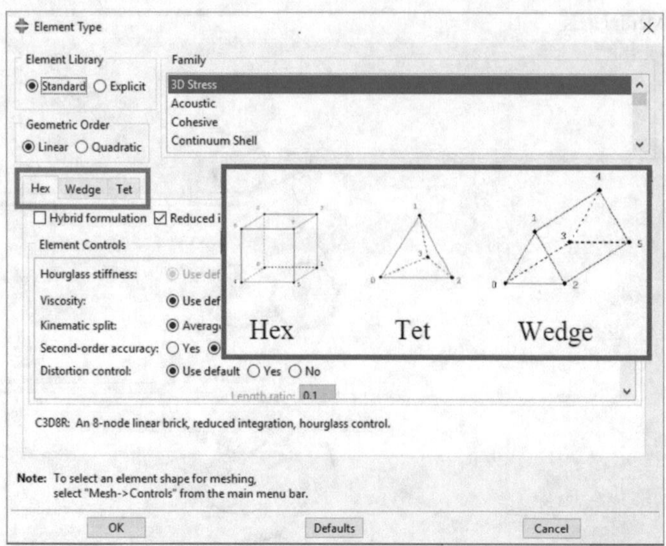

Fig. 6.26 3D element types

6.1.5.7 Unstructured Grids: 3D Elements

Figure 6.26 show the different element types that can be used to mesh a 3D solid structure. It is preferable to obtain Hex-Based meshing mainly equiangular Tets are not good for thin volumes and the model will need too many elements for reasonable resolutions.

Hex is used to maximize a volume covered per edge size with the maximum ratio of nodes per elements.
Hex/Wedges is clustering at the solid wall with high-quality elements.
Tets is used to get an automatic meshing of extremely complicated regions.

The benefits of a mesh mapping function of the element types are summarized in Table 6.1 and it can be clearly seen, with the exception of a complex geometric model, that having a structured mesh is the best option. This difficulty can be partially removed by using some smart partitioning strategies for the whole model, in order to balance the benefit performances of a structured mesh dealing with a complex geometry sometimes it is not possible to obtain a good balance and most of the time it is very difficult to figure it out.

6.1.5.8 Grid Quality

Quality measures are **NOT** absolute but should be considered in connection with solution schemes. The final accuracy of a procedure is **ALWAYS** a function of the

Table 6.1 Classification of the different mesh strategies as a function of the benefits of running analysis

Mesh mapping	Analysis benefits				
	Speed	Robustness	Quality and control	Complex geometry	Mesh sizes
Structured gridding (mapping)	Y[a]	Y	Y	N[b]	Y
Unstructured triangulation	N	Y	NS[c]	Y	N
Unstructured paving	N	N	NS	Y	Y
Unstructured coopering[d]	Y	N	NS	N	Y

[a]Y: Yes, it is recommended for use
[b]N: No, it is not recommended for use
[c]NS: Not Sure it will work correctly
[d]In general, the Cooper meshing scheme applies to volumes that demonstrate either of the two characteristics. First, at least one face is neither mappable nor sub-mappable. And second, all faces are mappable or sub-mappable, but the vertex types are specified such that the volume cannot be divided into mappable sub-volumes. Faces that meet either of the criteria outlined above, as well as those that are logically parallel to such faces, constitute source faces for the volume and the end caps of the corresponding logical cylinder

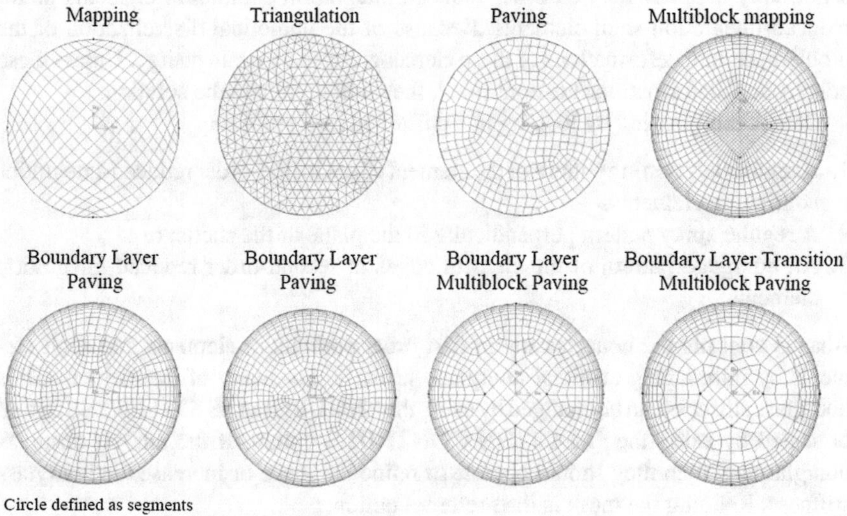

Fig. 6.27 Example of a mesh mapping used for a circle shape

grid quality. Figure 6.27 shows what kind of mesh makes is possible to get the function of the element types used or partitioning strategies used to help the mesher in meshing the region on the meshed structure and achieve as uniform a quadrilateral shape as possible, even though in theory whatever the mesh is there is still a compatibility equation of nodal fields. However, in practice, high element distortion on the loaded structure can lead to severe numerical difficulties. Therefore, the mesh quality can be a function of several geometrical measures which can be defined as follows:

- depending on the size of the elements,
- depending on the shape of the elements, and
- depending on relative dimensions of neighboring elements.

To give a better overview of meshing features, an example of a Pad eye design is given in Appendix A.5. This example includes all the different options available inside the mesh control panel as shown in Fig. 6.1, which presents the meshing of a 2D structure.

6.2 The Abaqus Model Meshed Has Changed into a Nonphysical Shape with a Regular Pattern

This can happen after a model has completed, but the deformed mesh has gone into a regular pattern that is clearly not physical, this is because the mesh may be experiencing a type of instability known as hourglassing. If hourglassing occurs, it normally happens for first-order reduced-integration continuum elements or for reduced-integration shell elements. Because of the numerical discretization of the problem, certain deformations of these elements cause no strain energy. Unless these nonphysical deformations are controlled, they will dominate the solution.

Detect hourglassing by looking in a deformed mesh plot for

1. A regular keystoning pattern of the element edges in first-order reduced-integration continuum elements;
2. A regular spiky pattern perpendicular to the plane of the shells; or
3. An hourglass pattern of the element edges in second-order reduced-integration elements.

Abaqus controls the hourglassing in first-order continuum elements and shell elements by introducing artificial hourglass stiffness. For many of the elements, this hourglass stiffness can be changed by using the ***HOURGLASS STIFFNESS** option or scaled by using the ***SECTION CONTROLS** option. If the analyst observes hourglassing, then they should operate to refine the mesh or increase the hourglass stiffness. Refining the mesh is the preferred option.

If the hourglass stiffness is too high, the model may become too stiff. Check the artificial energy **ALLAE** against internal energy **ALLIE**; **ALLAE** should normally be a small fraction, for instance, lesser than 1% of **ALLIE**. The greater the tendency to hourglass is, the higher the artificial energy.

If a model continues to hourglass, continue refining the mesh or use a different element type (first-order fully integrated or even a second-order element).

6.3 Excessive Element Distortion Warnings

As already described previously, if there are excessive element distortion warnings in the message file the mesh pattern has a mismatch with the design geometry. These messages are issued in the message (.msg) file, when the volume at an integration point of an element becomes negative. When parts and assemblies are used, error messages involving element and node numbers report these numbers in terms of the part instance.[3] When troubleshooting excessive element distortion warnings, consider the following:

- Check the deformed shape for the last converged increment. Is the mesh refinement reasonable considering the deformation in the element that is distorting? If not, try refining the mesh in the areas of large deformation.
- Are the elements hourglassing? Check Hourglassing to identify the causes and what to do about them.
- Check the model definition including boundary conditions, loading, and material properties. The problem could also be due to contact overclosures.
- If the error occurs in the first increment and the model contains contact pairs or tie constraints, have the slave nodes been adjusted to lie on the master surface? If nodes on the slave surface are adjusted onto the master surface by an amount that is significant compared to an element length, the elements may be highly distorted after the adjustment.

6.4 Compatibility Errors Printed to the Message File for a Model with Hybrid Elements

When the analyst runs an Abaqus standard analysis with hybrid elements, the message (.msg) file contains sections titled **COMPATIBILITY ERRORS**. What are these errors and can they be safely ignored?

The **COMPATIBILITY ERRORS** information printed to the message file when hybrid elements are used does not indicate that an error has occurred in the analysis. In the case of solid hybrid elements, volumetric strain is computed both from the displacement degrees of freedom and from the independent pressure degree of freedom. Abaqus checks that the difference between these two strain measures lies within a specific tolerance in order to ensure convergence. This is similar to the checks on force and moment equilibrium. In each iteration, information regarding the hybrid element compatibility is printed to the message file.

[3] See Abaqus Analysis User's Guide inside Sect. 2.10.1 Defining an assembly for more information.

For example, the following is from the last iteration of an increment using hybrid elements:

```
COMPATIBILITY ERRORS:
TYPE          NUMBER          MAXIMUM        IN ELEMENT
              EXCEEDING       ERROR
              TOLERANCE

VOLUMETRIC     0             -9.595E-10      49
```

In the case above, the maximum error (or difference) between the two strain values is well within the acceptable tolerance of 1.e-5. It is safe to ignore messages like these. A situation may arise when force and moment equilibrium have been satisfied but the hybrid element compatibility check fails. For instance, this happens in step 2, increment 4, iteration 3 of the bootseal.inp[4] example problem. In this analysis, the compatibility check is satisfied in iteration 4 and consequently the analysis proceeds and again the message can be ignored.

There may be situations where Abaqus has difficulty satisfying the compatibility check and the analysis fails to converge. Usually, this means that inappropriate elements are being used for the analysis, or that the material is completely incompressible. In these situations, it is generally necessary to change the model or the analysis to solve the problem.

6.5 User Element Subroutine

Abaqus standard has an interface that allows users to implement linear and nonlinear finite elements. A nonlinear finite element is implemented in user subroutine UEL.[5] The interface makes it possible to define any (proprietary) element of arbitrary complexity. If coded properly, user elements can be utilized with most analysis procedures in Abaqus standard. Multiple user elements can be implemented in a single UEL routine and can be utilized together. In this section, the implementation of nonlinear finite elements only will be discussed and illustrated with examples.

Abaqus standard is a versatile analysis tool with a large element library that allows analysis of the most complex structural problems. However, situations arise in which augmenting the Abaqus library with user-defined elements is useful for modeling nonstructural physical processes that are coupled to structural behavior, applying solution-dependent loads, and for modeling active control mechanisms. The advantages of implementing user elements in an analysis code such as Abaqus,

[4]See Abaqus Example Problems Guide inside Sect. 1.1.15 Analysis of an automotive boot seal for more information.

[5]See Abaqus Users Subroutines Reference Guide v6.14 in Sect. 1.1.28 UEL User subroutine to define an element.

instead of writing a complete analysis code, are obvious. Indeed, Abaqus offers a large selection of structural elements, analysis procedures, and modeling tools. Abaqus also offers pre- and post-processing. Many third-party vendors offer pre- and post-processors with interfaces to Abaqus. Moreover, maintaining and porting subroutines is much easier than maintaining and porting a complete finite-element program. An intensive work has been done by scientists around the world during many years in order to develop user element subroutine specifically matching with very special needs in many research fields. For example, in deformation processing [2], in reinforced concrete structures [3], and for crack propagation [4].

6.5.1 Guideline to Write a UEL

Before a UEL subroutine can be written, the following key characteristics of the element must be defined:

- The number of nodes on the element.
- The number of coordinates present at each node.
- The degrees of freedom active at each node.

In addition, the following properties must be determined:

- The number of element properties to be defined external to the UEL:
- The number of solution-dependent state variables (SDVs) to be stored per element.
- The number of (distributed) load types available for the element.

These items do not need to be determined immediately but they can be added easily after the basic UEL subroutine is completed.

The elements main contribution to the model during general analysis steps is to provide fluxes F^N at the nodes that depend on the values of the degrees of freedom u^N at the nodes.

F^N is defined as a residual quantity as shown in Eq. 6.1.

$$F^N = F^N_{ext} - F^N_{int} \tag{6.1}$$

F^N_{ext} is the external flux (due to applied distributed loads) and F^N_{int} is the internal flux (due to stresses, e.g.,) at node N.

If the degrees of freedom are displacements, the associated fluxes are the nodal forces. Similarly, rotations correspond to moments and temperatures to heat fluxes.

In nonlinear user elements, the fluxes per unit forces will often depend on the increments in the degrees of freedom Δu^N and the internal state variables H^α. The state variables must be updated in the user subroutine.

The solution of the (nonlinear) system of equations in general steps requires that the user to define the element Jacobian (stiffness matrix) is given by the Eq. 6.2.

$$K^{NM} = -\frac{dF^N}{du^M} \tag{6.2}$$

The Jacobian should include all direct and indirect dependencies of F^N on u^M, which includes terms of the form as shown in Eq. 6.3.

$$-\frac{\partial F^N}{\partial H^\alpha}\frac{\partial H^\alpha}{\partial u^M} \tag{6.3}$$

A more accurately defined Jacobian improves convergence in general steps.

The Jacobian (stiffness) determines the solution for linear perturbation steps, so it must be exact. The Jacobian can be symmetric or nonsymmetric.

The complexity of the formulation of a user element can vary greatly. Some simple elements can be developed to function as a control and a feedback mechanisms in an analysis that consists of regular elements. Some complex nonlinear structural elements often require significant effort in their development.

If the element is built out of a nonlinear material, the user should create a separate subroutine (or series of subroutines) to describe the material behavior. If the material model is implemented in user subroutine UMAT, a call to UMAT can be included in UEL. The integration issues discussed for UMAT also apply to the material models used in UEL.

6.5.1.1 Formulation and Calculation of Isoparametric Element

This section gives a short overview of the shape function and the formulation to calculate the Jacobian used in the finite-element method computation, but for further investigation it will be strongly recommended to read more about the finite-element procedures. For example, the finite-element procedures [5], the finite-element analysis of shells fundamentals [6], the inelastic analysis of solids and structures [7], the mechanics of solids and structures hierarchical modeling and the finite-element solution [8], and the modeling structures with the finite-element method Vol1: elastic solids [9], Vol2: beams and plates [10], Vol3: shells [11].

The basic concept in continuum elements is based on two aspects. First, the interpolate geometry and second the interpolation of displacements. For N, the number of nodes the interpolated geometry is given in Eq. 6.4 regarding the Cartesian coordinate system x, y, and z respectively.

$$x = \sum_{i=1}^{N} h_i x_i \qquad y = \sum_{i=1}^{N} h_i y_i \qquad z = \sum_{i=1}^{N} h_i z_i \tag{6.4}$$

Similarly, the interpolation of displacement is given in Eq. 6.5.

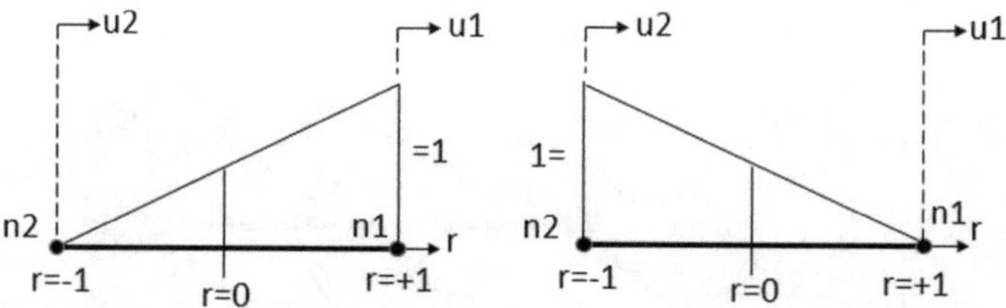

Fig. 6.28 The interpolation function for a 1D element with 2 nodes

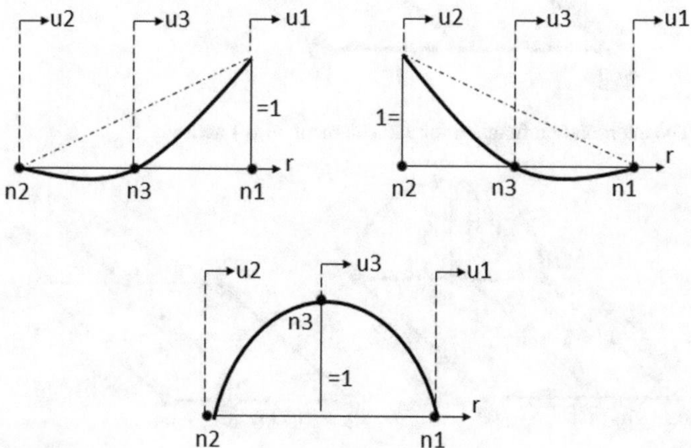

Fig. 6.29 The interpolation function for a 1D element with 3 nodes

$$u = \sum_{i=1}^{N} h_i u_i \qquad v = \sum_{i=1}^{N} h_i v_i \qquad w = \sum_{i=1}^{N} h_i w_i \tag{6.5}$$

After comparing Eqs. 6.4 and 6.5, the straightforward conclusion is that the geometry and the displacements are interpolated exactly in the same way with the interpolation function h_i. Therefore, this function is essential to build a finite element and need to be determined.

The interpolation function for a 1D element with two nodes is shown in Fig. 6.28.

Equation 6.6 gives the shape function as a function of the interpolation for each nodes according to the geometry and the displacements as shown in Fig. 6.28.

$$\begin{cases} h_1(r) = \frac{1}{2}(1 + r) \\ h_2(r) = \frac{1}{2}(1 - r) \end{cases} \tag{6.6}$$

The interpolation function for a 1D element with three nodes is shown in Fig. 6.29.

Equation 6.7 gives the shape function as a function of the interpolation for each node according to the geometry and the displacements as shown in Fig. 6.29.

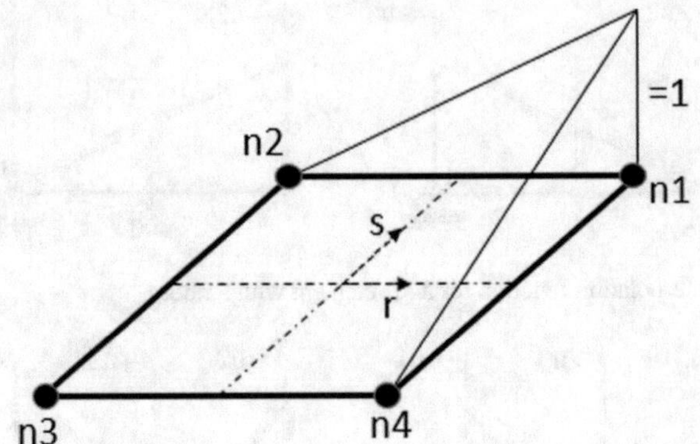

Fig. 6.30 The interpolation function for a 2D element with 4 nodes

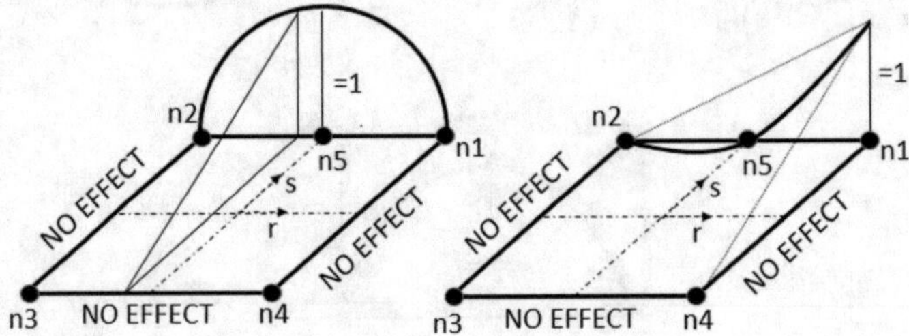

Fig. 6.31 The interpolation function for a 2D element with 5 nodes

$$\begin{cases} h_1(r) = \frac{1}{2}(1+r) - \frac{1}{2}\left(1-r^2\right) \\ h_2(r) = \frac{1}{2}(1-r) - \frac{1}{2}\left(1-r^2\right) \\ h_3(r) = \qquad 1-r^2 \end{cases} \tag{6.7}$$

The interpolation function for a 2D element with four nodes is shown in Fig. 6.30.

Equation 6.8 gives the shape function as a function of the interpolation for each node according to the geometry and the displacements as shown in Fig. 6.30.

$$\begin{cases} h_1(r,s) = \frac{1}{4}(1+r)(1+s) \\ h_2(r,s) = \frac{1}{4}(1-r)(1+s) \\ h_3(r,s) = \frac{1}{4}(1-r)(1-s) \\ h_4(r,s) = \frac{1}{4}(1+r)(1-s) \end{cases} \tag{6.8}$$

The interpolation function for a 2D element with five nodes is shown in Fig. 6.31.

Equation 6.9 gives the shape function as a function of the interpolation for each nodes according to the geometry and the displacements as shown in Fig. 6.31.

$$
\begin{cases}
h_1(r,s) = \frac{1}{4}(1+r)(1+s) - \frac{1}{2}h_5(r,s) \\
h_2(r,s) = \frac{1}{4}(1-r)(1+s) - \frac{1}{2}h_5(r,s) \\
h_3(r,s) = \quad \frac{1}{4}(1-r)(1-s) \\
h_4(r,s) = \quad \frac{1}{4}(1+r)(1-s) \\
h_5(r,s) = \quad \frac{1}{2}\left(1-r^2\right)(1+s)
\end{cases}
\tag{6.9}
$$

For a 4-node element in 2D as shown in Fig. 6.30, the interpolation of displacement is given in Eq. 6.10. A matrix-form Eq. 6.11 can be written from Eq. 6.10.

$$
u = \sum_{i=1}^{4} h_i(r,s)u_i \qquad v = \sum_{i=1}^{4} h_i(r,s)v_i \tag{6.10}
$$

$$
\begin{bmatrix} u(r,s) \\ v(r,s) \end{bmatrix} = \begin{bmatrix} h_1 & 0 & h_2 & 0 & h_3 & 0 & h_4 & 0 \\ 0 & h_1 & 0 & h_2 & 0 & h_3 & 0 & h_4 \end{bmatrix} \begin{bmatrix} u_1 \\ v_1 \\ u_2 \\ v_2 \\ u_3 \\ v_3 \\ u_4 \\ v_4 \end{bmatrix} = \underline{H}.\underline{u} \tag{6.11}
$$

The plane strain is given in Eq. 6.12. Similarly, a matrix-form 6.13 can be written from Eq. 6.12.

$$
\varepsilon_{rr} = \frac{\partial u}{\partial r} = \sum_{i=1}^{4} \frac{\partial h_i(r,s)}{\partial r} u_i \qquad \varepsilon_{ss} = \frac{\partial v}{\partial s} \qquad \varepsilon_{rs} = \frac{\partial u}{\partial s} + \frac{\partial v}{\partial r} \tag{6.12}
$$

$$
\begin{bmatrix} \varepsilon_{rr} \\ \varepsilon_{ss} \\ \varepsilon_{rs} \end{bmatrix} = \begin{bmatrix} \frac{\partial h_1}{\partial r} & 0 & \dots & \frac{\partial h_4}{\partial r} & 0 \\ 0 & \frac{\partial h_1}{\partial s} & \dots & 0 & \frac{\partial h_4}{\partial s} \\ \frac{\partial h_1}{\partial s} & \frac{\partial h_1}{\partial r} & \dots & \frac{\partial h_4}{\partial s} & \frac{\partial h_4}{\partial r} \end{bmatrix} \begin{bmatrix} u_1 \\ v_1 \\ u_2 \\ v_2 \\ u_3 \\ v_3 \\ u_4 \\ v_4 \end{bmatrix} = \underline{B}.\underline{u} \tag{6.13}
$$

In general, to know the displacement at an element point, it is very easy to determine thanks to the interpolated displacement function. However, regarding the strain, they are in respect of derivatives x and y, and **NOT** as a function of r and s anymore. Therefore, Eq. 6.10 has to be written with the Jacobian transformation system shown in Eq. 6.14. In this case, to transform each local coordinate system derivative for displacement $u(r,s)$ and $v(r,s)$ to a global coordinate system derivative $u(x,y)$ and $v(x,y)$ in the geometry model.

$$\begin{bmatrix} \frac{\partial}{\partial r} \\ \frac{\partial}{\partial s} \end{bmatrix} = \begin{bmatrix} \frac{\partial x}{\partial r} & \frac{\partial y}{\partial r} \\ \frac{\partial x}{\partial s} & \frac{\partial y}{\partial s} \end{bmatrix} \begin{bmatrix} \frac{\partial}{\partial x} \\ \frac{\partial}{\partial y} \end{bmatrix} = \underline{J} \begin{bmatrix} \frac{\partial}{\partial x} \\ \frac{\partial}{\partial y} \end{bmatrix} \tag{6.14}$$

The Jacobian components matrix are very easy to calculate. For example, the component $\frac{\partial x}{\partial r}$ is easy to calculate, because like the displacement and the geometry are interpolated with the same shape functions then Eq. 6.15 can be used as shown in Eq. 6.16 to calculate this Jacobian matrix component. Following this procedure, it is possible to rewrite the matrix $\underline{B}$ of general element coordinates in the geometry model.

$$x = \sum_{i=1}^{N} h_i(r, s) x_i \tag{6.15}$$

$$\frac{\partial x}{\partial r} = \sum_{i=1}^{N} \frac{\partial h_i(r, s)}{\partial r} x_i \tag{6.16}$$

In a 3D element, the $\underline{H}$ and $\underline{B}$ matrices are function of (r, s, t) locally and (x, y, z) in the geometry model. For example, the integration to calculate the stiffness matrix is given in Eq. 6.17 with the elementary element volume defined in Eq. 6.18.

$$\underline{K} = \int_V \underline{B}^T . \underline{C} . \underline{B} dV \tag{6.17}$$

$$dV = det(\underline{J}) drdsdt \tag{6.18}$$

6.5.1.2 UEL Interface

A user element is defined with the **USER ELEMENT** option. This option must appear in the input file before the user element is invoked with the ELEMENT option. The syntax for interfacing to UEL is as follows:

```
*USER ELEMENT, TYPE=Un, NODES=, COORDINATES=,
PROPERTIES=, I PROPERTIES=, VARIABLES=, UNSYMM
Data line(s)
*ELEMENT,TYPE=Un, ELSET=UEL
Data line(s)
*UEL PROPERTY, ELSET=UEL
Data line(s)
*USER SUBROUTINES, (INPUT=file_name)
```

A data line of the form dof_1, dof_2, and so on. Where dof_1 is the first degree of freedom active at the node and dof_2 is the second degree of freedom active at the node, and so on. It follows the **USER ELEMENT** option. If all nodes of the user element have the same active degrees of freedom, no further data are needed.

However, if some nodes have different active degrees of freedom, enter subsequent data lines of the form position, dof_1, dof_2, and so on. Where the position is the (local) node number (position) on the element, dof_1 is the first degree of freedom active at this and following nodes, and dof_2 is the second degree of freedom active at this and following nodes, and so on. The active degrees of freedom can be changed at any node in the element.

The dimensional units of a degree of freedom for a user element are the same as those for regular elements in Abaqus, as documented the indexes from 1 to 3 are displacements and from 4 to 6 are rotations. This correspondence is important for convergence controls in nonlinear analysis. It is also relevant for three-dimensional rotations in geometric nonlinear analysis, because of the nonlinear nature of finite rotations.

User elements can have some internal degrees of freedom in the sense that they belong to nodes that are not connected to other elements. The convergence will be checked for the internal degrees of freedom, so it is important to choose the internal degrees of freedom appropriately (i.e., an internal degree of freedom 1 should have the dimension of displacement). For efficiency reasons, the user should choose internal degree of freedom numbers that are present at external nodes on the element or elsewhere in the model.

Table 6.2 gives the definition of parameters used in UEL subroutine.

The maximum number of coordinates at any node of the element is specified with the COORDINATES parameter. The value of COORDINATES may be increased to match the highest displacement degree of freedom active on the element.

The total number of SDVs per element is set with the VARIABLES parameter. If the element is integrated numerically, VARIABLES should be set equal to the number of integration points times the number of SDVs per point. The solution-dependent state variables can be output with the identifiers SDV1, SDV2, and so on. The SDVs for any element can be printed only to the data (.dat), results (.fil), or output database (.odb) files and plotted as X-Y plots in Abaqus viewer.

The number of user element properties is given with the PROPERTIES and I PROPERTIES parameters shown in Table 6.2. PROPERTIES determines the number of floating-point property values. The I PROPERTIES determines the number of

Table 6.2 Parameter definition for UEL subroutine

Parameters	Definitions
TYPE	(User-defined) element type of the form Un, where n is a number
NODES	Number of nodes on the element
COORDINATES	Maximum number of coordinates at any node
PROPERTIES	Number of floating point properties
I PROPERTIES	Number of integer properties
VARIABLES	Number of SDVs
UNSYMM	Flag to indicate that the Jacobian is nonsymmetric

integer property values. The property values are given with the **UEL PROPERTY** option. The properties are assigned on an element set basis; hence, the same UEL subroutine can be used for user elements with different properties. Following this approach, hard-coding the property values in the user subroutine is not necessary.

Coding for the UEL is supplied in a separate file and invoked with the Abaqus execution procedure as follows:

```
abaqus job=... user=....
```

The user subroutine must be invoked in a restarted analysis because user subroutines are not saved on the restart file.

The distributed load and flux types can be applied with the **DLOAD** and **DFLUX** options by using load type keys **Un** and **UnNU**. In either case, the equivalent nodal load vector for the distributed load type must be defined in user subroutine UEL. If the load type key **Un** is used, the load magnitude is defined on the data line and can be varied in time with the **AMPLITUDE** option. If the load type key **UnNU** is used, all of the load definition is applied in user subroutine UEL: a time-dependent load magnitude vector must be coded. Finally, if the load depends on the solution variables, the corresponding load stiffness contributions matrix to the Jacobian should be included for best performance.

The following quantities are available in UEL:

- Coordinates; displacements; incremental displacements; and, for dynamics, velocities and accelerations.
- SDVs at the start of the increment.
- Total and incremental values of time, temperature, and user-defined field variables.
- User element properties.
- Load types as well as total and incremental load magnitudes.
- Element type and user-defined element number.
- Procedure-type flag and, for dynamics, integration operator values.
- Current step and increment numbers.

The following quantities must be defined, the right-hand-side vector (residual nodal fluxes or forces), the Jacobian (stiffness) matrix, and the solution-dependent state variables.

The following variables may be defined, the energies associated with the element (strain energy, plastic dissipation, kinetic energy, etc.) and the suggested new (reduced) time increment.

The solution variables (displacement, velocity, etc.) are arranged on a node/degree of freedom basis. The degrees of freedom of the first node are first, followed by the degrees of freedom of the second node, and so on. The flux vector and Jacobian matrix must be ordered in the same way. For instance, consider a planar beam that uses degrees of freedom 1, 2, and 6 (u_x, u_y, ϕ_z) at its first and second node and degrees of freedom 1 and 2 at its third (middle) node. The ordering is

Element variable	1	2	3	4	5	6	7	8
Node	1	1	1	2	2	2	3	3
Degree of freedom	1	2	6	1	2	6	1	2

The displacement, velocities, etc., passed into the UEL are in the global system, regardless of whether the TRANSFORM option is used at any of the nodes. The flux vector and Jacobian matrix must also be formulated in the global system.

The Jacobian must be formulated as a full matrix, even if it is symmetric. If the UNSYMM parameter is not used, Abaqus will symmetrize the Jacobian defined by the user.

For transient heat transfer and dynamic analysis, heat capacity and inertia contributions must be included in the flux vector.

At the start of a new increment, the increment in solution variable(s) is extrapolated from the previous increment. The flux vector and the Jacobian must be based on these extrapolated values. If the extrapolation is not desired, it can be switched off with **STEP, EXTRAPOLATION=NO**.

If the increment in solution variable(s) is too large, the variable **PNEWDT** can be used to suggest a new time increment. Abaqus will abandon the current time increment and will attempt the increment again with one that is a factor **PNEWDT** smaller.

Complex UELs may have many potential problem areas. Do not use a large model when trying to debug a UEL. Please, verify the UEL with a one-element input file.

1. Run tests using general steps in which all solution variables are prescribed to verify the resultant fluxes.
2. Run tests using linear perturbation steps in which all loads are prescribed to verify the element Jacobian (stiffness).
3. Run tests using general steps in which all loads are prescribed to verify the consistency of the Jacobian and the flux vector.

Gradually increase the complexity of the test problems, then compare the results with standard Abaqus elements, if possible.

6.6 UEL Subroutine Examples

Again as a good practice to debug the model, here the Fortran subroutine, is to insert at different location lines inside the user's code the instructions **PRINT** and **PAUSE** in that order. First to print the variable the user needs to track and then pause to monitor the sequence of the code executed by Abaqus. Second, this method allows the user to perform a simple step-by-step debug processing in the most efficient and effective way. Finally, the detection of unexpected behavior from the subroutine code through the FEA model will be easiest to identify in a complex program.

6.6.1 UEL Subroutine for Planar Beam with Nonlinear Cross Section

The frame is loaded to an extent where significant nonlinearity occurs in the concrete but the displacements are still small enough that geometric nonlinearity may be neglected.

In this example, the user develops a model that describes the nonlinear section behavior directly in terms of axial force and bending moment. Which is similar to the **BEAM GENERAL SECTION, SECTION=NONLINEAR GENERAL** option, but allows coupling between the axial and bending terms. The transverse shear deformation can be neglected.

The element is integrated numerically; hence, the following quantities require definition in the UEL:

- The element $\underline{B}$ matrix, which relates the axial strain, ε, and curvature, κ, to the element displacements, $\underline{u_e}$:

$$\begin{bmatrix} \varepsilon \\ \kappa \end{bmatrix} = \underline{B}.\underline{u_e} \tag{6.19}$$

- A constitutive law $\underline{D}$ relating axial force, F, and moment, M, to axial strain and curvature:

$$\begin{bmatrix} F \\ M \end{bmatrix} = \underline{D}.\begin{bmatrix} \varepsilon \\ \kappa \end{bmatrix} \tag{6.20}$$

- The element stiffness matrix:

$$\underline{K_e} = \int_0^L \underline{B}^T.\underline{D}.\underline{B}dl \tag{6.21}$$

- The element internal force vector:

$$\underline{F_e} = \int_0^L \underline{B}^T.\begin{bmatrix} F \\ M \end{bmatrix} dl \tag{6.22}$$

- The integration is done numerically:

$$\int_0^L A dl = \sum_{i=1}^{n} A_i.l_i \tag{6.23}$$

  where n is the number of integration points and l_i is the length associated with integration point i.

The element formulation is based on Euler–Bernoulli beam theory. The interpolation is described purely in terms of the displacements, which are continuous at the nodes.

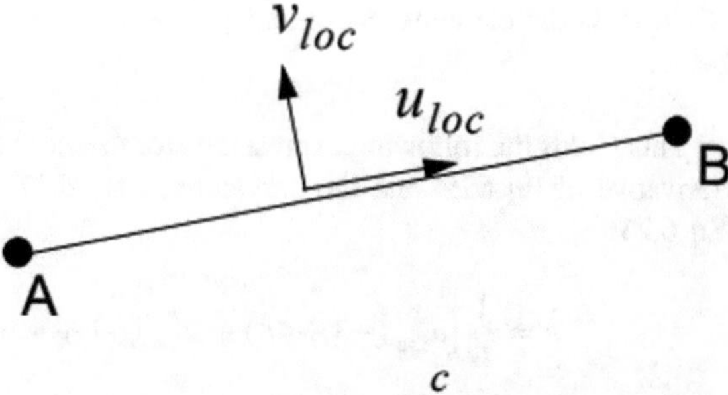

Fig. 6.32 Planar beam with nonlinear cross section model in UEL subroutine

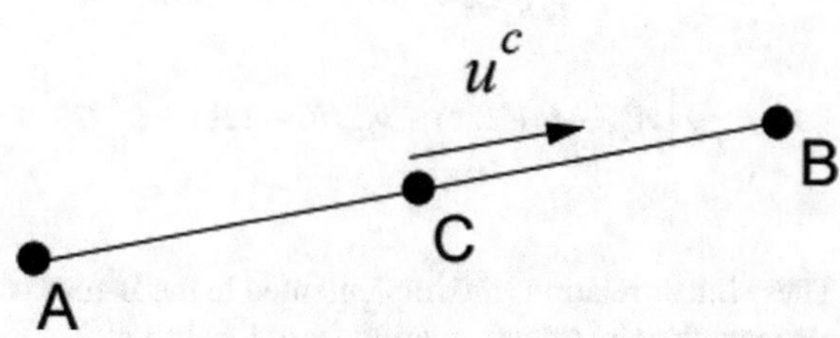

Fig. 6.33 Internal node added into the UEL planar beam with nonlinear cross section subroutine

The curvature is obtained as the second derivative of the displacement normal to the beam.

The simplest two-dimensional beam element has two nodes, with two displacements and one rotation (u_x, u_y, ϕ_z) at each node. The active degrees of freedom are indexed with 1, 2, and 6, respectively.

As shown in Fig. 6.32, in its basic form linear interpolation is used for the tangential displacement, u_{loc}, and cubic interpolation for the normal displacement, v_{loc}. The cubic interpolation for the normal displacement yields a linear variation for the curvature. The linear interpolation for the tangential displacement yields a constant axial strain.

The constant axial strain and linear curvature variation are inconsistent and may lead to excessive local axial forces if the axial and bending behaviors are coupled. Considering that the intent is to analyze nonlinear concrete behavior such coupling will be present. The excessive axial forces may lead to overly stiff behavior. In order to prevent this problem, an extra internal node is added to the element. The internal node has one degree of freedom: the tangential displacement as shown in Fig. 6.33.

Both the axial strain and the curvature now vary linearly. The interpolation functions are

$$u_{loc} = u^A_{loc}\left(1 - 3\xi + 2\xi^2\right) + u^B_{loc}\left(-\xi + 2\xi^2\right) + u^C 4\left(\xi - \xi^2\right) \tag{6.24}$$

$$v_{loc} = v^A_{loc}\left(1 - 3\xi^2 + 2\xi^3\right) + v^B_{loc}\left(3\xi^2 - 2\xi^3\right) + \phi^A L\left(\xi - 2\xi^2 + \xi^3\right) + \phi^B L\left(-\xi^2 + \xi^3\right) \tag{6.25}$$

where L is the element length and $\xi = \frac{s}{L}$ is the dimensionless position along the beam.

This yields the following expressions for the axial strain in Eq. 6.26 from the first derivative of Eq. 6.24 and the curvature in Eq. 6.27 from the second derivative of Eq. 6.25:

$$\varepsilon = \frac{1}{L}\left[u^A_{loc}\left(-3+4\xi\right)+u^B_{loc}\left(-1+4\xi\right)+u^C 4\left(1-2\xi\right)\right] \tag{6.26}$$

$$\kappa = \frac{1}{L^2}\left[v^A_{loc}\left(-6+12\xi\right)+v^B_{loc}\left(6-12\xi\right)+\phi^A L\left(-4+6\xi\right)+\phi^B L\left(-2+6\xi\right)\right] \tag{6.27}$$

These linear relations are implemented in the B-matrix of the element. The B-matrix also handles the transformation from local to global displacements at the nodes.

The element is integrated numerically with a two-point Gauss scheme.
The following data lines define the user element in the input file:

```
*user element,type=u1,nodes=3,coordinates=2,...
... properties=3,variables=8
1, 2, 6
3, 1
*element, type=u1, elset=one
1, 1, 2, 3
*uel property, elset=one
2., 1., 1000.
```

The user element name is U1, which is used in the **ELEMENT** option. There are eight state variables allocated, so four variables can be defined at each integration point. Three element properties are allocated: the section height, the section width, and Young's modulus. The element has three nodes: the third internal node is unique to each element.

This model can be coded as follows:

Listing 6.1 UEL1.f

```
c
c simple 2-d linear beam element with generalized
c section properties
c
      subroutine uel(rhs, amatrx, svars, energy, ndofel,
     1 nrhs, nsvars, props, nprops, coords, mcrd, nnode,
     2 u, du, v, a, jtype, time, dtime, kstep, kinc,
     3 jelem, params, ndload, jdltyp, adlmag, predef,
     4 npredf, lflags, mlvarx, ddlmag, mdload, pnewdt,
```

```
     5 jprops, njprop, period)
c
     include 'aba_param.inc'
c
     dimension rhs(mlvarx, *), amatrx(ndofel, ndofel),
     1 svars(*), props(*), energy(7), coords(mcrd,nnode),
     2 u(ndofel), du(mlvarx, *), v(ndofel), a(ndofel),
     3 time(2), params(*), jdltyp(mdload, *),
     4 adlmag(mdload, *), ddlmag(mdload, *),
     5 predef(2, npredf, nnode), lflags(4), jprops(*)
c
     dimension b(2, 7), gauss(2)
c
     parameter(zero=0.d0, one=1.d0, two=2.d0,
     1 three=3.d0, four=4.d0, six=6.d0, eight=8.d0,
     2 twelve=12.d0)
c
     data gauss/.211324865d0, .788675135d0/
c
c calculate length_and direction cosines
c
     dx=coords(1, 2)-coords(1, 1)
     dy=coords(2, 2)-coords(2, 1)
     dl2=dx**2+dy**2
     dl=sqrt(dl2)
     hdl=dl/two
     acos=dx/dl
     asin=dy/dl
c
c initialize rhs_and lhs
c
     do k1=1, 7
        rhs(k1, 1)= zero
        do k2=1, 7
           amatrx(k1, k2)= zero
        end do
     end do
c
     nsvint=nsvars/2
c
c loop over integration points
c
     do kintk=1, 2
        g=gauss(kintk)
c
```

```
c make b–matrix
c
      b(1,1)=(−three+four*g)*acos/dl
      b(1,2)=(−three+four*g)*asin/dl
      b(1,3)=zero
      b(1,4)=(−one+four*g)*acos/dl
      b(1,5)=(−one+four*g)*asin/dl
      b(1,6)=zero
      b(1,7)=(four−eight*g)/dl
      b(2,1)=(−six+twelve*g)*−asin/dl2
      b(2,2)=(−six+twelve*g)* acos/dl2
      b(2,3)=(−four+six*g)/dl
      b(2,4)= (six−twelve*g)*−asin/dl2
      b(2,5)= (six−twelve*g)* acos/dl2
      b(2,6)= (−two+six*g)/dl
      b(2,7)=zero
c
c calculate (incremental) strains_and curvatures
c
      eps=zero
      deps=zero
      cap=zero
      dcap=zero
      do k=1, 7
         eps = eps+b(1,k)*u(k)
         deps=deps+b(1,k)*du(k,1)
         cap = cap+b(2,k)*u(k)
         dcap=dcap+b(2,k)*du(k,1)
      end do
c
c_call constitutive routine ugenb
c
      isvint=1+(kintk−1)*nsvint
      bn=zero
      bm=zero
      daxial=zero
      dbend=zero
      dcoupl=zero
      call ugenb(bn, bm, daxial, dbend, dcoupl, eps,
      1 deps, cap, dcap, svars(isvint), nsvint,
      2 props, nprops)
c c assemble rhs_and lhs c
      do k1=1, 7
         rhs(k1,1)=rhs(k1,1)
      1            −hdl*(bn*b(1,k1)+bm*b(2,k1))
```

```
          bd1=hdl*(daxial*b(1,k1)+dcoupl*b(2,k1))
          bd2=hdl*(dcoupl*b(1,k1)+dbend *b(2,k1))
          do k2=1, 7
             amatrx(k1,k2)=amatrx(k1,k2)
          1                   +bd1*b(1,k2)+bd2*b(2,k2)
          end do
       end do
    end do
c
    return
    end
```

This UEL uses essentially the same formulation as the simple B23 element for geometrically linear analysis.

The routine can be used with and without the **TRANSFORM** option.

It would be relatively straightforward to generalize this routine for three-dimensional analyses. It is much more complicated to extend the routine to geometrically nonlinear analysis.

Even for linear analysis, this routine is called twice (for each element) during the first iteration of an increment: once for assembly and once for recovery. Subsequently, it is called once per iteration: assembly and recovery are combined.

6.6.2 Generalized Constitutive Behavior

At each integration point, the generalized constitutive behavior is processed in the user-created subroutine UGENB.

This subroutine is patterned after user subroutine UGENS, which allows the user to model the behavior of a shell.

The following quantities are passed into UGENB, a total and incremental axial strain and curvature, the state variables at the start of the increment, and the user element properties.

The user must define two things, first the axial force and bending moment, as well as the linearized force/moment-strain/curvature relations, and second the solution-dependent state variables.

A simple linear elastic subroutine UGENB follows:

Listing 6.2 UGENB.f

```
    subroutine ugenb(bn,bm,daxial,dbend,dcoupl,eps,
    1 deps,cap,dcap,svint,nsvint,props,nprops)
c
    include 'aba_param.inc'
c
    parameter(zero=0.d0,twelve=12.d0)
c
```

```
      dimension svint(*),props(*)
c
c variables_to be defined by the user
c
c bn – axial force
c bm – bending moment
c daxial – current tangent axial stiffness
c dbend – current tangent bending stiffness
c dcoupl – tangent coupling term
c
c variables that may be updated
c
c svint – state variables for this integration point
c
c variables passed_in for information
c
c eps – axial strain
c deps – incremental axial strain
c cap – curvature change
c dcap – incremental curvature change
c props – element properties
c nprops – # element properties
c nsvint – # state variables
c
c current assumption
c
c props(1) – section height
c props(2) – section width
c props(3) – Young s modulus
c
      h=props(1)
      w=props(2)
      E=props(3)
c
c formulate linear stiffness
c
      daxial=E*h*w
      dbend=E*w*h**3/twelve
      dcoupl=zero
c
c calculate axial force and moment
c
      bn=svint(1)+daxial*deps
      bm=svint(2)+dbend*dcap
c
```

```
c store internal variables
c
   svint(1)=bn
   svint(2)=bm
   svint(3)=eps
   svint(4)=cap
c
 return
 end
```

The coding in this routine is very similar in nature to what would be coded in the subroutines UMAT and UGENS.

The routine stores the axial strain, curvature, axial force, and bending moment at each Gauss point. For nonlinear material behavior, additional quantities would be stored.

The constitutive relation is written in incremental form for easy generalization to nonlinear section behavior.

UGENB and UEL must be combined in one external file.

6.6.3 UEL Subroutine for a Horizontal Truss and Heat Transfer Element

Both a structural and a heat transfer user element have been created to demonstrate the usage of subroutine UEL. These user-defined elements are applied in a number of analyses. The following excerpt is from the verification problem that invokes the structural user element in an implicit dynamics procedure:

```
*USER ELEMENT, NODES=2, TYPE=U1, PROPERTIES=4, ...
...COORDINATES=3, VARIABLES=12
1, 2, 3
*ELEMENT, TYPE=U1
101, 101, 102
*ELGEN, ELSET=UTRUSS
101, 5
*UEL PROPERTY, ELSET=UTRUSS
0.002, 2.1E11, 0.3, 7200.
```

The user element consists of two nodes that are assumed to lie parallel to the x-axis. The element behaves like a linear truss element. The supplied element properties are the cross-sectional area, Young's modulus, Poisson's ratio, and density, respectively.

The next excerpt shows the listing of the subroutine. The user subroutine has been coded for use in a perturbation static analysis; general static analysis, including

Riks analysis with load incrementation defined by the subroutine; eigenfrequency extraction analysis; and direct-integration dynamic analysis.[6] The flags passed in through the LFLAGS array are used to associate particular calculations with solution procedures.

During a modified Riks analysis, all force loads must be passed into UEL by means of distributed load definitions such that they are available for the definition of incremental load vectors; the load keys Un and UnNU must be used properly, as discussed in Sect. 6.5.1.2. The coding in subroutine UEL must distribute the loads into consistent equivalent nodal forces and account for them in the calculation of the RHS and ENERGY arrays.

The UEL code of this element follows:

Listing 6.3 UEL2.f

```
 SUBROUTINE UEL(RHS,AMATRX,SVARS,ENERGY,NDOFEL,NRHS,
    1 PROPS,NPROPS,COORDS,MCRD,NNODE,U,DU,V,A,JTYPE,
    2 DTIME,KSTEP,KINC,JELEM,PARAMS,NDLOAD,JDLTYP,
    3 PREDEF,NPREDF,LFLAGS,MLVARX,DDLMAG,MDLOAD,
    4 JPROPS,NJPROP,PERIOD,PNEWDT,ADLMAG,TIME,NSVARS)
C
      INCLUDE 'ABA_PARAM.INC'
      PARAMETER (ZERO=0.D0,HALF=0.5D0,ONE=1.D0)
C
    DIMENSION RHS(MLVARX,*),AMATRX(NDOFEL,NDOFEL),
    1 SVARS(NSVARS),ENERGY(8),PROPS(*),TIME(2),
    2 U(NDOFEL),DU(MLVARX,*),V(NDOFEL),A(NDOFEL),
    3 PARAMS(3),JDLTYP(MDLOAD,*),ADLMAG(MDLOAD,*),
    4 DDLMAG(MDLOAD,*),PREDEF(2,NPREDF,NNODE),
    5 JPROPS(*),COORDS(MCRD,NNODE),LFLAGS(*)
    DIMENSION SRESID(6)
C
C UEL_SUBROUTINE FOR A HORIZONTAL TRUSS ELEMENT
C
C SRESID - stores the static residual at time t+dt
C SVARS
c     - In_1-6, contains_the static residual at time t
C       upon entering the routine. SRESID is copied_to
C       SVARS(1-6) after the dynamic residual has been
C       calculated.
C     - For half-increment residual calculations:
C       In_7-12, contains_the static residual at the
C       beginning of the previous increment. SVARS(1-6)
C       are copied into SVARS(7-12) after the dynamic
C       residual has been calculated.
C
```

[6]See UEL, in Sect. 4.1.14 of the Abaqus Verification Guide v6.14.

```
      AREA = PROPS(1)
      E    = PROPS(2)
      ANU  = PROPS(3)
      RHO  = PROPS(4)
C
      ALEN = ABS(COORDS(1,2)-COORDS(1,1))
      AK   = AREA*E/ALEN
      AM   = HALF*AREA*RHO*ALEN
C
      DO K1 = 1, NDOFEL
        SRESID(K1) = ZERO
        DO KRHS = 1, NRHS
          RHS(K1,KRHS) = ZERO
        END DO
        DO K2 = 1, NDOFEL
          AMATRX(K2,K1) = ZERO
        END DO
      END DO
C
      IF (LFLAGS(3).EQ.1) THEN
C———Normal incrementation
        IF (LFLAGS(1).EQ.1 .OR. LFLAGS(1).EQ.2) THEN
C         *STATIC
          AMATRX(1,1) =  AK
          AMATRX(4,4) =  AK
          AMATRX(1,4) = -AK
          AMATRX(4,1) = -AK
          IF (LFLAGS(4).NE.0) THEN
            FORCE = AK*(U(4)-U(1))
            DFORCE = AK*(DU(4,1)-DU(1,1))
            SRESID(1) = -DFORCE
            SRESID(4) =  DFORCE
            RHS(1,1) = RHS(1,1)-SRESID(1)
            RHS(4,1) = RHS(4,1)-SRESID(4)
            ENERGY(2) = HALF*FORCE*(DU(4,1)-DU(1,1))
     *            + HALF*DFORCE*(U(4)-U(1))
     *            + HALF*DFORCE*(DU(4,1)-DU(1,1))
          ELSE
            FORCE = AK*(U(4)-U(1))
            SRESID(1) = -FORCE
            SRESID(4) =  FORCE
            RHS(1,1) = RHS(1,1)-SRESID(1)
            RHS(4,1) = RHS(4,1)-SRESID(4)
            DO KDLOAD = 1, NDLOAD
              IF (JDLTYP(KDLOAD,1).EQ.1001) THEN
```

```
            RHS(4,1)    = RHS(4,1)+ADLMAG(KDLOAD,1)
            ENERGY(8)  = ENERGY(8)+(ADLMAG(KDLOAD,1)
     *            - HALF*DDLMAG(KDLOAD,1))*DU(4,1)
            IF (NRHS.EQ.2) THEN
C              Riks
              RHS(4,2) = RHS(4,2)+DDLMAG(KDLOAD,1)
            END IF
          END IF
        END DO
        ENERGY(2) = HALF*FORCE*(U(4)-U(1))
      END IF
    ELSE IF(LFLAGS(1).EQ.11.OR.LFLAGS(1).EQ.12)THEN
C-----------DYNAMIC
      ALPHA = PARAMS(1)
      BETA  = PARAMS(2)
      GAMMA = PARAMS(3)

C
      DADU = ONE/(BETA*DTIME**2)
      DVDU = GAMMA/(BETA*DTIME)
C
      DO K1 = 1, NDOFEL
        AMATRX(K1,K1) = AM*DADU
        RHS(K1,1) = RHS(K1,1)-AM*A(K1)
      END DO
      AMATRX(1,1) = AMATRX(1,1)+(ONE+ALPHA)*AK
      AMATRX(4,4) = AMATRX(4,4)+(ONE+ALPHA)*AK
      AMATRX(1,4) = AMATRX(1,4)-(ONE+ALPHA)*AK
      AMATRX(4,1) = AMATRX(4,1)-(ONE+ALPHA)*AK
      FORCE = AK*(U(4)-U(1))
      SRESID(1) = -FORCE
      SRESID(4) =  FORCE
      RHS(1,1) = RHS(1,1) -
     *        ((ONE+ALPHA)*SRESID(1)-ALPHA*SVARS(1))
      RHS(4,1) = RHS(4,1) -
     *        ((ONE+ALPHA)*SRESID(4)-ALPHA*SVARS(4))
      ENERGY(1) = ZERO
      DO K1 = 1, NDOFEL
        SVARS(K1+6) = SVARS(k1)
        SVARS(K1)   = SRESID(K1)
        ENERGY(1)   = ENERGY(1)+HALF*V(K1)*AM*V(K1)
      END DO
      ENERGY(2) = HALF*FORCE*(U(4)-U(1))
    END IF
  ELSE IF (LFLAGS(3).EQ.2) THEN
```

```
C———Stiffness matrix
      AMATRX(1,1) =  AK
      AMATRX(4,4) =  AK
      AMATRX(1,4) = -AK
      AMATRX(4,1) = -AK
    ELSE IF (LFLAGS(3).EQ.4) THEN
C———Mass matrix
      DO K1 = 1, NDOFEL
        AMATRX(K1,K1) = AM
      END DO
    ELSE IF (LFLAGS(3).EQ.5) THEN
C———Half-increment residual calculation
      ALPHA = PARAMS(1)
      FORCE = AK*(U(4)-U(1))
      SRESID(1) = -FORCE
      SRESID(4) =  FORCE
      RHS(1,1)=RHS(1,1)-AM*A(1)-(ONE+ALPHA)*SRESID(1)
    *          + HALF*ALPHA*( SVARS(1)+SVARS(7) )
      RHS(4,1)=RHS(4,1)-AM*A(4)-(ONE+ALPHA)*SRESID(4)
    *          + HALF*ALPHA*( SVARS(4)+SVARS(10) )
    ELSE IF (LFLAGS(3).EQ.6) THEN
C———Initial acceleration calculation
      DO K1 = 1, NDOFEL
        AMATRX(K1,K1) = AM
      END DO
      FORCE = AK*(U(4)-U(1))
      SRESID(1) = -FORCE
      SRESID(4) =  FORCE
      RHS(1,1) = RHS(1,1)-SRESID(1)
      RHS(4,1) = RHS(4,1)-SRESID(4)
      ENERGY(1) = ZERO
      DO K1 = 1, NDOFEL
        SVARS(K1) = SRESID(K1)
        ENERGY(1) = ENERGY(1)+HALF*V(K1)*AM*V(K1)
      END DO
      ENERGY(2) = HALF*FORCE*(U(4)-U(1))
    ELSE IF (LFLAGS(3).EQ.100) THEN
C———Output for perturbations
      IF (LFLAGS(1).EQ.1 .OR. LFLAGS(1).EQ.2) THEN
```

```
C————STATIC
      FORCE   = AK*(U(4)-U(1))
      DFORCE = AK*(DU(4,1)-DU(1,1))
      SRESID(1) = -DFORCE
      SRESID(4) =  DFORCE
      RHS(1,1) = RHS(1,1)-SRESID(1)
      RHS(4,1) = RHS(4,1)-SRESID(4)
      ENERGY(2) = HALF*FORCE*(DU(4,1)-DU(1,1))
*        + HALF*DFORCE*(U(4)-U(1))
*        + HALF*DFORCE*(DU(4,1)-DU(1,1))
      DO KVAR = 1, NSVARS
        SVARS(KVAR) = ZERO
      END DO
      SVARS(1) = RHS(1,1)
      SVARS(4) = RHS(4,1)
    ELSE IF (LFLAGS(1).EQ.41) THEN
C————FREQUENCY
      DO KRHS = 1, NRHS
        DFORCE = AK*(DU(4,KRHS)-DU(1,KRHS))
        SRESID(1) = -DFORCE
        SRESID(4) =  DFORCE
        RHS(1,KRHS) = RHS(1,KRHS)-SRESID(1)
        RHS(4,KRHS) = RHS(4,KRHS)-SRESID(4)
      END DO
      DO KVAR = 1, NSVARS
        SVARS(KVAR) = ZERO
      END DO
      SVARS(1) = RHS(1,1)
      SVARS(4) = RHS(4,1)
    END IF
  END IF
C
  RETURN
  END
```

6.6.4 UELMAT Subroutine for 4 Nodes in Plane Strain

The user subroutine UELMAT differs from the subroutine UEL because UELMAT allows the user to define an element with access to Abaqus materials. Therefore, the user will not have to program the material law of behavior but only the structural element. For example, a structural user element with Abaqus isotropic linearly elastic material.

Both a structural and a heat transfer user element have been created to demonstrate the usage of subroutine UELMAT. These user-defined elements are applied in a number of analyses. The following excerpt illustrates how the linearly elastic isotropic material available in Abaqus can be accessed from user subroutine UELMAT:

```
...
*USER ELEMENT,TYPE=U1,NODES=4,COORDINATES=2,VAR=16,...
... INTEGRATION=4, TENSOR=PSTRAIN
1,2
*ELEMENT, TYPE=U1, ELSET=SOLID
1, 1,2,3,4
...
*UEL PROPERTY, ELSET=SOLID, MATERIAL=MAT
...
*MATERIAL, NAME=MAT
*ELASTIC
7.00E+010,  0.33
...
```

The user element defined above is a 4-node, fully integrated plane strain element, similar to the Abaqus CPE4 element. The next excerpt shows the listing of the user subroutine. Inside the subroutine, a loop over the integration points is performed. For each integration point, the utility routine MATERIAL_LIB_MECH is called, which returns stress and Jacobian at the integration point. These quantities are used to compute the right-hand-side vector and the element Jacobian.

The UELMAT code of this element follows:

Listing 6.4 UELMAT.f

```
      subroutine uelmat(rhs,amatrx,svars,energy,ndofel,
     1 nsvars,props,nprops,coords,mcrd,nnode,u,du,
     2 v,a,jtype,time,dtime,kstep,kinc,jelem,params,
     3 ndload,jdltyp,adlmag,predef,npredf,lflags,
     4 ddlmag,mdload,pnewdt,jprops,njpro,period,
     5 materiallib,nrhs,mlvarx)
c
      include 'aba_param.inc'
C
      dimension rhs(mlvarx,*), amatrx(ndofel, ndofel),
     1 svars(*), energy(*), coords(mcrd, nnode),
     2 du(mlvarx,*), v(ndofel), a(ndofel), time(2),
     3 jdltyp(mdload,*), adlmag(mdload,*),
     4 predef(2, npredf, nnode), lflags(*), jprops(*),
     5 props(*), u(ndofel), params(*), ddlmag(mdload,*)
```

```
c
      parameter (zero=0.d0, dmone=-1.0d0, one=1.d0,
      1 four=4.0d0, fourth=0.25d0,
      2 gaussCoord=0.577350269d0)
c
      parameter (ndim=2, ndof=2, nshr=1,nnodemax=4,
      1 ntens=4, ninpt=4, nsvint=4)
c
c ndim   the number_ of spatial dimensions
c ndof   the number_of degrees of freedom per node
c nshr   the number_of shear stress component
c ntens  the total _number_of stress tensor components
c        (=ndi+nshr)
c ninpt  the number_of integration points
c nsvint the number_of state variables per integration
c        point (strain)
c
      dimension stiff(ndof*nnodemax,ndof*nnodemax),
      1 force(ndof*nnodemax), shape(nnodemax),
      2 xjac(ndim,ndim),xjaci(ndim,ndim),
      3 statevLocal(nsvint),stress(ntens),
      4 stran(ntens), dstran(ntens), wght(ninpt),
      5 dshape(ndim,nnodemax), bmat(nnodemax*ndim),
      6 ddsdde(ntens, ntens)
c
      dimension predef_loc(npredf),dpredef_loc(npredf),
      1 defGrad(3,3),utmp(3),xdu(3),stiff_p(3,3),
      2 force_p(3)
c
      dimension coord24(2,4),coords_ip(3)
c
      data coord24 /dmone, dmone,
      2              one, dmone,
      3              one,   one,
      4            dmone,   one/
c
      data wght /one, one, one, one/
c
c-------------------------------------
c U1 = first-order, plane strain, full integration
c
c State variables: each integration point has nsvint
c SDVs
c
c isvinc=(npt-1)*nsvint    .integration point counter
```

```
c statev(1+isvinc          ) .strain
c---------------------------------
     if (lflags(3).eq.4) then
       do i=1, ndofel
         do j=1, ndofel
           amatrx(i,j) = zero
         end do
         amatrx(i,i) = one
       end do
       goto 999
     end if
c---PRELIMINARIES
     pnewdtLocal = pnewdt
     if(jtype.ne. 1) then
       write(7,*)'Incorrect element type'
       call xit
     endif
     if(nsvars .lt. ninpt*nsvint) then
       write(7,*)'Increase the number of SDVs to',
     1            ninpt*nsvint
       call xit
     endif
     thickness = 0.1d0
c---INITIALIZE RHS AND LHS
     do k1=1, ndof*nnode
       rhs(k1, 1)= zero
       do k2=1, ndof*nnode
         amatrx(k1, k2)= zero
       end do
     end do
c---LOOP OVER INTEGRATION POINTS
     do kintk = 1, ninpt

c-----EVALUATE SHAPE_FUNCTIONS AND THEIR DERIVATIVES
c-----determine (r,s) local coordinates
       r = coord24(1,kintk)*gaussCoord
       s = coord24(2,kintk)*gaussCoord
c-----shape_functions
       shape (1) = (one + r)*(one + s)/four;
       shape (2) = (one - r)*(one + s)/four;
       shape (3) = (one - r)*(one - s)/four;
       shape (4) = (one + r)*(one - s)/four;
c-----derivative d(Ni)/d(r)
       dshape (1,1) =  (one + s)/four;
       dshape (1,2) = -(one + s)/four;
```

```
      dshape (1,3) = -(one - s)/four;
      dshape (1,4) =  (one - s)/four;
c-----derivative d(Ni)/d(s)
      dshape (2,1) =  (one + r)/four;
      dshape (2,2) =  (one - r)/four;
      dshape (2,3) = -(one - r)/four;
      dshape (2,4) = -(one + r)/four;
c-----compute coordinates at the integration point
      do k1=1, 3
        coords_ip(k1) = zero
      end do
      do k1=1,nnode
        do k2=1,mcrd
          coords_ip(k2)=coords_ip(k2)+shape(k1)*
     1                  coords(k2,k1)
        end do
      end do
c-----INTERPOLATE FIELD VARIABLES
      if(npredf.gt.0) then
        do k1=1,npredf
          predef_loc(k1) = zero
          dpredef_loc(k1) = zero
          do k2=1,nnode
            predef_loc(k1) = predef_loc(k1)+
     1                    (predef(1,k1,k2)-
     2                    predef(2,k1,k2))*shape(k2)
            dpredef_loc(k1) = dpredef_loc(k1)+
     1                    predef(2,k1,k2)*shape(k2)
          end do
        end do
      end if
c-----FORM_B MATRIX
      djac = one
      do i = 1, ndim
        do j = 1, ndim
          xjac(i,j)  = zero
          xjaci(i,j) = zero
        end do
      end do
c
      do inod= 1, nnode
        do idim = 1, ndim
          do jdim = 1, ndim
            xjac(jdim,idim) = xjac(jdim,idim) +
     1            dshape(jdim,inod)*coords(idim,inod)
```

```
          end do
        end do
      end do
      djac = xjac(1,1)*xjac(2,2) - xjac(1,2)*xjac(2,1)
      if (djac .gt. zero) then
c-----jacobian is positive - o.k.
        xjaci(1,1) =  xjac(2,2)/djac
        xjaci(2,2) =  xjac(1,1)/djac
        xjaci(1,2) = -xjac(1,2)/djac
        xjaci(2,1) = -xjac(2,1)/djac
      else
c-----negative_or zero jacobian
        write(7,*)'WARNING: element',jelem,'has neg.
     1              Jacobian'
        pnewdt = fourth
      endif
      if (pnewdt .lt. pnewdtLocal) pnewdtLocal=pnewdt
      do i= 1, nnode*ndim
        bmat(i) = zero
      end do
      do inod = 1, nnode
        do ider = 1, ndim
          do idim = 1, ndim
            irow = idim + (inod - 1)*ndim
            bmat(irow) = bmat(irow) +
     1           xjaci(idim,ider)*dshape(ider,inod)
          end do
        end do
      end do
c-----CALCULATE INCREMENTAL STRAINS
      do i = 1, ntens
        dstran(i) = zero
      end do
c-----set deformation gradient to_Identity matrix
      do k1=1,3
        do k2=1,3
          defGrad(k1,k2) = zero
        end do
        defGrad(k1,k1) = one
      end do
c-----COMPUTE INCREMENTAL STRAINS
      do nodi = 1, nnode
        incr_row = (nodi - 1)*ndof
        do i = 1, ndof
          xdu(i)= du(i + incr_row,1)
```

```
            utmp(i) = u(i + incr_row)
          end do
          dNidx = bmat(1 + (nodi-1)*ndim)
          dNidy = bmat(2 + (nodi-1)*ndim)
          dstran(1) = dstran(1) + dNidx*xdu(1)
          dstran(2) = dstran(2) + dNidy*xdu(2)
          dstran(4) = dstran(4) + dNidy*xdu(1) +
     1                  dNidx*xdu(2)
c---------deformation gradient
          defGrad(1,1) = defGrad(1,1) + dNidx*utmp(1)
          defGrad(1,2) = defGrad(1,2) + dNidy*utmp(1)
          defGrad(2,1) = defGrad(2,1) + dNidx*utmp(2)
          defGrad(2,2) = defGrad(2,2) + dNidy*utmp(2)
        end do
c-------CALL_CONSTITUTIVE ROUTINE
c-------integration point increment
        isvinc= (kintk-1)*nsvint
c-------prepare arrays for entry_into material routines
        do i = 1, nsvint
          statevLocal(i)=svars(i+isvinc)
        end do
c-------state variables
        do k1=1,ntens
           stran(k1) = statevLocal(k1)
           stress(k1) = zero
        end do
        do i=1, ntens
          do j=1, ntens
            ddsdde(i,j) = zero
          end do
          ddsdde(i,j) = one
        enddo
c-------compute characteristic element length
        celent = sqrt(djac*dble(ninpt))
        dvmat  = djac*thickness
        dvdv0 = one
        call material_lib_mech(materiallib,stress,
     1  stran,dstran,kintk,dvdv0,dvmat,defGrad,ddsdde,
     2  predef_loc,dpredef_loc,npredf,celent,coords_ip)
c
        do k1=1,ntens
           statevLocal(k1) = stran(k1) + dstran(k1)
        end do
c-------integration point increment
        isvinc= (kintk-1)*nsvint
```

```
c———update element state variables
      do i = 1, nsvint
        svars(i+isvinc)=statevLocal(i)
      end do
c———form_stiffness matrix and_internal force vector
      dNjdx = zero
      dNjdy = zero
      do i = 1, ndof*nnode
        force(i) = zero
        do j = 1, ndof*nnode
          stiff(j,i) = zero
        end do
      end do
      dvol= wght(kintk)*djac
      do nodj = 1, nnode
        incr_col = (nodj - 1)*ndof
        dNjdx = bmat(1+(nodj-1)*ndim)
        dNjdy = bmat(2+(nodj-1)*ndim)
        force_p(1) = dNjdx*stress(1)+dNjdy*stress(4)
        force_p(2) = dNjdy*stress(2)+dNjdx*stress(4)
        do jdof = 1, ndof
          jcol = jdof + incr_col
          force(jcol)=force(jcol)+force_p(jdof)*dvol
        end do
        do nodi = 1, nnode
          incr_row = (nodi -1)*ndof
          dNidx = bmat(1+(nodi-1)*ndim)
          dNidy = bmat(2+(nodi-1)*ndim)
          stiff_p(1,1) = dNidx*ddsdde(1,1)*dNjdx
     1                 + dNidy*ddsdde(4,4)*dNjdy
     2                 + dNidx*ddsdde(1,4)*dNjdy
     3                 + dNidy*ddsdde(4,1)*dNjdx
          stiff_p(1,2) = dNidx*ddsdde(1,2)*dNjdy
     1                 + dNidy*ddsdde(4,4)*dNjdx
     2                 + dNidx*ddsdde(1,4)*dNjdx
     3                 + dNidy*ddsdde(4,2)*dNjdy
          stiff_p(2,1) = dNidy*ddsdde(2,1)*dNjdx
     1                 + dNidx*ddsdde(4,4)*dNjdy
     2                 + dNidy*ddsdde(2,4)*dNjdy
     3                 + dNidx*ddsdde(4,1)*dNjdx
          stiff_p(2,2) = dNidy*ddsdde(2,2)*dNjdy
     1                 + dNidx*ddsdde(4,4)*dNjdx
     2                 + dNidy*ddsdde(2,4)*dNjdx
     3                 + dNidx*ddsdde(4,2)*dNjdy
         do jdof = 1, ndof
```

```
                  icol = jdof + incr_col
                  do idof = 1, ndof
                    irow = idof + incr_row
                    stiff(irow,icol) = stiff(irow,icol) +
     1                            stiff_p(idof,jdof)*dvol
                  end do
                end do
              end do
            end do
c-------assemble rhs and_lhs
            do k1=1, ndof*nnode
                rhs(k1, 1) = rhs(k1, 1) - force(k1)
              do k2=1, ndof*nnode
                amatrx(k1,k2) = amatrx(k1,k2)+ stiff(k1,k2)
              end do
            end do
c-----end_ loop on material integration points
          end do
          pnewdt = pnewdtLocal
c
 999      continue
c
          return
          end
```

6.7 Using Nonlinear User Elements in Various Analysis Procedures

Nonlinear user elements can be utilized in most Abaqus standard analysis procedures. The **LFLAGS(1)** indicates which procedure type is used such as the **LFLAGS(1)=11** used for dynamic procedure with automatic time incrementation or the **LFLAGS(1)=12** used for dynamic procedure with fixed time incrementation.

Mainly, the usages described have applied only to **STATIC (LFLAGS(1)=1, 2)** in user subroutine.

The usage in many procedures is the same or similar to that for STATIC; VISCO; HEAT TRANSFER, STEADY STATE; COUPLED TEMPERATURE DISPLACEMENT, STEADY STATE; GEOSTATIC; SOILS, STEADY STATE or COUPLED THERMAL ELECTRICAL, STEADY STATE.

A special case of static analysis is STATIC, RIKS. An additional force vector containing only forces proportional to the applied loads, as well as the usual force vector and the Jacobian, must be supplied. These additional forces must include thermal expansion effects if any are present in the element.

If no forces are applied to the element, the usage is the same as that for a regular STATIC analysis.

The user elements can also be used in most linear perturbation procedures. For a static linear perturbation analysis (**STATIC, PERTURBATION**), a stiffness matrix and two force vectors must be returned by the UEL. The value of **LFLAGS(3)** denotes the matrix to be returned in a call. **LFLAGS(3)=1** is selected for an assemblyreturn the stiffness matrix for the base state and the force vector that contains only external perturbation loads. But a **LFLAGS(3)=100** is selected for a recoveryreturn, the force vector that contains the difference between external perturbation loads and internal perturbation forces is shown in Eq. 6.28. This force vector is used for the reaction force calculation.

$$F^N = \Delta P^N - K^{NM} \Delta u^M \tag{6.28}$$

For a **FREQUENCY** analysis, a stiffness and mass matrix must be returned by the UEL. The value of **LFLAGS(3)** denotes the matrix to be returned in a call. **LFLAGS(3)=2** is selected to return the stiffness matrix, **LFLAGS(3)=4** to return the mass matrix, and moreover there is no element output available for user elements utilized with the FREQUENCY option.

The eigenfrequencies and eigenvectors obtained with the FREQUENCY option can be used in all modal dynamics procedures MODAL DYNAMIC; STEADY STATE DYNAMICS; RESPONSE SPECTRUM or RANDOM RESPONSE.

The user elements limitation is that they cannot be used in the STEADY STATE DYNAMICS, DIRECT and BUCKLE procedures.

The first-order transient effects must be included in UELs that are used with the following procedures: HEAT TRANSFER (transient); SOILS, CONSOLIDATION; COUPLED TEMPERATURE DISPLACEMENT (transient) or COUPLED THERMAL ELECTRICAL, TRANSIENT. The heat (pore fluid) capacity term must be included in the flux vector and the Jacobian. If the user element has no heat (pore fluid) capacity, the user element usage is the same as in the corresponding steady-state analysis.

In transient heat transfer analysis, **LFLAGS(1)** indicates the transient heat transfer procedure type being used the **LFLAGS(1)=32** for transient heat transfer analysis with automatic time incrementation, or the **LFLAGS(1)=33** for transient heat transfer analysis with fixed time incrementation. It is important to outline that additional coding related to the transient terms in the equilibrium equation is required for the flux vector and Jacobian.

The flux vector must contain the externally applied fluxes, the fluxes due to conduction, and the fluxes due to changes in internal energy shown in Eq. 6.29, where $\Delta\theta^M$ is the temperature increment.

$$F^N = F^N_{ext} + F^N_{cond} + F^N_{cap} \tag{6.29}$$

If the heat capacity matrix C^{NM} is constant, the flux due to the heat capacity terms is given by Eq. 6.30.

$$F^N_{cap} = -C^{NM}\frac{\Delta\theta^M}{\Delta t} \tag{6.30}$$

If the heat capacity matrix varies with temperature (such as is the case during phase transformations), the flux vector must be calculated from the energy change vector shown in Eq. 6.31.

$$F^N_{cap} = -\frac{\Delta u^M}{\Delta t} \tag{6.31}$$

The Jacobian will contain contributions from the conductivity and heat capacity terms. If the heat capacity matrix is constant, the Jacobian has the form as shown in Eq. 6.32, where K^{NM} is the conductivity matrix.

$$K^{NM} + \frac{C^{NM}}{\Delta t} \tag{6.32}$$

If the heat capacity matrix is a function of temperature, the Newton algorithm requires the heat capacity at the temperature at the end of the increment as shown in Eq. 6.33. For cases in which the heat capacity varies strongly (such as in case of latent heat), convergence may be difficult.

$$C^{NM} = C^{NM}\left(\theta_{t+\Delta t}\right) \tag{6.33}$$

If the user element contains no heat capacity terms, the formulation for transient heat transfer is the same as for steady-state heat transfer.

The second-order transient (inertial) effects must be included in UELs that are used with direct-integration dynamic analysis DYNAMIC. **LFLAGS(1)** indicates the dynamics procedure type being used with **LFLAGS(1)=11** for dynamic procedure with automatic time incrementation, or with **LFLAGS(1)=12** for dynamic procedure with fixed time incrementation. It must be pointed out that additional coding related to transient terms, sudden changes in velocities or accelerations, and evaluation of the half-step residual (if automatic time incrementation is used) is required in the UEL.

The value of **LFLAGS(3)** indicates the coding being executed and the matrices to be returned. **LFLAGS(3)=1** is used for normal time increment. The user must specify the forces and Jacobian corresponding to the integration procedure used. The force vector has the form given in Eq. 6.34, where M^{NM} is the element mass matrix, G^N is the static force vector, and α is the Hughes–Hilbert–Taylor integration operator.

$$F^N = -M^{NM}\ddot{u}^M_{t+\Delta t} + (1+\alpha)\,G^M_{t+\Delta t} - \alpha G^N_t \tag{6.34}$$

The static force vector G^N must also contain the rate-dependent (damping) terms. The vector G^N_t must be stored as a set of state variables. The parameter α is passed into the subroutine as **PARAMS(1)**. The Jacobian has the form shown in Eq. 6.35, where C^{NM} is the element damping matrix and K^{NM} is the static tangent stiffness matrix.

$$M^{NM}\left(\frac{d\ddot{u}}{du}\right) + (1+\alpha)\,C^{NM}\left(\frac{d\dot{u}}{du}\right) + (1+\alpha)\,K^{NM} \tag{6.35}$$

$\left(\frac{d\ddot{u}}{du}\right)$ and $\left(\frac{d\dot{u}}{du}\right)$ follow from the integration operator. For the HHT operator as given in Eq. 6.36, where β and γ are the coefficients in the Newmark-β operator given in Eq. 6.37.

$$\left(\frac{d\ddot{u}}{du}\right) = \frac{1}{\beta\Delta t^2} \qquad \left(\frac{d\dot{u}}{du}\right) = \frac{\gamma}{\beta\Delta t} \tag{6.36}$$

$$\beta = \frac{1}{4}(1-\alpha)^2 \qquad \gamma = \frac{1}{2} - \alpha \tag{6.37}$$

The parameters β and γ are passed into the user subroutine UEL as **PARAMS(2)** and **PARAMS(3)**.

The coding is simplified considerably if the HHT parameter $\alpha = 0$. In particular, there is no need to store the static residual vector, G_t^N. The variable, α, can be set to zero with the ALPHA parameter on the DYNAMIC option.

LFLAGS(3)=4 is selected when a velocity jump calculation will be done at the start of each dynamic step and after contact changes. The purpose of this calculation is to make the velocities conform to constraints imposed by MPC, EQUATION, or contact conditions while preserving momentum. The Jacobian is equal to the mass matrix, and the force vector should be set to zero.

If the user element has no inertia or damping terms (i.e., if the force vector does not depend on the velocities and accelerations), the parameter can be ignored in the subroutine. If the user element includes viscous effects but no inertia terms, the same approach can be used for transient heat transfer analysis. The force vector then should contain the term shown in Eq. 6.38, and the term in Eq. 6.39 must be added to the stiffness, in that case the parameter can again be ignored.

$$-C^{NM}\frac{\Delta u^M}{\Delta t} \tag{6.38}$$

$$\frac{C^{NM}}{\Delta t} \tag{6.39}$$

LFLAGS(3)=5 is selected when an half-step residual calculation is needed only for automatic time incrementation. Only the force vector must be supplied, which has the form shown in Eq. 6.40.

$$F^N = M^{NM}\ddot{u}_{t+\frac{\Delta t}{2}} + (1+\alpha)\,G^N_{t+\frac{\Delta t}{2}} - \frac{\alpha}{2}\left(G_t^N + G_{t_0}^N\right) \tag{6.40}$$

where $G_{t_0}^N$ is the static residual at the beginning of the previous increment, it must be stored as a solution-dependent state vector. The vector $\ddot{u}_{t+\frac{\Delta t}{2}}$ is passed into the subroutine, $G^N_{t+\frac{\Delta t}{2}}$, and G^N_t must be calculated. It is obvious that this expression simplifies considerably if $\alpha = 0$.

LFLAGS(3)=6 is selected for an acceleration calculation, which will be done at the start of each dynamic step (unless INITIAL=NO on the DYNAMIC option) and after contact changes. The purpose of this calculation is to create dynamic equilibrium at the start of a step or after contact changes. The Jacobian is equal to the mass matrix, and the force vector should contain static and damping contributions only.

Implementation of a user element with inertia effects in a dynamic analysis is fairly complicated. Simplifications to the UEL can be realized if the ALPHA parameter on the DYNAMIC option is set to zero. Thus, there are no inertia effects included in the user element. The inertia effects can be included indirectly by overlaying standard Abaqus elements on top of user elements. In this case, the Abaqus elements should have negligible stiffness. For example, it is possible to overlay B23 elements on top of the beam elements with nonlinear section behavior shown in the example described in Sect. 6.6.1.

References

1. Lee DT, Schachter BJ (1980) Two algorithms for constructing a delaunay triangulation. Int J Comput Inf Sci 9(3)
2. Boyce MC (1992) Finite element simulations in mechanics of materials and deformation processing research. In: ABAQUS users' conference proceedings
3. Wenk T, Linde P, Bachmann H (1993) User elements developed for the nonlinear dynamic analysis of reinforced concrete structures. In: ABAQUS users' conference proceedings
4. Vitali R, Zanotelli GL (1994) User element for crack propagation in concrete-like materials. In: ABAQUS users' conference proceedings
5. Bathe KJ (2014) Finite element procedures, 1st edn. Prentice Hall, Upper Saddle River, 1996; 2nd edn. Watertown
6. Chapelle D, Bathe KJ (2011) The finite element analysis of shells fundamentals, 1st edn. Springer, Berlin, 2003; 2nd edn. Springer, Berlin
7. Kojic M, Bathe KJ (2005) Inelastic analysis of solids and structures. Springer, Berlin
8. Bucalem ML, Bathe KJ (2011) The mechanics of solids and structures hierarchical modeling and the finite element solution. Springer, Berlin
9. Batoz J-L, Dhatt G (1990) Modélisation des structures par éléments finis. Tome 1: solides élastiques (French). Hermes Science Publications, New Castle
10. Batoz J-L, Dhatt G (1990) Modélisation des structures par éléments finis. Tome 2: poutres et plaques (French). Hermes Science Publications, New Castle
11. Batoz J-L, Dhatt G (1992) Modélisation des structures par éléments finis. Tome 3: coques (French). Hermes Science Publications, New Castle

Chapter 7
Contact

7.1 Generalities

In general, what is contact? Simply put, it is a nonlinear spring set between two nodes. Indeed, a contact in FEA modeling, regardless of the software used, is just a simple mechanical element spring used between both parts in contact with each other with a linear or nonlinear stiffness behavior, as shown in Fig. 7.1. The contact is mainly set by defining slave and master surfaces. The contact stiffness is a function of the clearance between both surfaces with some tolerances in the penetration zone.

The cantilever beam shown in Fig. 7.1 is constrained at its free end to be in contact interaction with the ground. Therefore, the deformed shape deflection will be also constrained with the nonlinear spring zone. The nonlinear spring is always a better choice to simulate contact interaction in realistic matters, because the closer the deformed structure is at the end to the maximum clearance value. Consequently, the spring becomes more and more stiff to ensure a proper contact. If the springs were linear, the stiffness would increase even if the deformed shape is too far away to ensure a contact interaction with the master surface, which is fairly unusual and unrealistic.

To understand the principle, it is useful to provide a simple example according to the model in Fig. 7.1 and then try to calculate the theoretical solution, simulate this model with Abaqus using a normal contact behavior only and compare both deformed shapes. Thus, the first Euler–Bernoulli equilibrium equation will be used to calculate the exact solution without contact interaction.

$$EI.\frac{d^2}{dx^2}y\left(x\right) = -M_z\left(x\right) = F\left\langle\frac{L}{3} - x\right\rangle^1_{0\leq x\leq\frac{L}{3}} \tag{7.1}$$

As an example case, the total length of the beam is L, which equals to 45 mm with an elastic modulus E (200 000 MPa), I is the second moment of inertia taken for a beam cross-section circular in order to ensure a proper beam circumferential contact on a plane. The circular cross-sectional beam has a diameter ϕ equal to 5 mm according

R. J. Boulbes, *Troubleshooting Finite-Element Modeling with Abaqus*,
https://doi.org/10.1007/978-3-030-26740-7_7

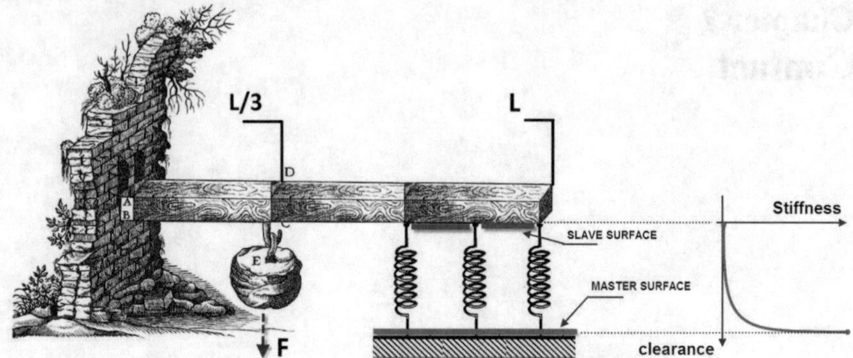

Fig. 7.1 Representation of a contact zone on a cantilever beam model

to formula $I = \frac{\pi}{64}\phi^4$; y is the deflection of deformed shape along the length at a location x, and M_z is the bending moment acting along the beam length. A force F is applied on the beam at a location $x = \frac{L}{3}$. The rotation equation can be written from Eq. 7.1 by an integration operation such as

$$EI.\theta\,(x) = -\frac{F}{2}\left\langle x - \frac{L}{3}\right\rangle^2_{0\leq x\leq \frac{L}{3}} + K_1 \tag{7.2}$$

The deflection of the deformed shape equation is given from Eq. 7.2 by another integration such as

$$EI.y\,(x) = \frac{F}{6}\left\langle \frac{L}{3} - x\right\rangle^3_{0\leq x\leq \frac{L}{3}} + K_1x + K_2 \tag{7.3}$$

The integration constants in Eq. 7.3 will be calculated thanks to the fixed boundary condition giving for $\theta\,(0) = 0$ and $y\,(0) = 0$ and the constant values for $K_1 = \frac{FL^2}{18}$ and $K_2 = -\frac{FL^3}{162}$, respectively. The theoretical solution of the model shown in Fig. 7.1 without the contact interaction is

$$EI.y\,(x) = \frac{F}{6}\left\langle \frac{L}{3} - x\right\rangle^3_{0\leq x\leq \frac{L}{3}} + \frac{FL^2}{18}x - \frac{FL^3}{162} \tag{7.4}$$

Now, let's put a distance h between the slave (from the beam neutral fiber) and the master surface, where h is measured at the total end of the beam, for instance

$$y\,(x = L) = h \tag{7.5}$$

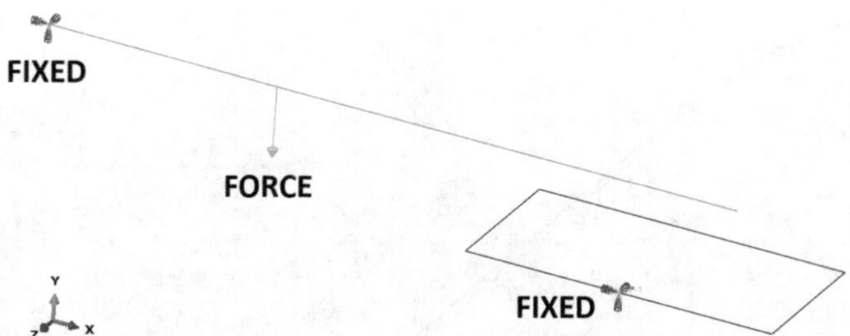

Fig. 7.2 An example of the cantilever beam to determine contact interaction behavior

By combining both Eqs. 7.4 and 7.5, it is possible to get a relation between the force applied on the structure and the maximum deflection of the deformed shape such as

$$F = \frac{81}{4}\frac{EIh}{L^3} \tag{7.6}$$

Equation 7.6 makes sense, if the maximum deflection h is equal to zero as there is no deformed shape and therefore, there is no force applied on the structure to deform it. If h is equal to the maximum clearance defined in contact interaction, then it is possible to drive the maximum deflection as a function of the force applied on the structure. Here, for instance, the distance clearance h is taken to be equal to 4.5 mm, so the force applied can be calculated according to Eq. 7.6 and is equal to 6 135.92 N.

An FEA beam modeling has been made from the configuration in Fig. 7.1 according to the input data above as shown in Fig. 7.2, with a fixed-free cantilever beam in contact interaction set as a hard contact normal behavior with a fixed underformable rectangular surface.

The solutions of the loaded beam with a force F equal to 6 135.92 N at x equal to 15 mm in Fig. 7.3 indicate the structural response in deflections for different contact settings. It can be observed that with the exception of the contact interaction set with a Surface to Surface with a small sliding, all other contact interaction settings give a response in accordance with the theoretical solution calculated in Eq. 7.4. Like the model set with a Node to Surface, contact interactions with small sliding also provide a correct solution. The first conclusion is actually something to think about because the master surface is always defined with a surface and therefore, the contact response is a function of the master surface selected. Here, the deflection response of a contact interaction Surface to Surface with small sliding gives a flat deflection for the last third length of the beam (for $2L/3 \leq x \leq L$) set in contact with a spring in series. This is shown in Fig. 7.1. The small sliding used with a Surface-to-Surface interaction must be set with caution, because it would affect the solution to cover the entire slave and master surfaces defined in contrast with other contact definitions as shown, for example, in Fig. 7.3.

The results discussed in relation to the plotted graphs in Fig. 7.3 give to the analysts a first clue about the contact interaction settings with contact pairs. To select

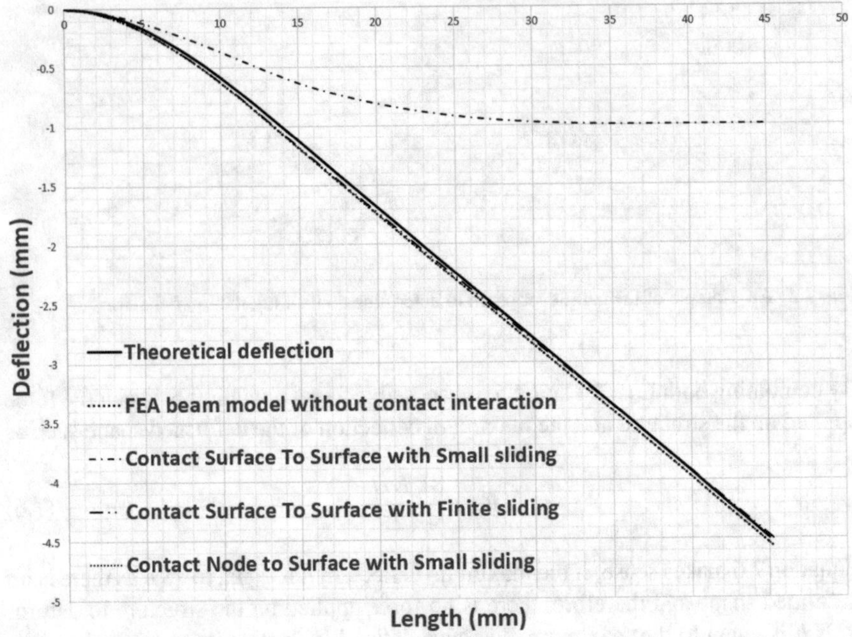

Fig. 7.3 Deflection results for the cantilever beam model

proper contact interaction, the analyst should carefully consider all aspects to make a prediction about the surfaces in contact with other elements and the type of contact sliding to define. This means the analyst must take into consideration the boundary and loading conditions set in the model assembly in order to anticipate the structural response. Quantification is not necessary because this will be done by the solver to return output data numbers but at least useful to qualify the structural response. It is often very difficult to figure out how components will deform each other, especially with a complex assembly in contact interaction with several components and with different contact interaction dependencies.

From Table 7.1, it is obvious that the best contact settings to use in this case after comparing with the theory is the contact Surface to Surface with a finite sliding. The next section will describe how to make a correct contact choice to set contact interaction without wasting time by performing all different options in the contact interaction features.

7.1.1 Understandings

A contact is fairly intuitive and numerically challenging. It is fairly intuitive because contact involves interactions between bodies, which would need to take into account

Table 7.1 Maximum deflection results function of contact settings used

Models	Maximum deflection (mm)	Maximum deviation between deflection versus theory (%)
Theory	4.5	0
Beam without contact interaction	4.568	1.51
Beam with contact STS[a] and small sliding	1	77.78
Beam with contact STS[a] and finite sliding	4.5	0
Beam with contact NTS[b] and small sliding	4.568	1.51
Beam with contact NTS[b] and finite sliding	NA[c]	NA[c]

[a]STS for Surface to Surface
[b]NTS for Node to Surface
[c]NA for Not Applicable because model returned the following warning message Whenever a translation (rotation) dof at a node is constrained by a kinematic coupling definition the translation (rotation) dofs for that node cannot be included in any other constraint including mpcs, rigid bodies, and so on. caused of the constraint set on the fixed undeformable rectangular surface shown in Fig. 7.2 which is not compatible with NTS contact algorithm

a contact pressure to resist penetration, a frictional stress to resist sliding, and if needed electrical or thermal interactions. It is also numerically challenging because contact includes severe nonlinearities, for instance, in inequality conditions resulting in discontinuous stiffness an initial gap distance $d_{gap} \geq 0$, a frictional stress $\tau = \mu\sigma$ with the friction coefficient (μ), and conductance properties suddenly change—even abrupt sometimes—when contact is established.

Various classifications of contact interactions can be considered like, for example, slender or bulky components. For bulky components, although typically many nodes are in contact at one time and the contact causes local deformation and shear, there is little bending. For slender components, there are often relatively few nodes in contact at one time, the contact causes bending and it is often more numerically challenging. However, common classifications used for contact interactions can be listed below:

- Slender or bulky components.
- Deformable or rigid surfaces.
- Degree of confinement and compressibility of components.
- Two-body contact or self-contact.
- Amount of relative motion (small or finite sliding).
- Amount of deformation.
- Underlying element type (first or second order).
- Interaction properties (friction, thermal, and so on.).
- The specific results that are of interest and importance (e.g., contact stresses).

There are many aspects which need to be explicitly specified to define contact interaction, so the following is an ingredients list to help analysts to determine what is required in contact settings:

1. The contact surfaces, the surfaces over bodies that may experience contact.
2. The contact interactions shall be defined by knowing which surfaces will interact with others surfaces.

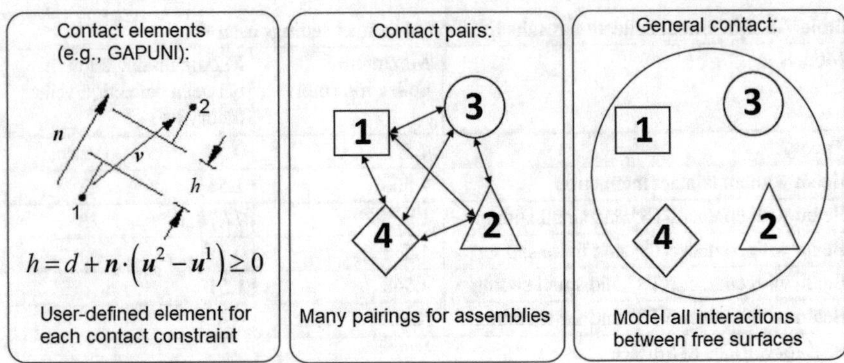

Fig. 7.4 Different contact interaction strategies in Abaqus

3. The surface property assignments, for example, contact thickness of a shell.
4. The contact property models, for example, the pressure versus overclosure relationship, friction coefficient, conduction coefficients, and so on.
5. The contact formulations shall be considered in case of a small sliding is assumed between surfaces in contact.
6. The algorithmic contact controls such as contact stabilization settings.

The main aim of contact modeling in a model is to improve usability, accuracy, and performance to ensure a greater focus on physical aspects rather than the idiosyncrasies of numerical algorithms and broad applicability, especially in large model assemblies. Figure 7.4 presents the main different options to set a contact interaction strategy with an assembly of four components directly from contact element to contact element with contact pairs to define each surface in contact with the others, or with general contact to define a zone in which all contact interactions will take place.

A physical master and slave entities must be defined properly in contact interaction otherwise a nonrealistic representation and therefore the solution will be computed as shown, for example, in Fig. 7.5. General contact algorithm can be defined with a contact domain that spans multiple bodies (both rigid and deformable), by default domain defined automatically via all-inclusive, element-based surface. The method geared toward models with multiple components and complex topology, which is easiest in defining contact model.

Be sure that all surfaces which must have a contact interaction will be defined, otherwise as shown in Fig. 7.6 some nonrealistic penetration between component might have occurred.

Transition to general contact is nearly complete for Abaqus explicit with most Abaqus explicit analyses using general contact is easy to use, robust and accurate, has good performance, and scalability is good or better than contact pairs. Transitioning to general contact in Abaqus standard shows a good feedback; it is an easier model to create than contact pairs, it has similar robustness and accuracy compared to contact

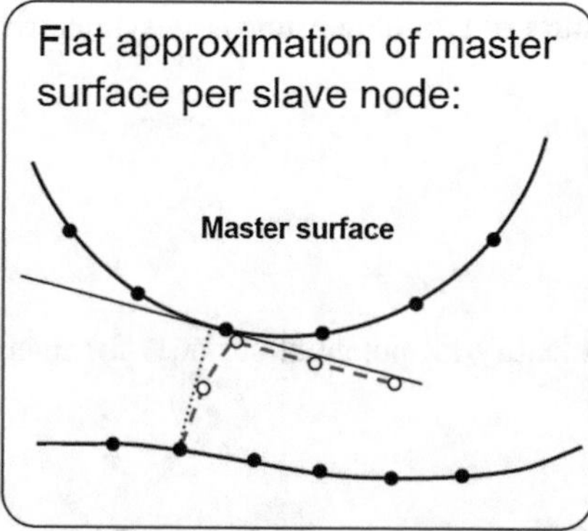

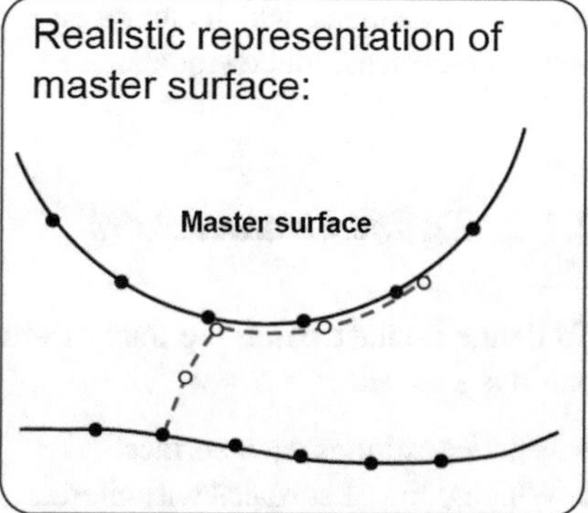

Fig. 7.5 Master surface in contact interaction

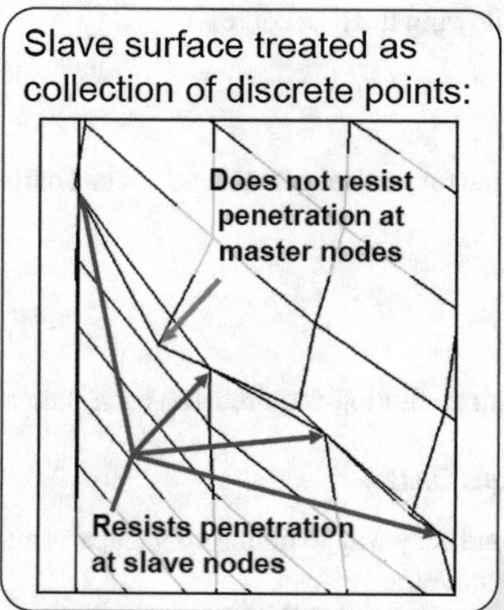

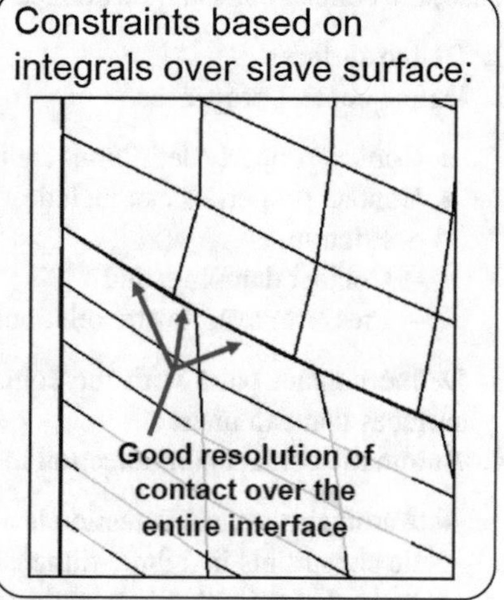

Fig. 7.6 Define all surfaces in contact

pairs with some extra contact tracking time, and so on. Contact pairs are required to access specific features not yet available with general contact like analytical rigid surfaces, node-based surfaces, or surfaces on 3D beams or small sliding formulation. General contact and contact pairs can be used together. The general contact algorithm automatically avoids processing interactions treated with contact pairs.

There are some surface restrictions—mostly context specific—that depend on which features use the surface. Restrictions on surfaces used in contact definitions depend on details of contact definition or trend toward fewer surface restrictions, for instance, with master surface connectivity requirements. For example, the contact formulation Node to Surface with finite sliding is not allowed for a discontinuous structure (or 3D faces joined at only one node) or for a T-intersection (more than two faces per edge). Another example of a general restriction on element-based surfaces is a parent element, which cannot be a mixture of two-dimensional, axisymmetric, and three-dimensional elements.

7.1.2 Define Contact Pairs

To define contact pairs, the analyst shall identify all potential surfaces for many pairings such as

- What constitutes each surface.
- Which pairs of surfaces will interact.
- Which surface is the master and which is the slave.
- Which surface interaction properties are relevant (e.g. friction)

Then, the contact pairs can be defined by following the procedure:

1. Define surfaces.
2. Define contact properties.
    - Contact property definitions are the same for general contact and contact pairs.
    - Contact properties can include
      - Friction,
      - Contact damping, and
      - Pressure–overclosure relationships.
3. Define contact pairs with the contact pair definition required for each pair of surfaces that can interact.
4. Automatic contact pair detection in Abaqus CAE:
    - Automatic contact detection is a fast and easy way to define contact pairs and tie constraints in a three-dimensional model.
    - Instead of individually selecting surfaces and defining the interactions between them, you can instruct Abaqus/CAE to automatically locate all surfaces in a model that are likely to interact based on initial proximity
    - It can be used to define contact with shells, membranes, and solids:
      - Including shell offset.
      - Native or orphan mesh parts.
    - Tabular display of candidate contact pairs is provided.
    - It includes various controls over selection criteria, and so on.

7.1.3 Define General Contact

The general contact user interface allows for a concise contact definition reflecting the physical description of the problem, for instance, a contact definition can be expanded in complexity, as needed. There is an independent specification of contact

Table 7.2 General contact between Abaqus explicit and standard solver

Characteristic	Abaqus/Explicit	Abaqus/Standard
Primary formulation	Node-to-surface	Surface-to-surface
Master–slave roles	Balanced master–slave	Pure master–slave
Secondary formulation	Edge-to-edge	Edge-to-surface
2-D and axisymmetric	Not available	Available
Most aspects of contact definition	Step-dependent	Model data

interaction domain, contact properties, and surface attributes permitted with minimal algorithmic controls required. The general contact user interface is very similar for Abaqus explicit and Abaqus standard analyses.

Table 7.2 gives an example of the differences between general contact in Abaqus explicit and Abaqus standard.

The procedure for setting a general contact is described below:

1. Begin the general contact definition.
2. Specify automatic contact for the entire model.
3. Assign global contact properties.

Each of these main steps can be subdivided into a sub-procedure as follows:

1. The contact definition can gradually become more detailed, as called for by the analysis with

   - Global/local friction coefficients and other contact properties can be defined.
   - Pair-wise specification of contact domain (instead of **ALL EXTERIOR**) is allowed for contact inclusions and contact exclusions.
   - User control of contact thickness (especially for shells) is provided with surface properties.
   - Contact initialization (initial adjustments, interference fits, and so on.)

2. Fine-tuning contact domain:

   - The general contact domain can be modified by including and/or excluding predefined surfaces.
   - For example, exclude consideration of contact between rigid surfaces in this example as it is not essential for this analysis (overlap between perpendicular surfaces is not resolved with the surface-to-surface contact formulation used by general contact)

3. Contact initialization:

   - The default behavior of general contact is to adjust small initial overclosures without strain.
   - Can instead be treated as **INTERFERENCE FIT**.

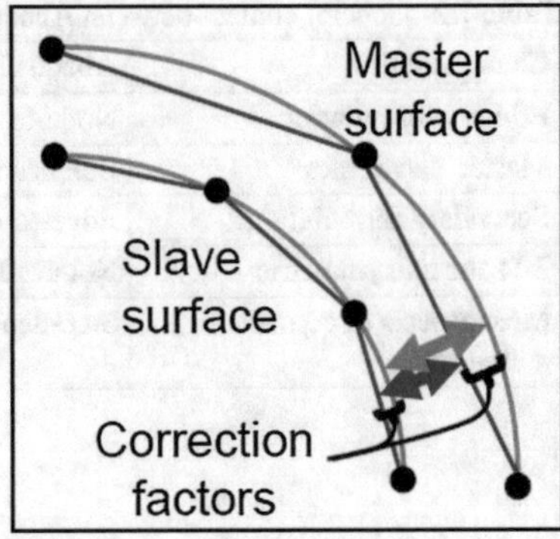

Fig. 7.7 Representation of curved slave and master surfaces. It is easy to observe the difference in contact between the design curved lines from the CAD model and the contact interaction with the mesh containing the straight lines

4. Contact properties:
   - These pertain to aspects such as
     – Contact pressure–overclosure relationship.
     – Friction.
     – Contact damping.
   - Defaults:
     – A hard pressure–overclosure relationship, where there is no contact pressure until nodes are in contact and unlimited contact pressure once contact has been established (enforced with a penalty method)
     – No friction.
     – No contact damping.
   - The user can override contact property defaults globally and locally with the last assignment. This applies to cases with conflicting assignments.

7.1.4 Representation of Curved Surfaces

As is evident from Fig. 7.7 having faceted representations of curved surfaces is sometimes detrimental to accuracy and convergence. Geometry corrections for the surface-to-surface contact formulation improve these aspects without degrading the per iteration performance and are available for near-axisymmetric and near-spherical surfaces.

Alternatively, surface-smoothing options for the node-to-surface contact formulation primarily target convergence issues associated with having discontinuous surface normals. However, they generally do not strive to represent exact initial geometry or details depend on whether surfaces are 2D or 3D, rigid or deformable.

There are some applicabilities of geometric corrections, for instance, obtaining a significant effect for small to moderate deformation. The effect is usually insignificant after large deformation. Or when using the Small or finite sliding, Surface to Surface in contact formulation.

Abaqus CAE automatically detects these surfaces in native geometry models and applies appropriate smoothing method in contact interactions. This improves accuracy, avoids the need for matched nodes across the contact interface, and reduces the iteration count (sometimes). As a rule of thumb, the fastest analysis time would be performed with a contact pairs definition followed by a general contact smooth definition; finally, the slowest setting definition should be made by the general contact faceted.

7.1.5 Contact Formulation Aspects

There are two main questions regarding the contact formulation. First, how are the constraints formed? With the discretization, for example, it is necessary to ascertain how to penetrate distances from nodal positions or how to work the Node-to-Surface, Surface-to-Surface, and Edge-to-Surface formulations. With the enforcement too, for example, a numerical method is used to resist penetrations or using direct (Lagrange multipliers) or a penalty. The second question is how do the constraints evolve upon sliding? The following is the evolution of the discretization, rigorous, nonlinear evolution (finite sliding) versus approximate (small sliding).

7.1.5.1 Contact Discretization

The different techniques used to define a contact discretization applicable for contact interaction surfaces as shown in Fig. 7.8 are listed below:

1. **Node-to-Surface** technique:
   - Nodes on one surface (the slave surface) contact the segments on the other surface (the master surface).
   - Contact is enforced at discrete points (slave nodes).
2. **Surface-to-Surface** technique:
   - Contact is enforced in an average sense over a region surrounding each slave node.

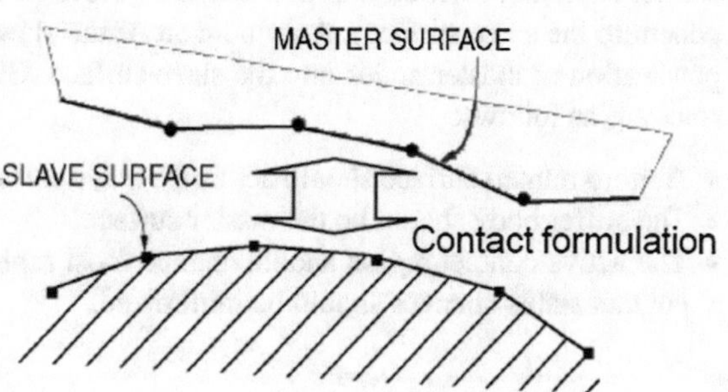

Fig. 7.8 Different techniques could be applied to establish such contact interaction between both slave and master surfaces

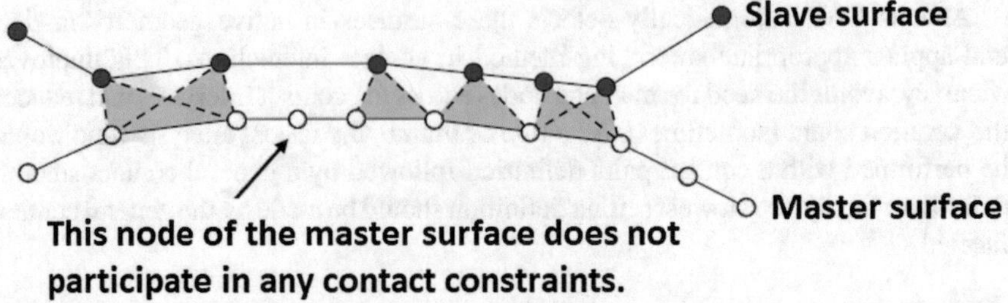

Fig. 7.9 Node-to-Surface technique is particularly mesh pattern dependent and if the quality of mesh between both slave and master surfaces is poor then some nodes cannot be included in the computed contact solution

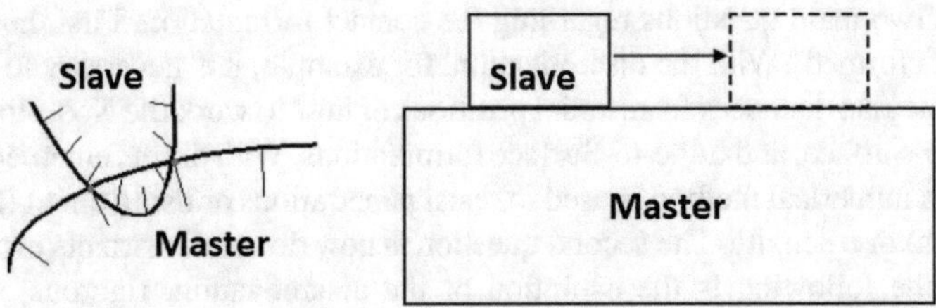

Fig. 7.10 Consequence of Node-to-Surface contact formulation

- The slave surface is much more than just a collection of nodes.
- This is fundamental to the development of general contact in Abaqus standard.

3. **Edge-to-Surface** technique:
   - Contact between a feature edge and a surface.
   - Enforced in an average sense over portions of feature edges.

Node to Surface

The Node-to-Surface contact discretization is a traditional point against the surface method, where each potential contact constraint with this formulation involves a slave node and a master facet as shown in Fig. 7.9.

As shown in Fig. 7.10, the key implications of node-to-surface formulation are that the slave nodes cannot penetrate master surface facets and the master nodes are not explicitly restricted from penetrating slave surface facets (and sometimes do penetrate the slave surface). A refinement of the slave surface helps to avoid gross penetration of master nodes into the slave surface. Guidelines for master and slave roles are as follows:

- A more refined surface should act as the slave surface,
- The stiffer body should be the master surface,
- The active contact region should change most rapidly on the master surface and contact status changes should be minimized.

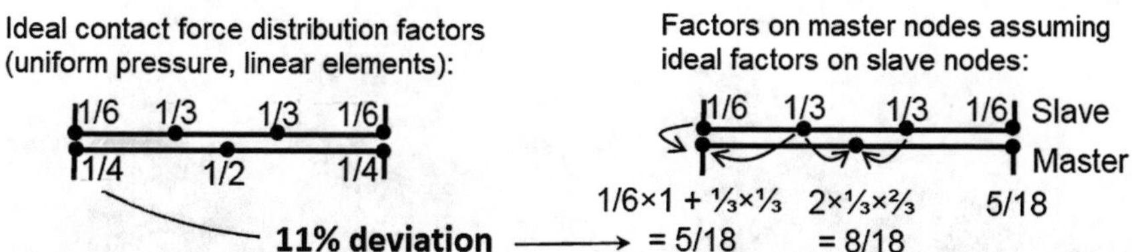

Fig. 7.11 Noise seen as a correction factor in solution deviation is produced by a misalignment of mesh pattern. Here, the deviation is 11%, which is equal to $1 - (5/18)/(1/4)$ at the pointed node

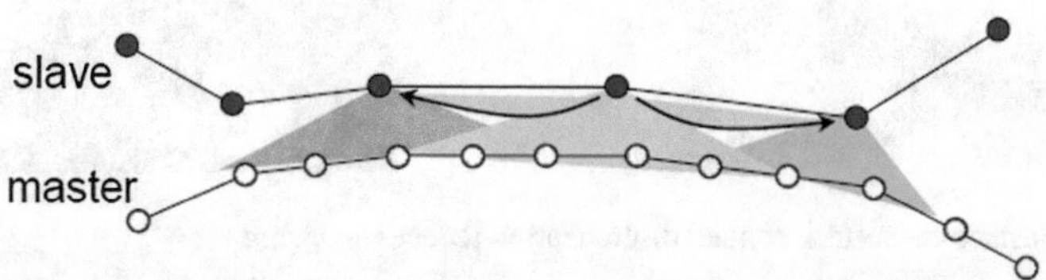

Fig. 7.12 Surface-to-Surface contact discretization with its working pattern

While refinement of slave surface leads to global accuracy, local contact stress oscillations may still be observed with Node to Surface. This noise, evaluated in percentage, acting on the slave surface in contact pressure **CPRESS** or stresses response can be evaluated as demonstrated in Fig. 7.11. Matching meshes across the contact interface can prevent this noise.

Surface to Surface

The Surface-to-Surface contact discretization shown in Fig. 7.12 is used for each contact constraint, which is formulated based on an integral over the region surrounding a slave node. It tends to involve more master nodes per constraint, especially if the master surface is more refined than the slave surface and also involves coupling among slave nodes. It is still ideal to have the more refined surface act as the slave in order to ensure better performance and accuracy. The benefits of the Surface-to-Surface approach are listed below:

- A reduced likelihood of large localized penetrations.
- A reduced sensitivity of results to master and slave roles.
- More accurate contact stresses even without a matching mesh.
- An inherent smoothing for better convergence.

A Surface-to-Surface discretization often improves the accuracy of contact stresses related to better distribution of contact forces among master nodes. For example, a classical Hertz contact problem will make the contact pressure contours much smoother and peak contact stress will have very close agreement with the analytical solution using the Surface-to-Surface approach.

As shown in Fig. 7.13, the Surface-to-Surface discretization reduces the likelihood of snagging. Indeed, the Node to Surface treats the slave surface as a collection of points, which can trigger snagging as the slave nodes to traverse a corner, whereas

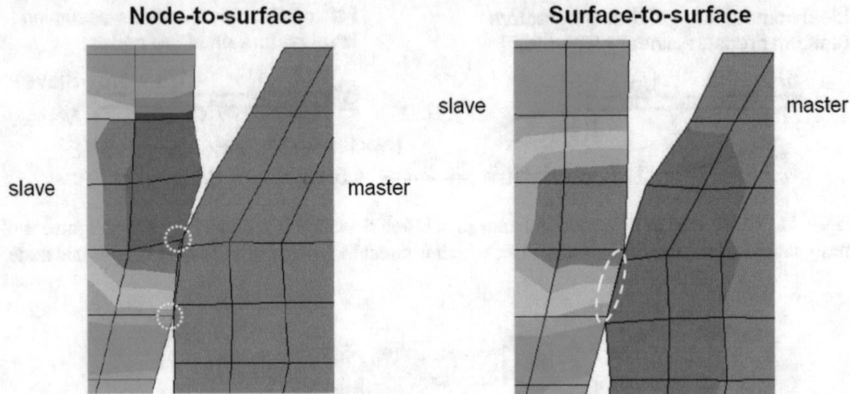

Fig. 7.13 Surface-to-Surface contact discretization reduces snagging

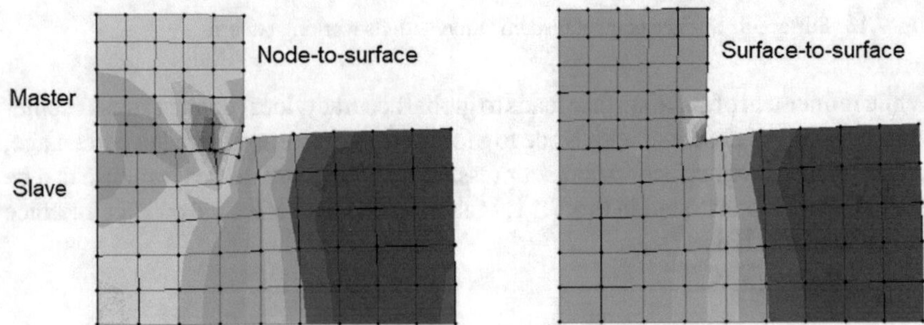

Fig. 7.14 Surface-to-Surface contact discretization prevents master node penetration

the Surface to Surface is computing the average penetration, and slips over finite regions; it has a smoothing effect that avoids snagging.

As demonstrated in Fig. 7.14, the Surface-to-Surface discretization reduces the likelihood of master nodes penetrating the slave surface. Indeed, some penetrations may be observed at individual nodes. However, large and undetected penetrations of master nodes into the slave surface do not occur.

The results with the Surface-to-Surface discretization are nearly independent of the master/slave roles in Fig. 7.15, where the two surfaces were switched with each other. A conclusion at this point is that choosing a slave surface to be finer mesh will still yield better results; choosing the master surface to be the more refined surface will tend to increase analysis costs (time and volume of output data).

A Surface-to-Surface discretization will generate multiple constraints at corners when appropriate. Indeed, a Node to Surface performs a single constraint in average normal direction at the corner, which is not stable and will lead to large penetrations and snagging. There is also a workaround with the two contact pairs. On the other hand, a Surface to Surface performs two constraints which are generated at the corner (even if one contact pair is used). This formulation is more accurate and stable without smoothing of surface normals.

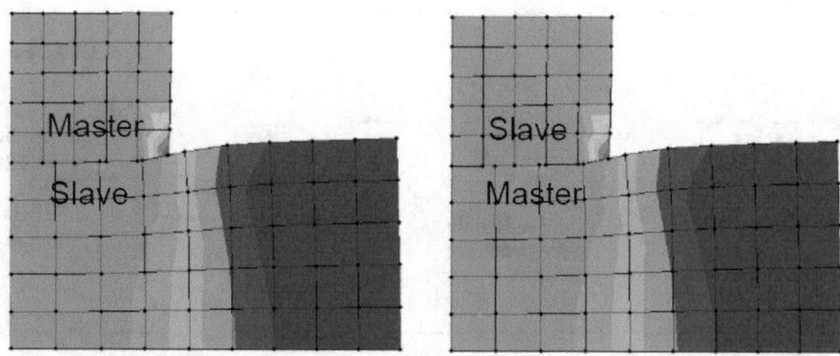

Fig. 7.15 Switched slave and master surfaces

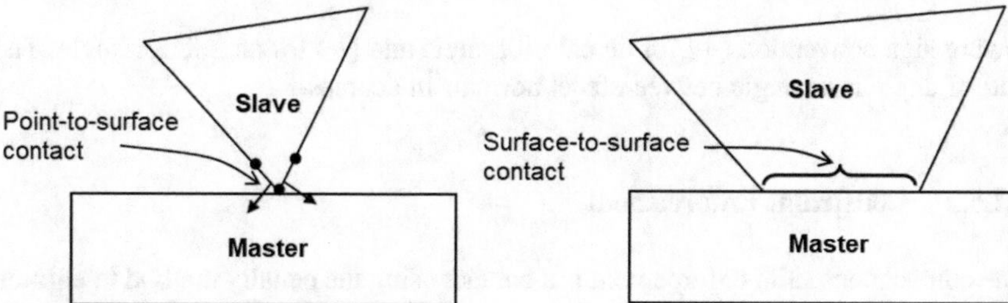

Fig. 7.16 Point-to-surface contact

A Surface-to-Surface discretization takes into consideration shell and membrane thicknesses when performing contact calculations, whereas a Node to Surface considers this effect only for the small sliding formulation.

A Surface-to-Surface discretization is fundamentally sound for situations in which quadratic elements underlie the slave surface, whereas a Node to Surface struggles with some quadratic element types related to a discrete treatment of the slave surface or a consistent force distribution for the element.

Moreover, a Surface-to-Surface discretization has a greater tendency to generate asymmetric stiffness terms, where the master and slave surfaces are not approximately parallel to each other. The usage of asymmetric solver ***STEP, UNSYMM=YES** is sometimes necessary to avoid convergence difficulties and is strongly recommended, especially when the friction coefficient is greater than 0.2.

Surface-to-Surface discretization works best when contacting surfaces have nearly opposing normals. It works well for many cases involving corners. However, it has difficulty resolving point-to-surface contact because its formulation is based on a penetration averaged over finite regions and a contact normal based on the slave surface normal (this is indicated in Fig. 7.16).

Edge to Surface

An Edge to Surface is good for enforcing certain contacts for which Surface-to-Surface formulation struggles. There are some limitations, for instance, when used with 3D solid edges only and in general contact only. It includes edges where the measured angle θ in non-deformed configurations such as $\theta \geq \theta_{cutoff}$ with the fol-

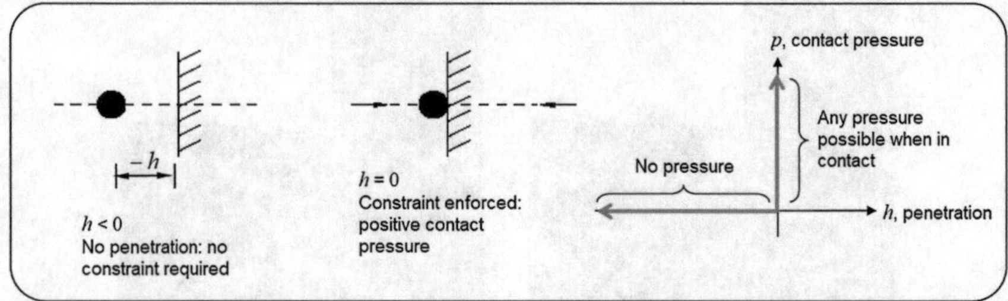

Fig. 7.17 Strict enforcement in contact formulation

lowing sign convention (+) for an exterior angle and (−) for an interior angle. The cut-off angle is an angle between facet normals in degrees.

7.1.5.2 Constraint Enforcement

The contact constraint enforcement is a contact using the penalty method to enforce the contact constraints by default. Contact pairs that use the finite sliding, node-to-surface formulation use a Lagrange multiplier method to enforce contact constraints by default in most cases.

The strict enforcement presented in Fig. 7.17 can be intuitively desirable for the user; it can be achieved with the Lagrange multiplier method in Abaqus standard with drawbacks. Drawbacks can make the solution processing very challenging for Newton iterations to converge as constraints are overlapped. This is problematic for the equation solver and the Lagrange multipliers added to the equation solver are an overanalysis cost because the size of the matrix system to solve will increase.

The direct enforcement using the Lagrange multiplier method constraints is presented in Eq. 7.7, which is the unconstrained system of equations where K is the stiffness matrix of the entire model, u is the nodal vector solution, and F the nodal force vector; the Lagrange multipliers are added to the system of equations in Eq. 7.8. λ is the vector of Lagrange multiplier degrees of freedom (constraint forces or pressures) with one per constraint. The matrix B^T is the unitless distribution coefficients for constraint force. The matrix C is the unitless constraint coefficients, and for a symmetric constraint $B = C$ is used.

$$(K)\{u\} = \{F\} \tag{7.7}$$

$$\begin{pmatrix} K & B^T \\ C & 0 \end{pmatrix} \begin{pmatrix} u \\ \lambda \end{pmatrix} = \begin{pmatrix} F \\ 0 \end{pmatrix} \tag{7.8}$$

The penalty method is a stiff approximation of hard contact as shown in Fig. 7.18, which is equivalent to replacing Eq. 7.7 with Eq. 7.9.

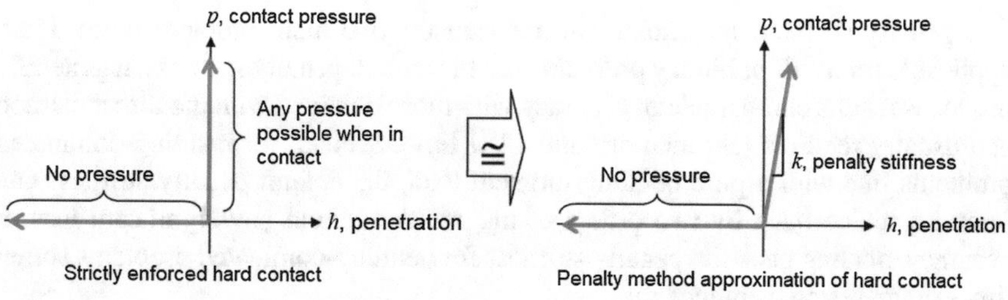

Fig. 7.18 The penalty method in contact formulation

$$\left(K + K_p\right)\{u\} = \{F\} \tag{7.9}$$

The advantages of using the penalty method are the improvement in the convergence rates, the better equation solver performance, and the effective treatment of overlapping constraints. There is also no Lagrange multiplier degree of freedom unless contact stiffness is very high. On the other hand, the main disadvantages are that there is a small amount of penetration which is typically insignificant and it may be necessary to adjust penalty stiffness relative to the default setting in some cases.

The default penalty stiffness K_p in Eq. 7.9 is calculated as follows: Abaqus tries to find a happy medium between the penalty stiffness which is too low, causing excessive penetrations, and the penalty stiffness which is too high in Abaqus standard based on a convergence rates degrading with Lagrange multiplier degrees of freedom needed to avoid an ill-conditioning; the penalty stiffness which is too high in Abaqus explicit causes a significant reduction in the stable time increment. As shown below, the scale factor is step dependent with ***CONTACT CONTROLS** and it is a multiplicative parameter.

```
*SURFACE INTERACTION
*SURFACE BEHAVIOR, PENALTY
penalty stiffness, clearance offset, scale factor
:
*STEP
:
*CONTACT CONTROLS, STIFFNESS SCALE FACTOR=value
```

The default penalty stiffness is based on the representative stiffness of the underlying elements with a scale factor applied to this representative stiffness to set the default penalty stiffness. The magnitude is higher in Abaqus standard than in Abaqus explicit. The options to scale the penalty stiffness are available:

- for cases in which default penalty stiffness is not suitable,
- when order-of-magnitude changes are recommended,
- if the scale factor is greater than 100, Abaqus will automatically invoke a variant of a method that uses Lagrange multipliers to avoid ill-conditioning issues.

The penalty stiffness magnitude can be seen for two main problem types. First, applicable for a stiff or blocky problem with the default penalty stiffness, it generally produces results comparable in accuracy with those obtained with the direct method and usually requires less memory and CPU time. Second, for bending-dominated problems like with a pure bending moment load, the default penalty stiffness can often be scaled back by two orders of magnitude without any significant loss of accuracy. Scaling back the penalty stiffness for bending-dominated problems sometimes increases convergence rate.

7.1.5.3 Evolution of Discretization

Abaqus offers finite and small sliding versions of Surface-to-Surface and Node-to-Surface contact formulations.

- The finite sliding formulation is applicable in general for a point in interaction on an updated master surface using a true representation of the master surface.
- The small sliding formulation is an approximation intended to reduce solution costs; it has limited applicability for a planar representation of the master surface per slave node based on initial configuration, only available for contact pairs (and not self-contact or general contact).

Small Sliding Approximation
Every slave node interacts with its own local slide plane; in 2D or axisymmetric it is depicted as a line. It also assumes that relative motion per slave node remains small compared to the local curvature of the master surface and facet sizes of the master surface.

```
*CONTACT PAIR, SMALL SLIDING
```

- The main advantage is dealing with less nonlinearity, reducing cost per iteration, and finding a converged solution in fewer iterations.
- The disadvantage is mainly that the results can be nonphysical if relative tangential motion does not remain small. Therefore, it is the user's responsibility to ensure that the assumption is not infringed.

7.1.5.4 Static Instabilities

The contact interaction can lead to three main types of static instabilities:

1. Unconstrained rigid body modes;
2. Geometric instabilities (snap through, and so on.);
3. Material instabilities (softening).

Figures 7.19 and 7.23 show the static instability for an unconstrained rigid body mode and geometric or material instabilities, respectively.

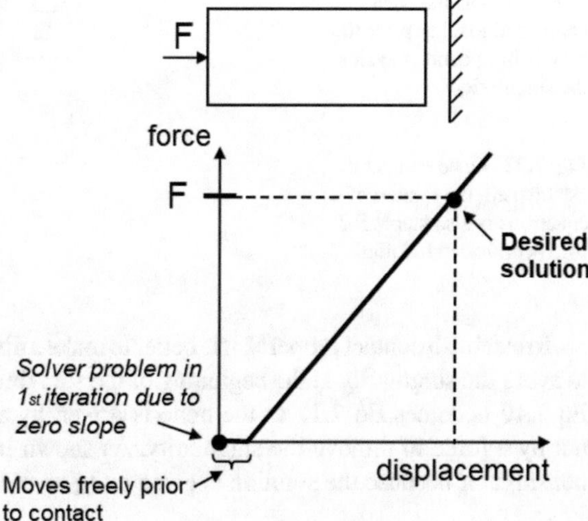

Fig. 7.19 Unconstrained rigid body modes

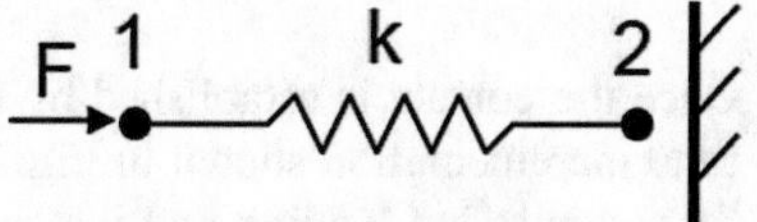

Fig. 7.20 Singular system of equations prior to establishing contact

For an unconstrained rigid body motion as shown in Fig. 7.19, for instance, in many mechanical assemblies, there is a reliance on contact between bodies to prevent unconstrained rigid body motion. Often it is impractical or impossible to model such systems with contact initially established without user intervention. Abaqus may report solver singularities in the message (.msg) file, which also often leads to slow or no convergence.

```
***WARNING: SOLVER PROBLEM.
NUMERICAL SINGULARITY WHEN PROCESSING NODE 17
D.O.F. 2 RATIO = 3.93046E+16
```

Indeed, in Fig. 7.19, there is a contact interaction issue; at the beginning of the time increment a singular system of equations prior to establishing contact makes a move freely prior to establishing contact. It is possible to write such a system of equations based on Eq. 7.7, with the spring stiffness k modeling the contact interaction as follows:

$$\begin{pmatrix} k & -k \\ -k & k \end{pmatrix} \begin{pmatrix} u_1 \\ u_2 \end{pmatrix} = \begin{pmatrix} F \\ 0 \end{pmatrix} \tag{7.10}$$

The reason for the instability is obvious because, as shown in Eq. 7.10, the determinant of the stiffness matrix is zero and this means there is no solution. This is called a singularity in the system (Fig. 7.20).

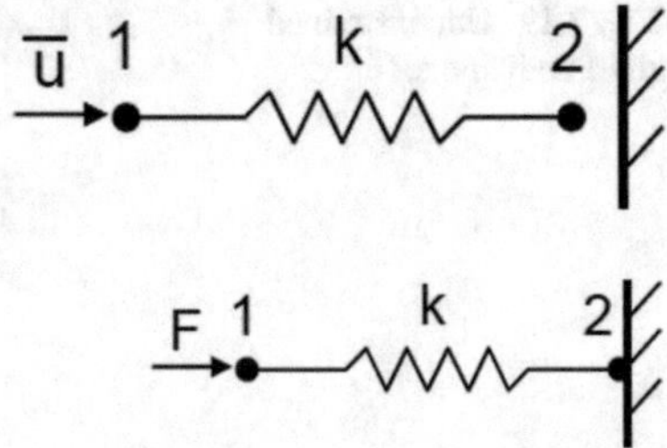

Fig. 7.21 Displacement-controlled loading prior to establishing contact avoids the singularity

Fig. 7.22 Once contact is established, the system of equations is also stable for force-controlled loading

To establish contact properly, it is better to make a displacement-controlled loading to avoid the singularity at the beginning of the step time incrementation. In this case, Eq. 7.10 becomes Eq. 7.11 as the node is driven by a controlled displacement and not by a force to remove the singularity. As shown in Fig. 7.21, the system is now nonsingular because the solution is given by $u_2 = u_1 = \bar{u}$ (Fig. 7.22).

$$(k)\,\{u_2\} = \{ku_1\} \tag{7.11}$$

Once the contact is established by removing the flat curve at the beginning of the time incrementation shown in Fig. 7.19, the system of equations is also stable for force-controlled loading and the system of equation is finally written in Eq. 7.12, which is a nonsingular system with a solution $u_1 = F/k$, $u_2 = 0$ and $\lambda = F$.

$$\begin{pmatrix} k & -k & 0 \\ -k & k & 1 \\ 0 & 1 & 0 \end{pmatrix} \begin{pmatrix} u_1 \\ u_2 \\ \lambda \end{pmatrix} = \begin{pmatrix} F \\ 0 \\ 0 \end{pmatrix} \tag{7.12}$$

This is also true with the penalty enforcement method, where the system of equations written in Eq. 7.10 does not use a Lagrange multiplier but an additional stiffness as shown in Fig. 7.18 with a new system of Eq. 7.13 for a solution equal to $u_1 = F\left(\frac{1}{k} + \frac{1}{k_p}\right)$ and $u_2 = \frac{F}{k_p}$. It can be observed that the more k_p tends to infinity the closer this solution becomes to that presented in Eq. 7.12.

$$\begin{pmatrix} k & -k \\ -k & k + k_p \end{pmatrix} \begin{pmatrix} u_1 \\ u_2 \end{pmatrix} = \begin{pmatrix} F \\ 0 \end{pmatrix} \tag{7.13}$$

Figure 7.23 shows a typical negative eigenvalues issue caused by a nonlinear system, which often experience temporary instabilities associated with a negative tangent stiffness for a particular incremental deformation mode. This is a typical issue that can be identified as a geometric instability (snap through) or a material instability (softening). Without intervention, Abaqus will report negative eigenvalues in the message (.msg) file and often leads to slow or no convergence.

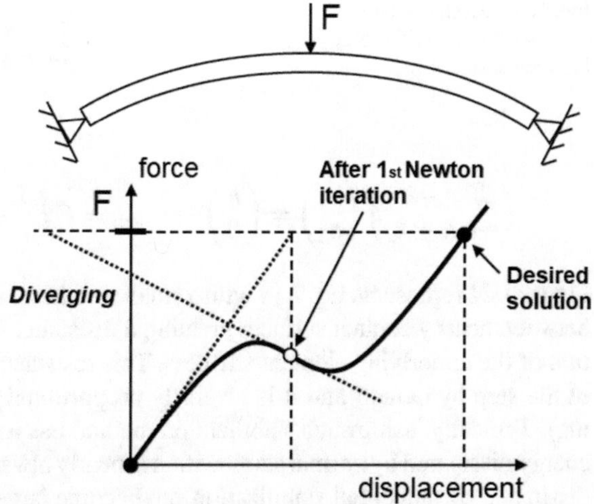

Fig. 7.23 Geometric and/or material instabilities

In both cases shown in Figs. 7.19 and 7.23, an intervention procedure is used to

- Add boundary conditions (e.g., displacement-controlled loading);
- Adjust the initial contact state;
- Add stabilization stiffness (damping);
- Consider inertia effects (dynamic analysis).

7.1.5.5 Stabilization Methods

If the stabilization method is used to fix convergence issues caused by numerical difficulties with contact interaction there are two main options: the first is to use an artificial stiffness to produce a kind of damping effect, and the second—and preferred—option is to define a contact-based stabilization or a volume-based stabilization.

A contact-based stabilization adds a small resistance to relative motion between nearby surfaces while contact constraints are inactive, which is quite effective for stabilizing initial rigid body modes prior to establishing contact.

A volume-based stabilization will create an adaptive stabilization throughout bodies, which is quite effective in terms of overcoming temporary instabilities that sometimes occur during mid-analysis.

Contact-Based Stabilization

Here the principle identical to Eq. 7.13 is used for the penalty enforcement method except that now it is used not to define a contact but to stabilize a contact. Equation 7.13 is, therefore, replaced by the nonsingular system of equations in Eq. 7.14 with a constant stiffness stabilization k_s, where u_2 is used to trigger a contact status change for the next iteration.

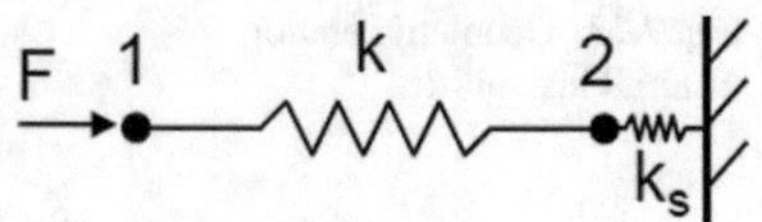

Fig. 7.24 Primarily targets cases with small initial play between surfaces

$$\begin{pmatrix} k & -k \\ -k & k+k_s \end{pmatrix}\begin{pmatrix} u_1 \\ u_2 \end{pmatrix} = \begin{pmatrix} F \\ 0 \end{pmatrix} \rightarrow \quad u_1 = F\left(\frac{1}{k}+\frac{1}{k_s}\right) \quad u_2 = \frac{F}{k_s} \tag{7.14}$$

Figure 7.24 represents Eq. 7.14 with a small resistance to incremental relative motion between nearby contact surfaces creating a resistance (stiffness) that is a small fraction of the underlying element stiffness. This resistance is ramped to zero at the end of the step by default and it is inversely proportional to the increment size (damping). Typically, using such stabilization method has a minimal effect on results; the energy dissipated by normal stabilization is nearly always insignificant but the energy dissipated by tangential stabilization can become large if large sliding occurs.

The analyst can define contact stabilization using different options with a different parameter to set such as

- ***CONTACT CONTROLS, STABILIZE**, use the default damping coefficient.
- ***CONTACT CONTROLS, STABILIZE=factor**, to scale the default damping coefficient.
- ***CONTACT CONTROLS, STABILIZE, damping factor**, to specify the damping coefficient directly.
- ***CONTACT CONTROLS, STABILIZE, ramp-down factor** to specify a non-default ramp- down factor.
- ***CONTACT CONTROLS, STABILIZE, TANGENT FRACTION=value** to decrease or increase the tangential damping or set it to zero.
- ***CONTACT STABILIZATION**, to specify local or global contact stabilization controls. This is the first step-dependent suboption of ***CONTACT** for Abaqus standard. It is not active by default but when activated the built-in settings target temporary, initial unconstrained rigid body modes. These built-in settings do not have tangential stabilization and stabilization is aggressively ramped down over increments.

A special case is presented in Fig. 7.25 as an example of an initially touching surfaces for Surface-to-Surface discretization by considering an average gap greater than zero for each slave node; thus, Surface-to-Surface contact constraints are initially inactive and the initial system of equations would have no resistance to the applied load. Then, the stabilization stiffness is automatically added for such cases (even if the point of touching does not correspond to a node). The stabilization stiffness is zero by the end of the step and is inversely proportional to the increment size.

This special form of automatic stabilization is on by default for the **finite sliding**, **Surface-to- Surface** formulation which cannot be applied to other formulations.

If contact pairs is used then

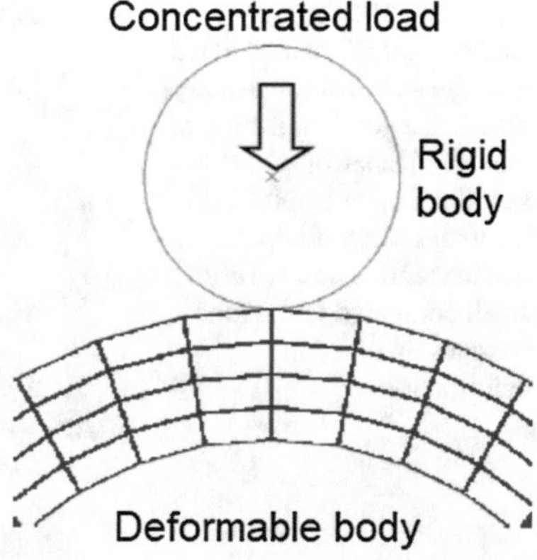

Fig. 7.25 An example of contact stabilization

```
*CONTACT PAIR, TYPE=SURFACE TO SURFACE,
MINIMUM DISTANCE = [YES(DEFAULT)/NO]
```

If general contact is used then

```
*CONTACT INITIALIZATION DATA, NAME=xyz,
MINIMUM DISTANCE = [YES(DEFAULT)/NO]

*CONTACT
*CONTACT INCLUSIONS
*CONTACT INITIALIZATION ASSIGNMENT
, , xyz
```

Volume-Based Stabilization

***STATIC, STABILIZE**, also referred to as static stabilization, with a volume proportional damping targeting local dynamic instabilities. It is applicable to the following quasi-static procedures in static, visco, coupled temperature displacement, and soils consolidation. The damping term used in the equilibrium equation is shown in Eq. 7.15, where $\dot{u}$ is the quasi velocity, M^* is the mass matrix with unit density, and c is the damping factor.

$$cM^*\dot{u} + I(u) = P \tag{7.15}$$

The effect of Eq. 7.15 on the equations solved in each Newton–Raphson iteration can be written as follows in Eq. 7.16:

$$\left[K_t + \left(\frac{c}{\Delta t}M^*\right)\right]\partial u = R - \left(cM^*\frac{\Delta u}{\Delta t}\right) \tag{7.16}$$

With an automatic selection of the damping factor, Abaqus automatically calculates the damping factor c. This can vary in space and with time and can be adaptive based on convergence history and the ratio of energy dissipated by viscous damping to

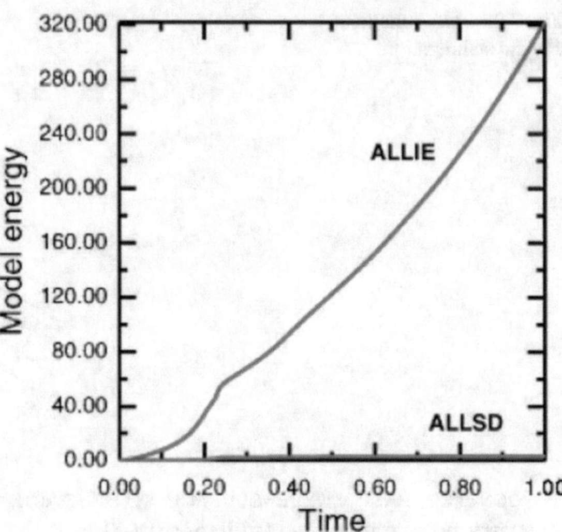

Fig. 7.26 The amount of energy dissipation associated with the stabilization usually provides a good indication of the significance of stabilization on results. Here, the total energy dissipated due to stabilization is very small compared to the total energies involved in deformation

the total energy. Possessing an initial damping factor that is based on the following premises, the model response in the first increment of a step to which damping is applied is stable but not particularly effective for stabilizing unconstrained rigid body modes at the beginning of an analysis. Under stable circumstances, using a damping factor produces an energy ratio where the amount of dissipated energy **ALLSD** is a minimal fraction of the total energy **ALLIE** as shown in Fig. 7.26.

Another approach for overcoming static instabilities is to use a dynamic procedure, where inertia is inherently stabilizing with the equation of motion, where Eq. 4.4 becomes Eq. 7.17. Abaqus provides implicit and explicit dynamics procedures as already explained in Sect. 4.8 specifically for contact in Sect. 4.8.5, or for materials in Sect. 4.8.6.

$$M\ddot{u} + C\dot{u} + I\,(u) = P \tag{7.17}$$

7.1.5.6 Overconstraints

A direct consequence of overconstraining a model implies the Lagrange multipliers that impose contact constraints are indeterminate when the node is overconstrained; therefore, analyses will typically fail in such cases. This situation occurs when multiple kinematic (boundary condition, contact, or MPC) constraints act in the same direction on the same node and may be caused by a single slave node interacting with a number of different master surfaces from different contact pairs. An example of this situation is given in Fig. 7.27.

There are numerous situations when contact interactions in combination with other constraint types may lead to overconstraints, one is shown in Fig. 7.27. Since contact status typically changes during the analysis, it is not possible to detect redundant constraints associated with the contact in the model preprocessor. Instead, these

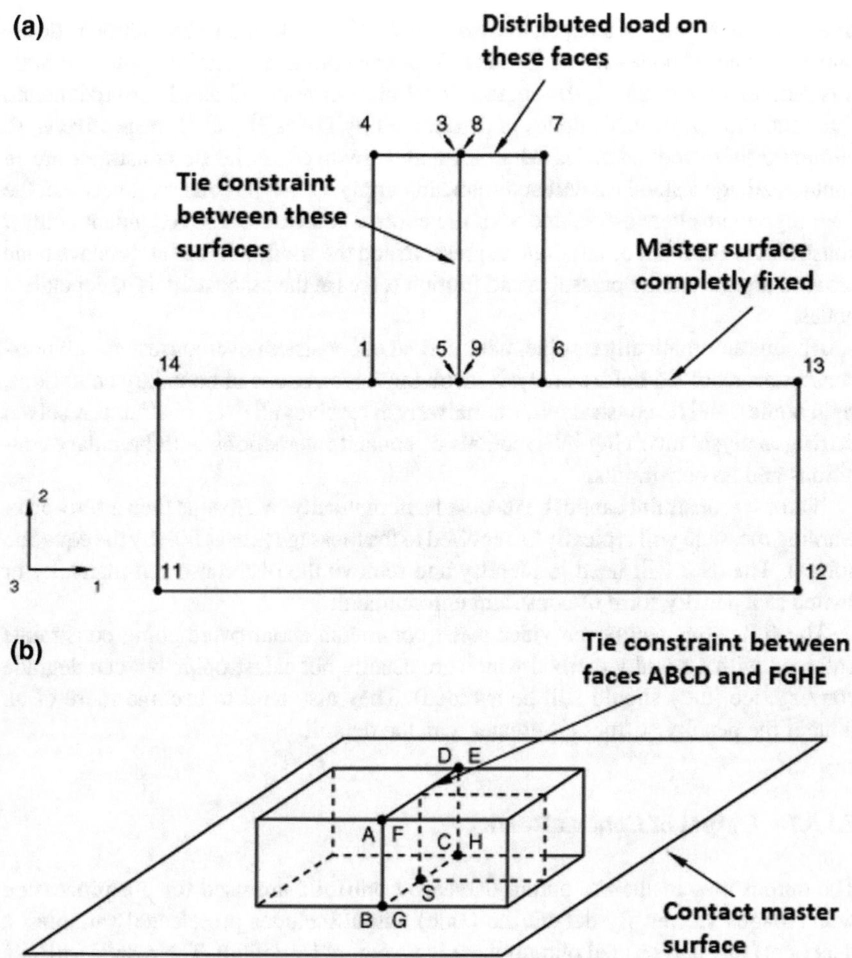

Fig. 7.27 Overconstraints by a single slave node in contact interaction, and with redundant constraints arising from contact interactions and tie constraints. Figure used by permission ©Dassault Systemes Simulia Corp

checks are performed during the analysis. Due to the complexities associated with contact interactions, only a limited number of redundant constraint cases are resolved automatically.

Redundant constraints are common in cases when slave nodes used in surface-based tie constraints are also slave nodes in contact as illustrated in Fig. 7.27. In Fig. 7.27a, nodes 5 and 9 are connected with a tie constraint, and both are in contact with a master surface. Since the two nodes are tied together, one of the contact constraints is redundant. A similar situation is presented in Fig. 7.27b: two mismatched solid meshes are connected with a tie constraint, and contact is defined with a flat rigid surface. Node S is a dependent node in the tie constraint, so its motion is determined by that of nodes B and C. Therefore, any contact constraint applied at node S is redundant. Moreover, the contact constraints at nodes G and H are redundant, since the motion of these nodes is determined by nodes B and C, respectively. To eliminate these redundancies when all nodes involved in the tie constraint are in contact, Abaqus/Standard will automatically apply a tie-type constraint between the Lagrange multipliers associated with the contact constraint. The redundant contact constraint is eliminated. The contact pressure and the friction forces at the slave node are recovered from the pressures and friction forces at the associated tie-independent nodes.

Abaqus automatically resolves a limited set of consistent overconstraints, all overconstraints resolved **before** analysis involving intersections of boundary conditions, rigid bodies, and tie constraints. Alternatively, it resolves all overconstraints resolved **during** analysis involving intersections of contact interactions with boundary conditions and tie constraints.

If an overconstraint cannot be resolved automatically by Abaqus then a zero-pivot warning message will typically be reported to the message (.msg) file (by the equation solver). The user will need to identify and remove the overconstraint manually or switch to a penalty form of constraint enforcement.

The following section provides some comments about overlapping constraints enforced with a penalty method which are usually not catastrophic but can degrade convergence (they should still be avoided). They also tend to become more of an issue if the penalty stiffness is greater than the default.

7.1.5.7 Output of Contact Results

The output files inside the output database (.odb) file are used for postprocessing with Abaqus viewer. By default the (.odb) output includes preselected variables, a data (.dat) file, and printed output; there is no output by default. The results (.fil) file is used for postprocessing with third-party postprocessors, with no output by default. The output variable types stored nodal variables with all surface variables.

The nodal output to the (.odb) file is set by default with nodal contact output to (.odb) file and includes the following variables the contact stresses **CSTRESS**, the contact pressure **CPRESS**, the frictional shear stresses **CSHEAR1** and **CSHEAR2**, the contact displacements **CDISP**, the contact openings **COPEN**, and the accumulated relative tangential motions with both **CSLIP1** and **CSLIP2**. **CSHEAR2** and **CSLIP2** are provided only in three- dimensional problems. All outputs are available as both field and history data.

Additional nodal output to the (.odb) file include the contact nodal force vectors **CFORCE** with **CNORMF** and **CHEARF**, the nodal areas associated with active contact constraints **CNAREA**, and the contact status **CSTATUS**, which enables contour plots of sticking, slipping, or open status.

The self-contact results have values of **CPRESS**, **CSHEAR**, **CNORMF**, and **CSHEARF** in the output database file and represent net quantities with the contributions while a node acts as a slave in some constraints and as a master in other constraints for a given self-contact definition.

The small sliding is the contact area always based on **reference configuration**, regardless of whether or not geometrically nonlinear effects are considered.

The finite sliding is the contact area always based on the **current configuration**, regardless of whether or not geometrically nonlinear effects are considered.

There are two options available for generating a printed output that is relevant to the contact analyses:

1. ***PREPRINT, CONTACT=YES**, controls output to the printed output (.dat) file during the preprocessing phase and gives details of internally generated contact elements.
2. ***PRINT, CONTACT=YES**, controls output to the message (.msg) file during the analysis phase and gives details of the iteration process.

With the contact stress error indicator **CSTRESSERI** which is a nodal variable similar to **CSTRESS** in the field variable output and which cannot be used to drive adaptive remeshing then the Art of interpreting error indicators can be defined with the following warning:

- Error indicator output variables are approximate and do not represent an accurate or conservative estimate of the solution error. The quality of an error indicator can be particularly poor if the user's mesh is coarse. The error indicator quality improves as the mesh is refined; however, these variables should never be interpreted as indicating what the value of a solution variable would be upon further refinement of the mesh.
- Error indicators do not replace the need to perform mesh refinement studies or other ways that analysts can gain confidence in modeling practices.

7.1.5.8 Best Practices for Treating Initial Overclosures

First of all, the analyst should check any initial overclosure in the model in order to determine whether the initial overclosure intended is seen as an interference fits or unintended.

Indeed, the common causes of initial overclosure intended include modeling interference fit in Abaqus standard or unintended issues such as the shell thickness not being accounted for in the preprocessor, preprocessor errors or a discretization of curved surfaces without geometry corrections as shown in Fig. 7.28.

To avoid such unintended initial gaps like in Fig. 7.28, there is a technique known in contact pairs. The adjustment zone in contact settings is used to define a zone with the **ADJUST** parameter (a) to set a zone between the master and slave surface where the slave nodes will lie down on the master surface. This is demonstrated in Fig. 7.29.

```
*CONTACT PAIR, INTERACTION=myProp,
ADJUST=a
```

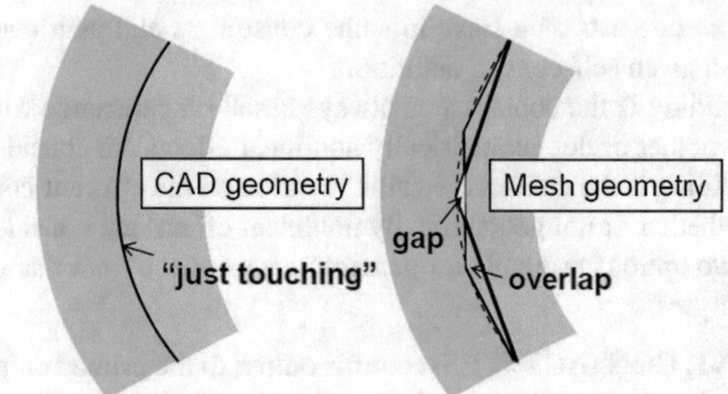

Fig. 7.28 Discretization of curved surfaces creates an initial overclosures

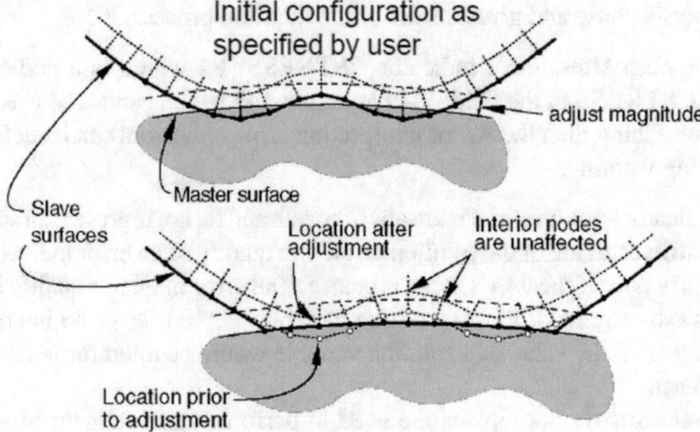

Fig. 7.29 Configuration after adjustment and prior to start of analysis: slave nodes outside adjust bands are unaffected. There are some exceptions for a Surface-to-Surface formulation

The user can also define the same adjustment zone in the contact interface for general contact in Abaqus standard using the command **SEARCH ABOVE**.

```
*Contact Initialization Data, name=adjust-1,
SEARCH ABOVE=1.E-5
*Contact Initialization Assignment
allHeads, myPart.outer, adjust-1
```

General contact in Abaqus standard and Abaqus explicit treats an initial overclosures (within a given tolerance) with a strain-free adjustment by default. All overclosures greater than the specified tolerance are ignored and, alternatively, in Abaqus standard overclosures can be treated as interference fits that are gradually resolved over the first step.

For contact pairs in Abaqus standard, an initial overclosures is treated as interference fits by default. It resolves all interference in the first (i.e., a single) increment but

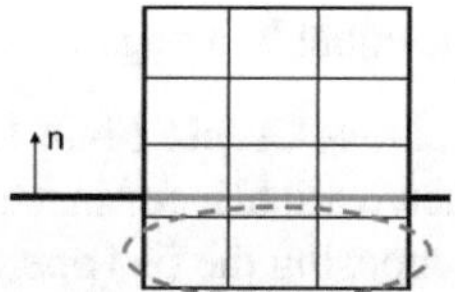

Fig. 7.30 Slave surface nodes with element inversion where a negative volume will occur after strain- free adjustments. **n** is the normal vector to the master surface in contact interaction

can cause convergence difficulties because the loading does not scale with the increment size. Alternatively, overclosures can be resolved gradually or via strain-free adjustments.

Strain-Free Adjustments

The output variable **STRAINFREE** contains nodal vectors representing initial strain-free adjustments. By default, this output variable is written to the output database (.odb) file for the original field output frame at zero time if any strain-free adjustments are made by Abaqus standard.

General contact in Abaqus standard includes a contact initialization by default, which removes small initial overclosures via strain-free adjustments. The default tolerance is based on the size of underlying element facets. The initial gaps remain unchanged by default adjustments. It is optional for large initial overclosures and initial gaps which can also be adjusted by specifying search distances above and below the surfaces a search above to close gaps (as discussed previously), or even a search below to increase the default overclosure tolerance as shown in the command lines below.

```
*Contact Initialization Data,
name=Init-1,
SEARCH ABOVE=distance,
SEARCH BELOW=distance
*Contact Initialization Assignment
, , Init-1
```

A typical warning message involving only slave surface nodes which are relocated for gross (large) adjustments can severely distort initial element shapes. In cases such as the example shown in Fig. 7.30, the user should rely only on strain-free adjustments to resolve small initial overclosures (relative to element dimensions).

The nodal output variable called **STRAINFREE** is provided to visualize strain-free adjustments in Abaqus standard. The output variable is written by default if any initial strain-free adjustments are made and the variable is only available for the initial output frame at $t = 0$. The following procedure is given to visualize strain-free adjustment:

1. Create a field output variable equal to a negative STRAINFREE:
   - Go to Abaqus viewer **tools** → **Create Field Output** → **From Fields**.
   - Choose a name for the new variable, for instance, negStrainfree here.
   - Enter the expression by choosing the (−) operator and **STRAINFREE** in the output variable.
2. View the deformed plot based on this variable.
   - Abaqus viewer **Result** → **Step/Frame** → Choose the **Session Step**.
   - Create a deformed plot with the new variable driving the displacements.

The configuration that appears in the plot is in accordance with Eq. 7.18 as the sum of the configuration with strain-free adjustments x_0 plus a net effect to subtract strain-free adjustments with the negStrainfree.

$$x = x_0 + \text{negStrainfree} \tag{7.18}$$

The following inconsistency exists between Abaqus standard and Abaqus explicit with respect to strain-free adjustments between the standard adjustment x_0 and the explicit adjusts u according to Eq. 7.19 and given in Table 7.3.

$$x = x_0 + u \tag{7.19}$$

Interference Fit

In general contact set with Abaqus standard, the general contact algorithm can treat initial overclosures as interference fits which use a shrink fit method to resolve the interference gradually over the course of the first analysis step to generate stresses and strains.

```
*Contact Initialization Data, name=Fit-1,
INTERFERENCE FIT
*Contact initialization Assignment
Surface_BUMPER-EXT, Surface_SHAFT, Fit-1
```

Table 7.3 Technique in Abaqus viewer between the difference with strain-free adjustment used in Abaqus standard or in Abaqus explicit models

Desired aspect to visualize	Abaqus/Standard	Abaqus/Explicit
Nodal adjustment vectors	Symbol plot of STRAINFREE at $t = 0$	Symbol plot of U[a] at $t = 0$
Nodal adjustment magnitudes	Contour plot of STRAINFREE at $t = 0$	Contour plot of U[a] at $t = 0$
Adjusted configuration	Undeformed shape or deformed shape at $t = 0$	Deformed shape at $t = 0$
Configuration prior to adjustments	Substitute STRAINFREE for U[a] in deformed plot $t = 0$	Undeformed shape

[a]Displacement U

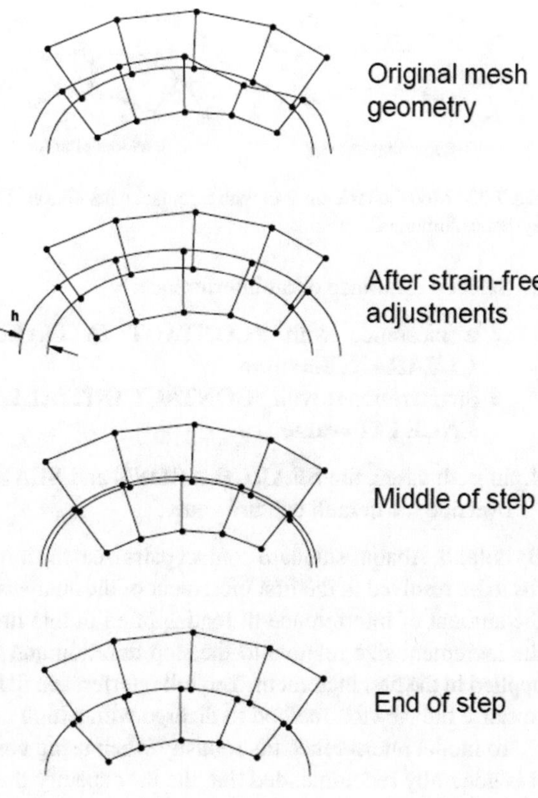

Fig. 7.31 User-specified interference and clearance distance for a general contact in Abaqus standard between two bodies, where h is the desired interference fit distance. Figure used by permission ©Dassault Systemes Simulia Corp

The user-specified interference and clearance distance for a general contact in Abaqus standard with no equivalent to the procedure is described below and shown in Fig. 7.31 for contact pairs.

1. The original mesh required does not reflect the desired interference or clearance distance.
2. Strain-free adjustments are used to achieve the user-specified interference clearance distance. The user is aware that a large adjustment may cause element distortion problems.
3. This is followed by a shrink fit during the first step to resolve interference and therefore generate stress and strain.
4. Both surfaces which are in the interference fit will appear compliant at the end of the first step (aside from penalty penetration).

Using the keyword interface, the user shall:

1. Assign a contact initialisation method using the ***CONTACT INITIALIZATION ASSIGNMENT** option,
2. Specify the clearance or interference distance with the ***CONTACT INITIALIZATION DATA** option,

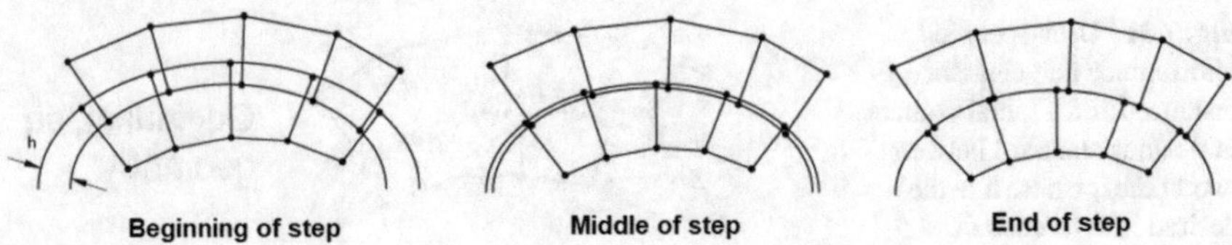

Fig. 7.32 Model interference fit with a contact pairs option. Figure used by permission ©Dassault Systemes Simulia Corp

3. Select a clearance or an interference:
   - a clearance with ***CONTACT INITIALIZATION DATA, INITIAL CLEARANCE=value**,
   - an interference with ***CONTACT INITIALIZATION DATA, INTERFERENCE FIT=value**
4. In both cases, the **SEARCH ABOVE** and **SEARCH BELOW** parameters can override the default capture zone.

By default, Abaqus standard contact pairs treat the initial overclosures as interference fits to be resolved in the first increment of the analysis. However, with this approach, the amount of interference fit load applied in this first increment is independent of the increment size relative to the step duration and the full interference fit load is applied in the first increment. The full interference fit load is sometimes large enough to cause the Newton method to diverge with a high nonlinear response.

To model interference fits robustly when using contact pairs in Abaqus standard, it is generally recommended that the user specify the shrink fit option such that the interference fit can be resolved over multiple increments within the first step to obtain the configuration demonstrated in Fig. 7.32.

```
*CONTACT INTERFERENCE, SHRINK
slave, master
```

Modeling an interference distance that differs from the initial mesh overclosure with contact pairs is a tricky combination of options; it is awkward, confusing, and not as accurate compared to the method used in general contact. Indeed, in general contact the process is as simple as

1. Strain-free adjustments to zero penetration using the ADJUST parameter.
2. Ramp allowed interference from 0.0 to h in the first step, where h is the desired interference fit distance. Using the contact interference option, it will appear as if a gap of distance h exists between the surfaces at the end of the first step even though contact constraints are active.

The interference fits and the surface-to-surface contact discretization mean that normal constraints are applied along the directions of slave surface normals. If penetration is deeper than the element size, the user may need to use the node-to-surface formulation instead.

7.1.5.9 Summary

Good formulation characteristics regarding accuracy, robustness, and generality can be summarized as

- Accurate representation of surface geometry
  - Slave surface: Not just a collection of points. **Surface to Surface**
  - Master surface: Not approximated as flat per slave node. **Finite sliding**
  - Geometric corrections to reduce discretization error. **Surface to Surface**
- The distribution of nodal forces is consistent with underlying element formulation, which is the ability to satisfy patch tests for contact. **Surface to Surface**
- Continuity in contact forces upon sliding. **Surface to Surface**
- Individual constraint stresses should oppose penetration (and sliding), This is a nontrivial aspect for some quadratic element types. **Surface to Surface**
- Avoid overconstraints and under constraints. Generally, the number of contact constraints in an active contact region should equal the number of nodes of the more refined surface in that region. **Master and Slave roles**
- Small amount of numerical softening. **Penalty method**
- Robust contact search algorithm to avoid missing contacts, and so on. **Finite sliding**
- Special treatment of feature edges. **Edge to Surface**.

Table 7.4 lists the available formulations for general contact and contact pairs in Abaqus standard.

The most common issues when converting contact are listed as follows:

- Most issues are related to **initial overclosures**.
- General contact accounts for shell/membrane thickness and the finite sliding. The Node-to-Surface contact pairs do not make an initial penetration if shell thickness is considered.
- General contact typically considers all exposed surfaces, and the contact pairs may not be defined on some penetrated regions.
- Different default treatment of initial overclosures:

Table 7.4 Modeling approach

Formulation aspect	General contact	Contact pairs
Contact discretization	Primary: Surface to Surface Suppl.: Edge to Surface	Default: Node to Surface Optional: Surface to Surface
Contact enforcement	Default: penalty Optional: direct	Node to Surface default: direct Surface to Surface default: penalty
Constraint evolution upon sliding	Finite sliding	Default: finite sliding Optional: small sliding approx.

  - The contact pairs make initial overclosures, which are treated as interference fits by default.
  - The general contact resolves small initial overclosures with strain-free adjustments or large initial overclosures are assumed to be nonphysical/unintended.

- The initial overclosure is an issue determined by the user:

  - The user is responsible for directing the treatment of initial overclosures by choosing whether to resolve them with or without strains.
  - Common characteristics of interference fits, where the overclosure distance may be large or limited to specific interfaces. This requires pair-wise attention from the user.
  - The strain-free adjustments are intended to resolve small overclosures (e.g., due to faceted representation of curved surfaces). For small overclosures, the automated algorithm can determine which nodes to move and where to move them.

Some convergence issues are related to the following topics:

- Newton iterations, radius of convergence, and incrementation.
- Diagnostics output, which is helpful for determining the location and cause of convergence problems.
- Changes in contact status (open/closed and slip/stick) are characterized as severe discontinuities by the iteration-control algorithm:

  - Strict enforcement with a change from no-contact stiffness to an infinite stiffness.
  - Penalty enforcement with change from no contact stiffness to finite stiffness (less severe).

- Smooth contact formulation characteristics enhance convergence (e.g., continuity in nodal contact forces upon sliding), with Surface-to-Surface contact discretization being smoother than Node-to-Surface contact discretization.
- Smooth (and more accurate) representation of curved surfaces and accounting for nonsymmetric stiffness terms in the equation solver is also helpful for convergence.

Unintentionally having this bad combination of features is quite common, so the following suggestions are provided to prevent this problematic combination of features:

- Surface-to-surface enforcement and penalty method are generally recommended.
- Somewhat neutral on element-type recommendation, but, for example, C3D10(I) gives a more accurate representation of curved surfaces than C3D10M.
- Default penalty stiffness is a factor of 10–100 higher in Abaqus standard:

  - Increasing penalty stiffness tends to reduce the time increment size in Abaqus explicit.
  - Increasing penalty stiffness tends to degrade convergence behavior in Abaqus standard.

- With surface thickness reductions, Abaqus may automatically reduce the contact thickness associated with structural elements to avoid issues of self-intersection.

If the thickness is reduced, a warning is produced in the status file along with element set **WarnElemGContThickReduce**. Reducing the contact thickness of a surface may mean that contact occurs later than expected, for instance, with a pinched shell. Use output variable **CTHICK** to contour the actual shell thickness used for general contact.

- With surface erosion,[1] the nodes attached only to eroded elements are by default treated as point masses that can experience contact with intact facets, meaning that some additional momentum is transferred and they do not interact with other such nodes. Alternatively, the user can specify ***CONTACT CONTROLS ASSIGNMENT, NODAL EROSION=YES**. In this case, nodes are excluded from contact interaction. The output variable **STATUS** indicates whether or not an element has failed; (STATUS = 0) for failed elements or (STATUS = 1) for active elements. Subsequently, Abaqus viewer will automatically remove failed elements when the output database file includes STATUS.
- Abaqus explicit is not well suited for modeling interference fits so Abaqus standard is preferable.
- Contact overclosures present in the first step are resolved with strain-free adjustments by default; adjustments are made to nodal displacements in Abaqus explicit.

Some comments are made here about diagnostic interpretation in relation to contact issues:

- A resolution of an initial overclosures can be identified in Abaqus viewer with the symbol (vector) plots of displacements (U) at time=0.0 used in contour plots of displacements (U) at time=0.0. To identify the automatically generated node sets regarding an adjusted node **InfoNodeOverclosureAdjust**, or for some nodes with unresolved initial overclosures **InfoNodeUnresolvInitOver**.
- Initial crossed surfaces generally indicate the geometry is wrong. The diagnostic output provides a view element set **WarnElemSurfaceIntersect** for use in the Display Group dialog box. Any wrong geometries in the FEA model should be manually avoided otherwise the surfaces will remain locked together for the duration of analysis.

Abaqus provides two alternatives regarding contact behavior. First, a physical pressure versus overclosure, it can be defined in the FEA model a softened contact (Exponential, Linear, or Tabular) which is motivated by physically based (surface coatings) or a numerical based to improve converge. Second, a contact without separation. There are other features influencing overall contact constitutive behavior

[1]In Abaqus explicit, the user can define the facets of a surface on the interior of a solid element mesh. The faces of the specified elements that are not on the exterior (free) surface of the model will be included in the surface definition. For example, interior surfaces are used with the general contact algorithm in Abaqus explicit for modeling surface erosion due to element failure. The automatic generation of an interior surface is equivalent to constructing a surface consisting of all faces of the elements and then subtracting the free surfaces of those elements. Shell elements, beam elements, pipe elements, membrane elements, and so on are ignored since they do not have any interior faces by definition. Multi-point constraints are not taken into account when generating interior surfaces. This can result in faces that are on the interior of a body being excluded from the surface definition.

like a breakable bond, surface-based cohesive behavior, or crack propagation along a contact interface including normal and tangential behavior. Another aspect is the usage of specific behavior with a user subroutine **UINTER**, which also controls tangential behavior programmed by the user.

A contact without separation is useful for modeling adhesives. This feature causes surfaces to be bonded for the duration of analysis once contact is established only normal contact is affected with relative sliding still allowed and it is often used with the rough friction option (no sliding either). This option is sometimes used for numerical motivation to improve convergence.

```
*SURFACE INTERACTION
*SURFACE BEHAVIOR, NO SEPARATION
```

7.2 Friction

Friction is the force resisting the relative motion of solid surfaces, fluid layers, and material elements sliding against each other. There are several types of friction:

- **Dry friction** is a force that opposes the relative lateral motion of two solid surfaces in contact. Dry friction is subdivided into static friction (stiction) between non-moving surfaces and kinetic friction between moving surfaces. With the exception of atomic or molecular friction, dry friction generally arises from the interaction of surface features known as asperities.
- **Fluid friction** describes the friction between layers of a viscous fluid that are moving relative to each other.
- **Lubricated friction** is a case of fluid friction where a lubricant fluid separates two solid surfaces.
- **Skin friction** is a component of drag, the force resisting the motion of a fluid across the surface of a body.
- **Internal friction** is the force resisting motion between the elements which constitute a solid material while it undergoes deformation.

There are three types of friction model available in Abaqus:

1. Coulomb friction
    - Isotropic or anisotropic;
    - Optional friction coefficient dependence on slip rate, pressure, temperature, and field variables; with linear interpolation of tabular data or exponential dependence on slip rate or even with a customized user subroutine **FRIC_COEF**.
2. Rough friction, which is sticking regardless of contact pressure as long as the normal contact constraint is active.
3. User defined through user subroutine **FRIC** or **UINTER**

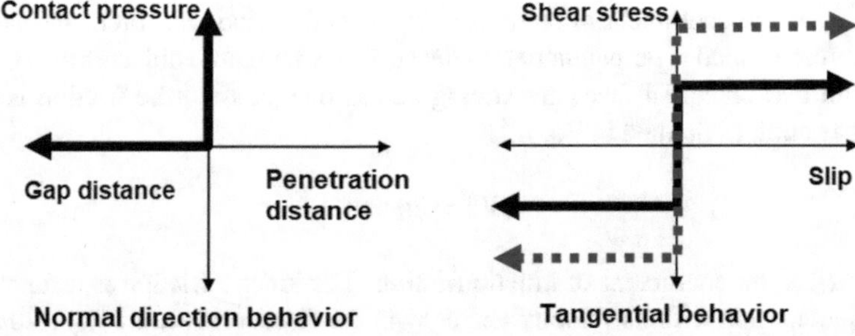

Fig. 7.33 Constraints enforced with the Lagrange multiplier method

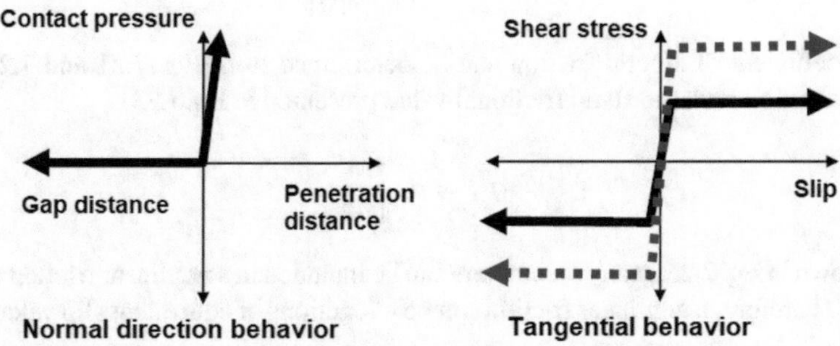

Fig. 7.34 Constraints enforced with the penalty method. The dashed line shows the tangential stress response when the contact pressure increases in a normal direction behavior

The stick or slip discontinuity for friction is similar to open or closed discontinuity in the normal direction. Figures 7.33 and 7.34 depict both normal and tangential friction behavior for two different types of constraints with a Lagrange multiplier and the penalty method, respectively.

The shear stress function of slip has in both cases a dependency on contact pressure (the dotted line) with the penalty method used by default called here the stick stiffness. The shear stress τ, the yielded function of the normal stress σ, and the friction coefficient μ such as $\tau = \mu\sigma$.

The Lagrange multiplier method can cause overconstraint problems, for instance, at junctions. Overlapping strict constraints cause problems for the equation solver.

7.2.1 Static and Kinetic Friction

When two object surfaces are not sliding across one to another, the friction is called static friction as defined in Eq. 7.20.

$$\|\mathbf{F}\| = \mu_s \|\mathbf{N}\| \tag{7.20}$$

where μ_s is the coefficient of static friction, F is the frictional force, and N is the normal force which is perpendicular to the surface where both objects are in contact.

When two object surfaces are sliding across one another, the friction is called kinetic friction as defined in Eq. 7.21.

$$\|\mathbf{F}\| = \mu_k \|\mathbf{N}\| \tag{7.21}$$

where μ_k is the coefficient of kinetic friction. The kinetic friction is a function of the object motion with its velocity vector $\mathbf{v_t}$ in the direction vector $\mathbf{t}$ Eq. 7.20 can be rewritten as

$$\|\mathbf{F_t}\| = \mu_s \|\mathbf{N}\| \frac{\mathbf{v_t}}{\|\mathbf{v_t}\|} \tag{7.22}$$

The coefficient of kinetic friction can be determined from Eqs. 7.21 and 7.22 as a yield function with the static frictional value presented in Eq. 7.23.

$$\mu_k = \mu_s \frac{\mathbf{v_t}.\mathbf{t}}{\|\mathbf{v_t}\|} \tag{7.23}$$

As shown in Eq. 7.23, many parameters can be included in a nonlinear friction coefficient. Therefore, a nonlinear friction can be a functions of equivalent slip velocity, in plane $\dot{\gamma}_{eq} = \sqrt{\dot{\gamma}_1^2 + \dot{\gamma}_2^2}$; of the contact pressure, p; of an average surface temperature, $\bar{T} = \frac{1}{2}(T_A + T_B)$; and a function of an average field variable value, $\bar{f}_i$.

For linear interpolation of tabular data, if μ is a function of field variables, the dependencies parameter must be used on the ***FRICTION** option to specify the number of field variable dependencies.

User subroutines **FRIC_COEF** and **VFRIC_COEF** allow the user to specify an expression for the friction coefficient in Abaqus standard and also provide expressions for derivatives. For example, if the friction coefficient is user defined such as $\mu = A\left(1 + B\dot{\gamma} + C\dot{\gamma}^2\right)(1 + Dp)$ then the user subroutine coded in Fortran is given below with the FRIC.f file.

Listing 7.1 FRIC.f

```
subroutine fric_coef(fCoef,fCoefDeriv,
* nBlock,nProps,nTemp,nFields,jFlags,rData,
* surfInt,surfSlv,surfMst,props,slipRate,
* pressure,tempAvg,fieldAvg)

include "aba_param.inc"
dimension fCoefDeriv(3)
parameter (one=1.d0,two=2.d0)

fs=one+props(2)*slipRate+props(3)*slipRate**2
fp=one+props(4)*pressure

fCoef=props(1)*fs*fp
```

```
      fCoefDeriv(1)=props(1)*(props(2)+
     1              two*props(3)*slipRate)*fp
      fCoefDeriv(2)=props(1)*fs*props(4)
      fCoefDeriv(3)=zero

      return
      end
```

where :

fCoef is the coefficient of friction,
fCoefDeriv(1) is the first derivative $\partial \mu / \partial \dot{\gamma}$,
fCoefDeriv(2) is the first derivative $\partial \mu / \partial p$,
and **fCoefDeriv(3)** is the first derivative $\partial \mu / \partial T$. This derivative is equal to zero and it indicates the friction is not dependent on the temperature, which remains constant whatever the friction value,
props(1), **props(2)**, **props(3)**, and **props(4)** are constants A,B,C and D respectively.

Once the user subroutine is coded to call it into the FEA model, the user will have to use the following command lines:

```
*SURFACE INTERACTION, NAME=name
*FRICTION, USER=COEFFICIENT, PROPERTIES=4
A, B, C, D (substitute real numbers)
```

The kinetic friction model is a specific form of friction coefficient versus slip rate in Abaqus an exponential transition from a static friction coefficient μ_s to a kinetic friction coefficient μ_k can be made according to Eq. 7.24, where d_c is the decay coefficient.

$$\mu = \mu_k + (\mu_s - \mu_k)\, e^{-d_c \dot{\gamma}_{eq}} \tag{7.24}$$

There are two methods for defining this model. First, the user should provide the static, kinetic, and decay coefficients directly with the following command lines:

```
*SURFACE INTERACTION
*FRICTION, EXPONENTIAL DECAY
```

Or use test data to fit the exponential model.

The Rough friction is an optional behavior in which sticking conditions are always enforced while surfaces are in contact (i.e., while normal constraints are active). It is similar to Coulomb friction with an infinite friction coefficient. However, if **NO SEPARATION** behavior is also specified, it will resist relative motion even if normal contact forces are tensile. Although a rough friction idealized model has zero slip while in contact, a small amount of slipping may occur due to numerical softening (for penalty enforcement of the sticking condition). Motivation for using rough friction may be physical or numerical (avoid convergence problems).

```
*SURFACE INTERACTION, NAME=name
*FRICTION, ROUGH
```

7.2.2 Change Friction Properties During an Analysis

If the analyst needs to change friction properties during an analysis, then the updated friction will depend on what solver is defined.

Abaqus explicit will assign a different named grouping of contact properties. The friction model is one part of a contact property grouping. The library of named groupings of contact properties is then constructed, for instance, for **Property grouping i** to **Property grouping j**; now, in Step 1 **Surface pairing k** is defined and in Step 2 **Surface pairing k**.

The Abaqus explicit solver will make an assignment of contact property grouping (or surface interaction), that is, step dependent, meaning that in Step 1 **Surface pairing k** will be assigned to **Property grouping i** and in Step 2 **Surface pairing k** will be assigned to **Property grouping j**.

Abaqus standard will modify the contact property grouping already assigned. In the model definition. **Property grouping i** is defined with **Surface pairing k**. In the Step module, there are no contact changes for Step 1 but contact is changed in Step 2; next, in Step 2 the solver will modify the friction model in **Property grouping i**. There is a very limited step dependence per contact property grouping (surface interaction) in Abaqus standard. To make such changes with friction properties, the user needs to use the following command lines:

```
*CHANGE FRICTION, INTERACTION=name
*FRICTION
```

The following is an example of what can be changed with a friction coefficient defined (most common) and gradually ramped from the old value to the new value over the step increments for most step types. In the next step, there is a slip tolerance associated with penalty enforcement of stick conditions (uncommon). With the altered friction, in most cases, the slip tolerance transition uses the same ramping behavior as friction coefficient transitions.

7.2.3 Classic Friction Values

Without any information or specification documents about the state of the surface in contact with or without lubricant, it is always difficult to set a correct coefficient of friction value. Of course, the most conservative value is zero but this is not at all

Table 7.5 Classic static and kinematic friction values function of the type of materials in contact [2]

Material 1	Material 2	μ_s DC[a]	μ_s L[b]	μ_k DC[a]	μ_k L[b]
Aluminum	Steel	0.61		0.47	
Aluminum	Aluminum			1.5	
BAM[c]	TiB_2[f]	0.04–0.05	0.02		
Brass	Steel	0.35–0.51	0.19	0.44	
Cast iron	Copper	1.05		0.29	
Cast iron	Zinc	0.85		0.21	
Concrete	Rubber	1.0	0.30[g]	0.6–0.85	0.45–0.75[g]
Concrete	Wood	0.62			
Copper	Glass	0.68			
Copper	Steel	0.53		0.36	
Glass	Glass	0.9–1.0		0.4	
HSF[d]	Cartilage		0.01		0.003
Ice	Ice	0.02–0.09			
Polyethene	Steel	0.2	0.2		
PTFE[e]	PTFE[e]	0.04	0.04		0.04
Steel	Ice	0.03			
Steel	PTFE[e]	0.04–0.2	0.04		0.04
Steel	Steel	0.74–0.80	0.16	0.42–0.62	
Wood	Metal	0.2–0.6	0.2[g]		
Wood	Wood	0.25–0.5	0.2[g]		

[a]DC for Dry and Clean
[b]L for Lubricated
[c]Ceramic alloy $AlMgB_{14}$
[d]HSF for an Human Synovial Fluid
[e]Teflon
[f]Titanium boride
[g]For a wet state of surface

realistic even in the case of two ice cubes in contact the friction is not zero. Under certain conditions, some materials have very low friction coefficients. An example is (highly ordered pyrolytic) graphite, which can have a friction coefficient below 0.01 [1]. This ultralow friction regime is called super lubricity (Table 7.5).

7.3 Hard or Soft Contact

An hard contact uses penalty constraint enforcement and a soft contact uses linear and exponential pressure–overclosure relations. Hard contact has been defined in order to optimize the reaction force response function of the gap as shown in Fig. 7.35.

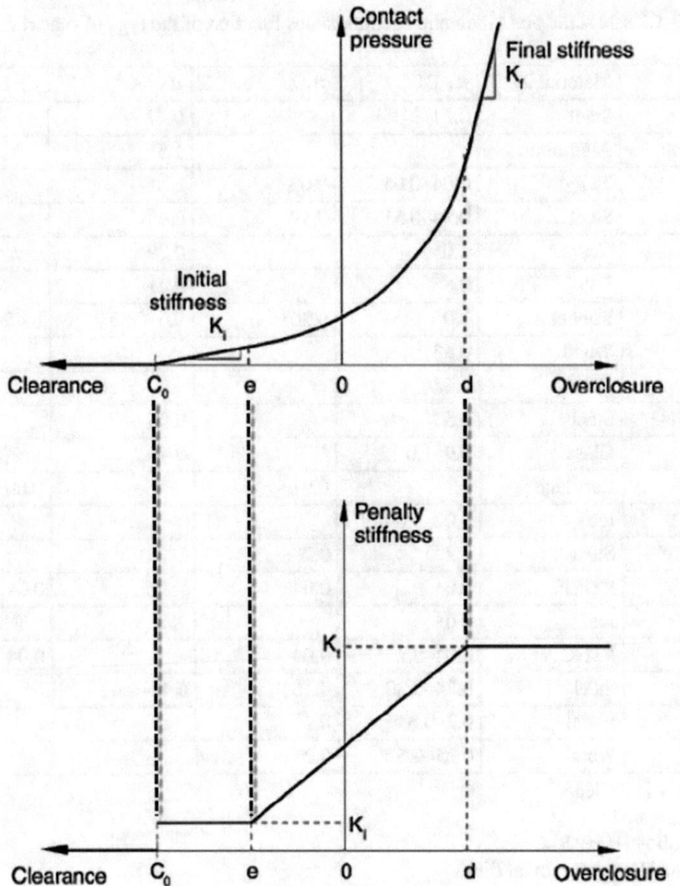

Fig. 7.35 Nonlinear contact pressure and linear contact stiffness. Figure used by permission ©Dassault Systemes Simulia Corp

The penalty method approximates hard pressure–overclosure behavior. With this method the contact force is proportional to the penetration distance, so some degree of penetration will occur. Advantages of the penalty method include

- Numerical softening associated with the penalty method can mitigate overconstraint issues and reduce the number of iterations required in an analysis.
- The penalty method can be implemented such that no Lagrange multipliers are used, which allows for improved solver efficiency.

The parameters shown in Fig. 7.36 are used to define such type of contact behavior with Abaqus shown in Fig. 7.35, and are listed below.

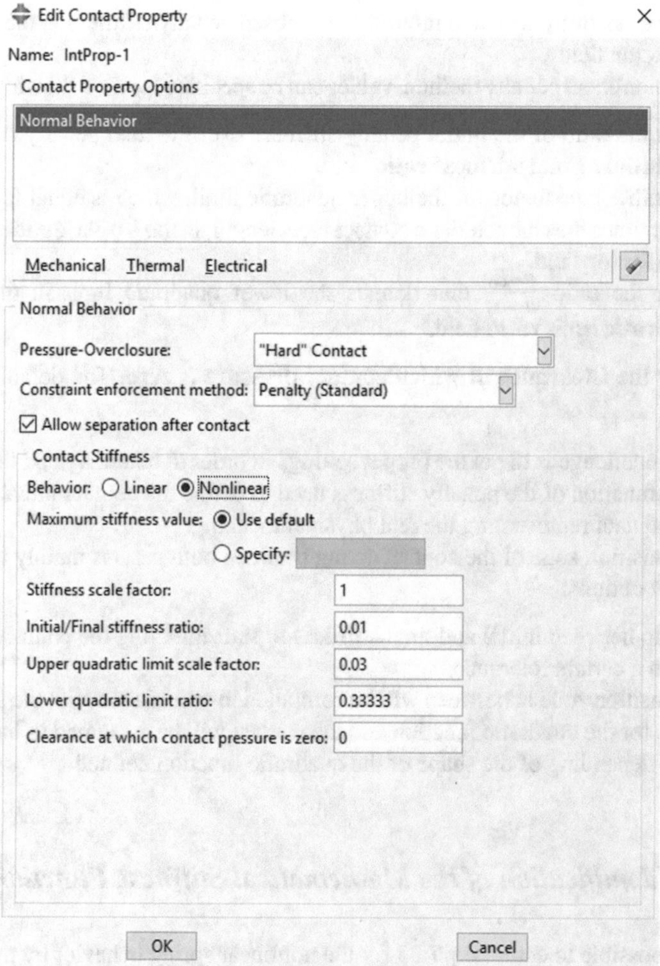

Fig. 7.36 Edit contact properties for a nonlinear contact pressure

- Specify the contact stiffness.
  - For the linear penalty method, specify the contact stiffness in the **Stiffness value** field. Use default can be selected to have Abaqus calculate the penalty contact stiffness automatically or can select Specify and enter a positive value for the linear penalty stiffness.
  - For the nonlinear penalty method, specify the contact stiffness in the **Maximum stiffness value** field. Use default can be selected to have Abaqus calculate the penalty contact stiffness automatically or can select Specify and enter a positive value for the final nonlinear penalty stiffness.

- Specify a factor by which to multiply the chosen penalty stiffness in the **Stiffness scale factor** field.
- For the nonlinear penalty method, values can be specified for the following options:
  - Enter the ratio of the initial penalty stiffness over the final penalty stiffness in the **Initial / Final stiffness ratio** field.
  - Enter the scale factor for the upper quadratic limit, which is equal to the scale factor times the characteristic contact facet length, in the **Upper quadratic limit scale factor** field.
  - Enter the ratio $\frac{e-c_0}{d-c_0}$ that defines the lower quadratic limit in the **Lower quadratic limit ratio** field.
- Specify the **Clearance at which contact pressure is zero**. The default value is zero.

The main challenge is to get the proper settings in order to obtain a good fit between the linear function of the penalty stiffness used to define the contact interaction and the hard contact representing the real physical meaning.

The transition zone of the contact acting between both parts is mainly a function of two key options:

- The ratio between initial and final stiffness to start initiating the contact and stop it within a certain tolerance.
- The transition zone in between will be computed in accordance with the parameter settings for the quadratic function and the contact will be asssumed as hard or soft contact depending of the shape of the quadratic function defined.

7.3.1 *Identification of the Mathematical Stiffness Function*

It is now possible to define Eq. 7.25 for the nonlinear spring behavior by the penalty stiffness curve in Fig. 7.35, the function of the clearance plus an overclosure u_0.

$$k(x) = \begin{cases} k_1 & c_0 \le x \le e \\ \Omega x + \Gamma & e \le x \le d \\ k_2 & d \le x \le u_0 \end{cases} \tag{7.25}$$

The continuity conditions are given by the system of Eq. 7.26.

$$\begin{cases} k(e) = k_1 \\ k(d) = k_2 \end{cases} \tag{7.26}$$

The coefficients Ω and Γ for the linear stiffness used in Eq. 7.25 can be calculated in Eq. 7.27 with the equations of continuity in Eq. 7.26.

$$\Omega = \frac{k_1-k_2}{e-d} \qquad \Gamma = k_1 - \frac{k_1-k_2}{e-d}e \tag{7.27}$$

Finally, the equation of the multi-linear stiffness for the spring element in contact interaction is given by Eq. 7.28.

$$k(x) = \begin{cases} k_1 & c_0 \leq x \leq e \\ \frac{k_1-k_2}{e-d}(x-e) + k_1 & e \leq x \leq d \\ k_2 & d \leq x \leq u_0 \end{cases} \tag{7.28}$$

The relation between the stiffness Eq. 7.28 and the reaction force used for the contact interaction is given by Eq. 7.29

$$\frac{\mathrm{d}}{\mathrm{d}x} f(x) = k(x) \tag{7.29}$$

According to the general form of the stiffness Eq. 7.25, the reaction force function based on Eq. 7.29 will be given by Eq. 7.30

$$f(x) = \begin{cases} a_1 x + a_2 & c_0 \leq x \leq e \\ a_3 x^2 + a_4 x + a_5 & e \leq x \leq d \\ a_6 x + a_7 & d \leq x \leq u_0 \end{cases} \tag{7.30}$$

From Eq. 7.29, it is, therefore, possible to identify some coefficients given by Eq. 7.31

$$\begin{cases} a_1 & c_0 \leq x \leq e \\ 2a_3 x + a_4 & e \leq x \leq d \\ a_6 & d \leq x \leq u_0 \end{cases} = \begin{cases} k_1 & c_0 \leq x \leq e \\ \Omega x + \Gamma & e \leq x \leq d \\ k_2 & d \leq x \leq u_0 \end{cases} \tag{7.31}$$

The integral constants in Eq. 7.30 will be calculated with the continuity equations in Eq. 7.32.

$$\begin{cases} a_1 c_0 + a_2 & = 0 \\ a_1 e + a_2 & = a_3 e^2 + a_4 e + a_5 \\ a_3 d^2 + a_4 d + a_5 & = a_6 d + a_7 \end{cases} \tag{7.32}$$

All integral constants are given by the system of Eq. 7.33

$$\begin{cases} a_2 = & -c_0 a_1 \\ a_5 = & -a_3 e^2 + (a_1 - a_4)e + a_2 \\ a_7 = & a_3 d^2 + (a_4 - a_6)d + a_5 \end{cases} \tag{7.33}$$

All coefficients used to determine the contact pressure based on general reaction force equations given in Eq. 7.30 are listed as functions of the stiffness, clearance, and overclosure parameters in the list of equations presented in Eq. 7.34.

Parameters	Equations
$a_1 =$	k_1
$a_2 =$	$-c_0 k_1$
$a_3 =$	$\frac{1}{2}\frac{k_1-k_2}{e-d}$
$a_4 =$	$k_1 - \frac{k_1-k_2}{e-d} e$
$a_5 =$	$\frac{1}{2}\frac{k_1-k_2}{e-d} e^2 - c_0 k_1$
$a_6 =$	k_2
$a_7 =$	$\frac{1}{2}(k_1 - k_2)(e + d) - c_0 k_1$

(7.34)

It is now possible to locally recalculate the theoretical contact force pressure based on the parameters given in Fig. 7.35 and therefore, determine how the contact pressure is acting between both surfaces in contact.

The contact behavior is highly sensitive regarding the clearance parameter and how the contact starts being defined. Moreover, both the clearance c_0 and lower quadratic limit $R_{\text{Lower Quadratic Limit}}$ parameters are linked in Eq. 7.35 as functions of the minimum clearance e and the maximum overclosure d as shown in Fig. 7.35.

$$R_{\text{Lower Quadratic Limit}} = \frac{e - c_0}{d - c_0} \tag{7.35}$$

The challenge with contact interaction is to deal with an equation and two unknown parameters, which are the initial clearance and the lower quadratic limit to ensure a proper adjustment to define the contact behavior. The most logical choice is to keep the default lower quadratic limit ratio and figure out the best initial clearance to set despite the imperfect reflection of the hard normal contact behavior of both surfaces in contact. Otherwise, the user will simply need to calculate the lower quadratic limit ratio in accordance with Eq. 7.35 by knowing the initial clearance parameter.

In cases where an initial contact interaction is defined without any contact openings, the normal behavior can be assumed to be linear with a low stiffness value instead of setting nonlinear contact; in order to save computational time and avoid numerical difficulties with convergence problems.

There is a relation between the initial/final stiffness ratio shown in Fig. 7.36 and the lower quadratic limit ratio defined in Eq. 7.35 thanks to the model developed in Eq. 7.30. The analyst is now able to build a customized contact pressure behavior, as shown in Fig. 7.35, by taking five points between the gap space of two parts in contact interaction. The five coordinate points (c_0, e, d, u_0) plus one point between the points e and d, will identify the coefficients a_n $(1 \le n \le 7)$ in Eq. 7.30. The clearance e and the overclosure d can be written as a function of coefficients a_n $(1 \le n \le 7)$ according to the definition given in Eq. 7.30, and shown in Eq. (7.36) and Eq. (7.37) respectively.

$$e = \frac{(a_1 - a_6)^2 + 2\,(a_7 - a_2)}{4a_3\,(a_1 - a_6)} \tag{7.36}$$

$$d = \frac{2\left(a_7 - a_2\right) - \left(a_1 - a_6\right)^2}{4a_3\left(a_1 - a_6\right)} \tag{7.37}$$

The Eqs. (7.36) and (7.37) show a forbidden value such as a1 must not be equal to a6, meaning that the initial stiffness k_1 must not be equal to the final stiffness k_2. Which is consistent with the penalty stiffness behavior shown in Fig. 7.35, otherwise the curve will be a constant function.

The lower quadratic limit ratio can now be written as a function of coefficients a_n $(1 \leq n \leq 7)$ in Eq. (7.30), as given below in Eq. (7.38).

$$R\left(X\right) = \frac{AX^2 + BX + C}{-AX^2 + BX + C} \tag{7.38}$$

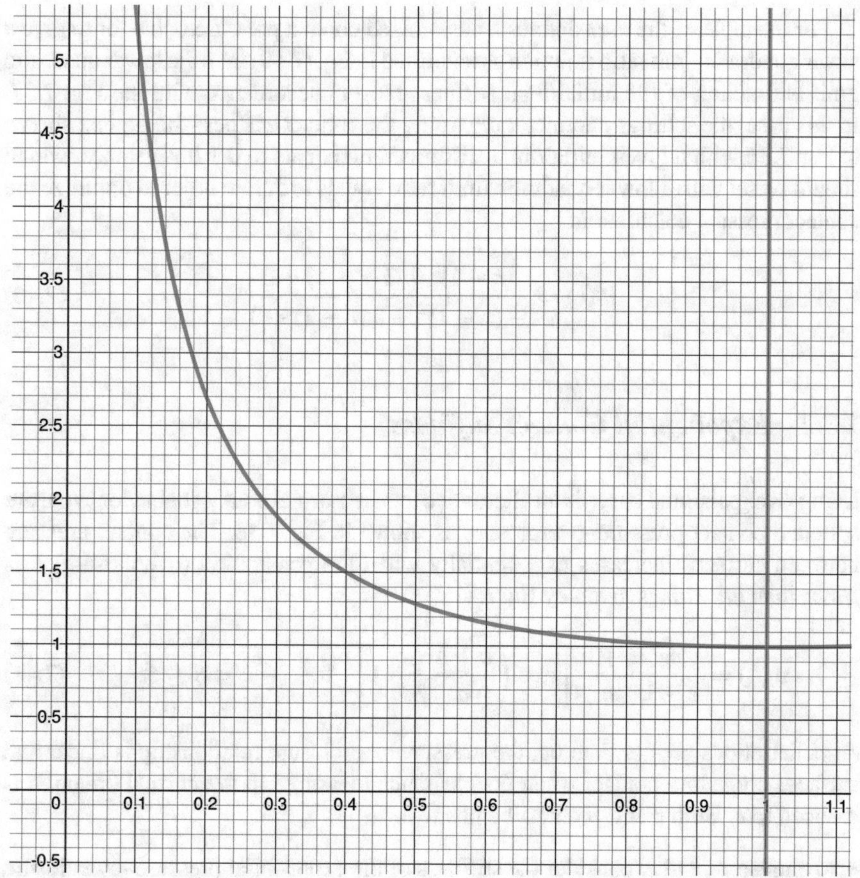

Fig. 7.37 Example of a function to set contact stiffness parameters, where the lower quadratic limit ratio parameter is given as a function of the initial/final stiffness ratio parameter

Where, $A = a_1 a_6^2$, $B = 4a_2 a_3 a_6$, $C = 2a_1 (a_7 - a_2)$, and $X = (a_1 - a_6)/a_6$ According to Fig. 7.35, the penalty stiffness behavior is positive and must be defined such as the final stiffness k_2 must be greater than the initial stiffness k_1, respectively ($a_6 > a_1$). Therefore, the X value is always negative. The X variable is also equal to $(a_1/a_6) - 1$, where the ratio (a_1/a_6) is the initial/final stiffness ratio. When X tends to zero, the ratio r tends to one meaning the value of k2 tends to k_1. The value of the ratio r varies between zero to one, when the ratio r tends to zero the X tends to minus one meaning that the value k_1 tends to zero or k_2 tends to infinity. It is now possible to establish the lower quadratic limit ratio R which is function of the initial/final stiffness ratio r, as given below in Eq. (7.39).

$$R\,(0 < r < 1) = \frac{A\,(r-1)^2 + B\,(r-1) + C}{-A\,(r-1)^2 + B\,(r-1) + C} \tag{7.39}$$

For example, if the user defines a customized contact pressure behavior to give a lower quadratic limit ratio function as identified in Eq. (7.40) then the lower quadratic limit is function of the initial/final stiffness ratio in accordance with the Fig. 7.37. In this case, the settings parameter regarding the contact stiffness dialog box shown in Fig. 7.36 with a given initial/final stiffness ratio equals to 0.1 for instance, which should be set with a lower quadratic limit ratio equals to 5.2, in order to optimize the contact pressure calculation.

$$R\,(r) = \frac{(r-1)^2 + 2\,(r-1) + 3}{-\,(r-1)^2 + 2\,(r-1) + 3} \tag{7.40}$$

7.3.2 *Exponential Contact Stiffness*

Using an exponential curve is an alternative to hard normal contact behavior to ensure softened contact pressure–overclosure as shown in Fig. 7.38. The contact pressure between surfaces increases exponentially when the penetration (overclosure) h is lesser than $6c$. This is given by Eq. 7.41.

$$p\,(h) = \frac{p_0}{e-1}\frac{c+h}{c}\left[exp\left(\frac{c+h}{c}\right) - 1\right] \quad \text{for} \quad -c < h \leq 6c \tag{7.41}$$

At penetrations greater than $6c$, the pressure–overclosure relationship is linear and surfaces come into contact when the clearance measured in the normal direction is reduced to c. Both c and p_0 must be positive.

- to choose p_0 and c, consider the stiffness of the exponential pressure–overclosure relationship by setting p_0 and c to match the stiffness of the surface being modeled.
- If only a single value k is available, the user will need to approximate the pressure clearance after using a linear relationship for positive clearance as follows:

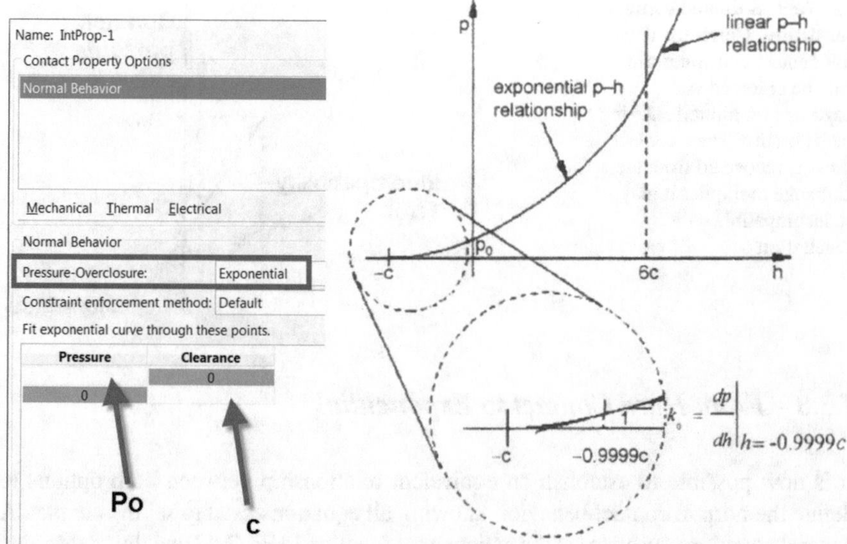

Fig. 7.38 Softened contact for pressure–overclosure with an exponential function

- Set p_0 for the expected contact pressure p. If this value is unknown, then use an average value of expected stress acting on the structure.
- Set c so that the contact stiffness k is given by the equation: $c = p_0/k$

- Always check for overclosure at closed contact points. If the overclosure is too severe, the pressure at zero clearance p_0 needs to be changed and the analysis must be resubmitted.
  - to decrease the amount of overclosure, increase the expected contact pressure at zero clearance p_0.
  - to increase the amount of overclosure, decrease the expected contact pressure at zero clearance p_0.

- The difference in clearance is expressed in terms of an incompatibility error.
  - This is **NOT** a runtime error and its role in establishing the correct contact state is analogous to that played by a force residual for determining force equilibrium.
  - The figure above shows an example using the exponential format.

- The default tolerance for convergence: incompatibility error $\leq 0.005c$ for $p > p_0$
- The tolerance for $0 \leq p \leq p_0$ is linearly interpolated between $0.005c$ at $p = p_0$ and $0.1c$ at $p = 0$. The tolerance at $p = 0$ can be modified with the ***CONTROLS** option (Fig. 7.39).

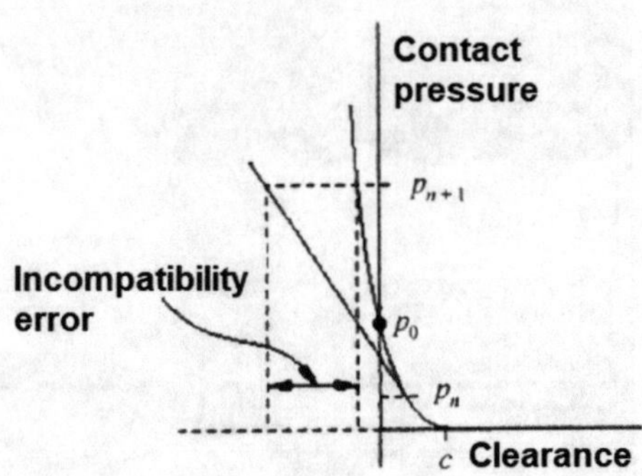

Fig. 7.39 Softened contact constraint. Currently, the soft contact constraint can only be enforced via Lagrange multipliers. During each iteration, since contact stress is recovered from the Lagrange multiplier it may be incompatible with penetration

7.3.3 *From Hard Contact to Exponential*

It is now possible to establish an equivalent relationship between both options to define the normal contact behavior, knowing all equations used to set the contact. A general equation is used to set the exponential function in Eq. 7.42 and thus determine what must be a softer contact behavior using an exponential with an equivalent starting point of the contact function of the initial clearance x to minimize the risk of numerical difficulties.

$$e \le x \le d : \begin{cases} f(x) = \quad b_1 e^{b_2 x} + b_3 \\ k(x) = f'(x) = b_1 b_2 e^{b_2 x} \end{cases} \tag{7.42}$$

As shown in Fig. 7.38, the exponential function is defined with two coordinates, then both ends of the exponential function behave with the linear stiffness slope identified for the hard contact. The boundary conditions described in Eq. 7.34 are used to identify the coefficient b_1 and b_2 in Eq. 7.43.

$$\begin{cases} b_1 e^{b_2 e} = a_1 \\ b_1 e^{b_2 d} = a_6 \end{cases} \tag{7.43}$$

The identification of coefficients are, therefore, given in Eq. 7.44.

$$\begin{cases} b_1 = \quad a_6 e^{-b_2 d} \\ b_2 = (e-d)^{-1} \, ln \left| \frac{a_1}{a_6} \right| \end{cases} \tag{7.44}$$

The last coefficient b_3 will be identified in Eq. 7.46 such as the contact should start at the same value of clearance between the hard contact and exponential contact.

$$\begin{cases} e-d \le x-d \le 0 : f(e-d) \le f(x-d) \le f(0) : 0 \le f(x-d) \le P_0 \\ \qquad f(x) = b_1 e^{b_2 (x-d)} + b_3 \end{cases} \tag{7.45}$$

$$\begin{cases} b_3 = a_1 e + a_2 - b_1 e^{b_2(e-d)} \\ f(x) = b_1 \left[e^{b_2 x} - e^{b_2(e-d)} + a_1 e + a_2 \right] \end{cases} \tag{7.46}$$

It is obvious to make an equivalent mathematical model between a hard contact and an exponential soft contact to provide the analyst with an idea of how to determine the contact behavior between two surfaces on a local basis in order to properly interpret the differences between both contact interactions in order to be physically consistent with real contact interaction. A mathematical model can also be used in reverse to determine the input parameter from a known set of output data like a certain contact pressure limit and/or a maximum contact stiffness value.

According to Fig. 7.38 parameters p_0 and c need to be defined for an exponential pressure over-closure selected in the contact interaction property settings box. Similarly to Fig. 7.35, the Fig 7.36 shows a linear p-h relationship at a clearance e equals to $-0.9999\ c_0$ and after an overclosure d equals to $6c_0$, assuming c_0 equals to one. The linear $p-h$ relationship correlate the initial and final stiffness respectively. To find the values of p_0 and c the system of equations in Eq. (7.47) will be resolved, using the pressure overclosure equation in Eq. (7.41) as shown in Eq (7.47).

$$\begin{cases} \left(\frac{dp}{dh}\right)_{(h=-0.9999)} = k\,(-0.9999) = k_1 \\ \left(\frac{dp}{dh}\right)_{(h=d)} = k\,(d) = k_2 \end{cases} \tag{7.47}$$

$$\begin{cases} \frac{p_0}{c(e-1)} \left[\left(2 + \frac{-0.9999}{c}\right) exp\left(1 + \frac{-0.9999}{c}\right) - 1 \right] = k_1 \\ \frac{p_0}{c(e-1)} \left[\left(2 + \frac{d}{c}\right) exp\left(1 + \frac{d}{c}\right) - 1 \right] = k_2 \end{cases} \tag{7.48}$$

The solution for the parameters c and p_0 are shown in Eqs. (7.49) and (7.50), respectively.

$$\frac{\left[\left(2 + \frac{-0.9999}{c}\right) exp\left(1 + \frac{-0.9999}{c}\right) - 1 \right]}{\left[\left(2 + \frac{d}{c}\right) exp\left(1 + \frac{d}{c}\right) - 1 \right]} = \frac{k_1}{k_2} \tag{7.49}$$

$$p_0 = \frac{c k_2\,(e-1)}{\left[\left(2 + \frac{d}{c}\right) exp\left(1 + \frac{d}{c}\right) - 1 \right]} \tag{7.50}$$

To determine the value of the clearance c in Eq. (7.50) the user will interpolate the value of the initial/final stiffness ratio $r = k_1/k_2$ with the curve defined as a function of the overclosure d as shown for example in Fig. 7.40. According to Fig. 7.40, more the overclosure increases more the response tends to zero. When the ratio r tends to zero, k_2 tends to infinity like a very hard contact behavior, or when the ratio tends to one, k_2 tends to k_1 like a soft contact behavior.

Once the value of the clearance c will be determined from the Fig. 7.40, knowing the ratio r, then the pressure overclosure p_0 can be determined from Eq. (7.50).

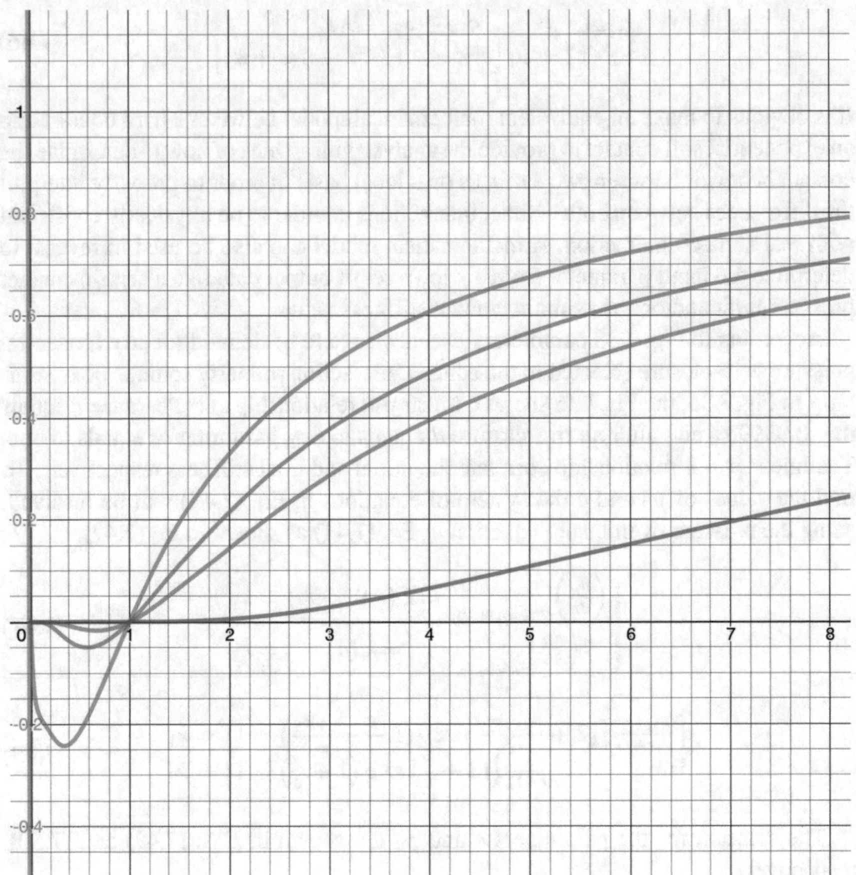

Fig. 7.40 Example of initial/final stiffness ratio versus the clearance parameter c for different values of an overclosure d defined by the user to establish the contact interaction. In red, the overclosure is equals to 0.01. In blue, the overclosure is equals to 0.5. In green, the overclosure is equals to 1. And in purple, the overclosure is equals to 6.

7.4 Obtain a Converged Contact Solution

This section provides the analyst with a procedure to use when an Abaqus standard contact analysis is having convergence difficulties and how to check it. First, a detailed output of the contact state should be requested:

- with Abaqus CAE in Step Module: **Output** → **Diagnostic Print ...** → **Contact**.
- with keyword: ***PRINT, CONTACT=YES**

Contact state information will be written in the message (.msg) file. The analyst should also write the necessary contact output variables to the data (.dat) and the output database (.odb) files frequently enough so that the user can diagnose the problem.

The following steps should be taken:

1. Check that the contact surfaces are properly defined and correct any errors. The surfaces can be viewed in Abaqus viewer. If the surface has been created in the input file by automatically using the free surface of an element set, check whether the TRIM parameter should have been used.
2. Check the contact direction defined for each surface. The normals to a surface can be viewed in Abaqus viewer. If the normal directions are wrong, analysts will frequently get large overclosures that may lead to convergence difficulties.
3. Frictional contact problems are generally more difficult to solve than frictionless contact problems. Try solving the problem without friction to see if the difficulty is due to friction or if it is due to something else. If the difficulty is indeed due to friction, consider the following:
   - Reexamine the choice of the friction coefficient: larger coefficients of friction are generally more difficult to use.
   - Examine the allowable elastic slip (low values can cause convergence problems, although too high a value can give physically incorrect solutions).
   - Refine the mesh so that more points come into contact at the same time.
4. Rough friction with intermittent contact generally produces convergence problems. If appropriate for the analysis, when using rough friction do not allow the contacting surfaces to separate after contact is established:
   - with Abaqus CAE in Interaction Module: **Interaction** → **Property** → **Create** → **Contact** → **Mechanical** → **Normal Behavior** → deselect Allow separation after contact
   - with keyword: ***SURFACE BEHAVIOR, NO SEPARATION**
5. Three-dimensional finite sliding contact with highly curved faceted master surfaces generates a highly nonsymmetric tangent stiffness matrix. Use the asymmetric solver even if the coefficient of friction is less than 0.2.
6. If the model has sharp corners on the contact surfaces, the comments from the previous point apply. In addition, try the following:
   - Smooth the surface. Nodes on the slave surface can be caught in folds in the master surface, causing convergence difficulties when the surrounding elements deform to take this into account. The elements making up the slave surface should be small enough to be able to resolve the geometry. A rough guideline is to use ten elements around a 90-deg corner; the analyst must use their judgment to decide whether this is adequate or too fine.
   - If the physical problem has a sharp concave fold, use two separate surface definitions.
   - Sharp convex folds cannot be modeled with a reasonable finite-element mesh. Smooth the fold with a radius larger than the element size on the slave surfaces. A rough guideline is to use ten elements around a 90-deg corner; obviously, the analyst must use their judgment to decide whether this is adequate or too fine.

7.5 Convergence Difficulty in the First Increment

This is a situation in which an Abaqus standard contact analysis terminates in the first increment after too many severe discontinuity iterations. To determine the cause of such troubleshooting, let's investigate the following areas.

One or more of the contact pairs in the analysis may have an excessive initial overclosure. The overclosure may be a feature of the model, or it may be the result of a modeling error. In either case, an excessive overclosure may cause the analysis to terminate prematurely. In this section, some common sources of excessive initial overclosures are discussed.

1. In the case of an overclosure that is a feature of the model, as with an interference fit analysis, the overclosure may be too large to resolve in one increment. to help with the diagnosis, print information about the overclosures during the datacheck and solution phases:
   - with Abaqus CAE in Job Module go to **Job Manager** and then **Edit** → **General** → **Preprocessor printout** → **Print contact constraint data**.
   - In Step Module go to **Output** → **Diagnostic Print...** → **Contact**.
   - with keyword enter command lines ***PREPRINT, CONTACT=YES** and ***PRINT, CONTACT=YES**.
2. If the user is using General Contact, the solution, in this case, is to use the ***CONTACT INITIALIZATION DATA, INTERFERENCE FIT** option to remove the overclosure. If the user is using Contact Pairs method, ***CONTACT INTERFERENCE** is already the default behavior to treat overclosures as an interference fit. Sometimes resolving the overclosure over several increments is necessary. Adjusting the solution controls to increase the maximum number of severe discontinuity iterations allowed in an increment may also be needed to overcome this difficulty. Sometimes small overclosures (relative to element dimensions) can be removed by adjusting the slave nodes in the contact pair:
   - with Abaqus CAE in Interaction Module go to Interaction Manager and then **Edit** → **Slave node adjustment**.
   - with keyword use command ***CONTACT PAIR, ADJUST=[Node set label | Adjustment value]**
3. The master surface normals may be pointing away from the slave surface. Use Abaqus viewer to check the normals on the master surface. The solution in this case is to redefine the master surface so the normals point toward the slave surface, assuming the surfaces are open.
4. The model may be permitting a rigid body motion with very large overclosures detected; these are sometimes on the order of 10–15, typically due to loading a body with forces when there are insufficient boundary conditions or insufficient active contact constraints to remove rigid body motion. In most cases, the analyst will also see messages about NUMERICAL SINGULARITIES in the message (.msg) file. Print overclosure information during the solution (as described in Item 1) to diagnose the problem. This case can be resolved with different approaches:

- Use boundary conditions to move the bodies until they are just in contact (do this in a dummy step). In the next step, remove these boundary conditions and replace them with forces that maintain the contact (this is the recommended technique).
- Add soft springs to the model in the directions of the rigid body motion. The degrees of freedom (dofs) requiring springs can be seen from the dofs associated with the numerical singularity messages. The stiffness of the springs must be sufficiently small that the forces in the springs are negligible compared to typical forces in the problem.
- Use contact stabilization ***CONTACT CONTROLS, APPROACH** or ***CONTACT CONTROLS, STABILIZE** to stabilize the motion (using damping effects) when, at some point in the analysis, contact is meant to prevent rigid body motion. If this is not done carefully, the viscous forces can dominate the solution. Print out the forces due to viscous effects applied with these options using the **VF** output variable. Additionally, output variable **ALLSD** measures the energy dissipated by viscous damping; the ratio of this quantity to the elastic strain energy or other appropriate general energy measures should be small.

  If the analysis converges in an iteration during which NUMERICAL SINGULARITIES appear, check the solution carefully. Abaqus standard attempts to fix the solution, but sometimes this fix will not give a correct solution. The analyst should remove the cause of the NUMERICAL SINGULARITIES instead. If numerical singularities only occur in iterations prior to the iteration which converges, the solution should be correct; however, the analyst should always check that this is so.
- Contact chattering is when the contact state changes from one iteration to the next so that the maximum number of SDIs (severe discontinuity iterations) is reached.

7.6 Causes and Resolutions of Contact Chattering

Contact chattering occurs when Abaqus has difficulty resolving a contact constraint. The contact status of a particular slave node may sometimes repeatedly change from open to closed between each severe discontinuity iteration (SDI); this is known as contact chattering. It can lead to cutbacks in the time increment if the contact state cannot be resolved within a certain number of severe discontinuity iterations. If several cutbacks are required, contact chattering may ultimately lead to a loss of convergence.

As outlined below, there are many possible causes of contact chattering. Invoking the automatic overclosure tolerances very often alleviates the difficulty. These tolerances are introduced as follows:

- with Abaqus CAE in Interaction Module go to **Interaction → Contact Controls → Create → Continue → check Automatic overclosure tolerances**
- with keyword use command line ***CONTACT CONTROLS, AUTOMATIC TOLERANCES**

With the automatic tolerances, Abaqus calculates an alternative set of tolerances that are intended for those problems where the standard controls do not provide cost-effective solutions. These problems often require several iterations at the start of the analysis to establish the correct contact state. Thus, the automated tolerances increase the allowable penetration of a slave node to twice the maximum displacement correction, and allow a tensile contact pressure equal to ten times the maximum allowable force residual divided by the contact area of a node during the first two iterations.

If convergence should occur in the first two iterations with these modified tolerances, at least one more additional iteration is made with the separation tolerance set equal to the largest allowable residual.

If the use of the automatic contact controls does not resolve the chattering, some additional causes and suggested resolutions are

1. If a slave node is sliding off the master surface (when using finite sliding), then check and extend the master surface if necessary.
2. If only a few nodes are in contact, then refine the underlying mesh of the slave surface or use softened contact (exponential, tabular, or linear pressure–overclosure relationship) to distribute the contact over more nodes.
3. If the size of the region in contact is changing rapidly make sure that the application of friction is not being delayed to the increment after contact occurs (i.e., make sure that ***CONTACT CONTROLS, FRICTION ONSET=DELAYED** is not being used). Indeed, immediate onset of friction is the default behavior. The model has long, flexible parts with small contact pressures then use softened contact.
4. If the master surface is not sufficiently smooth and kinks in the master surface get caught between two slave surface nodes then smooth the master surface by refining the underlying mesh or defining the contact directions with ***NORMAL, TYPE=CONTACT SURFACE**. If possible, use analytical rigid surfaces instead of rigid surfaces defined with elements.
5. If rigid elements must be used, smooth the rigid surface with ***CONTACT PAIR, SMOOTH**.
6. Make sure analytical rigid surfaces are smooth between segments then use the ***SURFACE, FILLET RADIUS** option. It is the user's responsibility to make sure rigid surfaces are sufficiently smooth.
7. If contact can be established but Abaqus standard has difficulty (changing the contact status from closed to open), try viscous damping to control chattering:

    - with Abaqus CAE in Interaction Module go to **Interaction → Contact Controls → Create → Stabilization** tab → click **Stabilization coefficient → Specify damping parameters**.

- with keyword enter command line ***CONTACT DAMPING, DEFINITION=DAMPING COEFFICIENT**

8. If none of the above cases seem to apply, try using automatic stabilization in a static, coupled temperature displacement, soils, or quasi-static step:
   - with Abaqus CAE in Step Module go to **Step** → **Create...** → select **Step type** → select **Use Stabilization with**.
   - with keyword enter command line to include the STABILIZE parameter on the ***STATIC, *COUPLED TEMPERATURE-DISPLACEMENT, *SOILS** or ***VISCO** options.

   Automatic stabilization is normally used to stabilize globally unstable problems but can sometimes help prevent contact chatter problems from occurring.
   In some cases, using dashpots on elements on and near the contact region can help stabilize the contact. If the analyst uses this technique, dashpots should be applied to all translation degrees of freedom until experience indicates the user can apply the dashpots only in a particular direction or at a limited number of nodes.
9. For extremely difficult situations, and only as a last resort, the analyst can allow some points to violate contact conditions by setting the maximum number of points permitted to violate contact and the maximum value of tensile stress allowed to be transmitted at a contact point:
   - with Abaqus CAE in Interaction Module go to **Interaction** → **Contact Controls** → **Create** → **General** tab.
   - with keyword enter command line ***CONTACT CONTROLS, [MAXCHP | PERRMX]**.

   These parameters can be reset in a subsequent step.
   WARNING: These controls are intended for experienced analysts and should be used with great care.

It can also be noted that the ***CONTACT CONTROLS, APPROACH** and ***CONTACT CONTROLS, STABILIZE** options are not designed to solve contact chatter problems, so they will probably not help in these situations. These options are meant to prevent excessive overclosures due to unconstrained rigid body motion.

7.7 Understand Finite Sliding with Surface-to-Surface Contact

A frequent question regarding contact is: how does the finite sliding surface-to-surface contact constraint enforcement method differ from the finite sliding node-to-surface method? An answer is, therefore, discussed here to complement the overview given in Sect. 7.1.5.1.

In the node-to-surface formulation, the contact condition is enforced at discrete points—i.e., the slave nodes. In the surface-to-surface formulation, the contact condition is enforced in an average sense over the slave surface, rather than at discrete points.

The basic premise behind the surface-to-surface formulation is that the negative effects associated with surface discretization can be reduced if the contact constraint enforcement is based upon the average penetration over a finite surface region rather than the penetration at a single node. The inherent smoothing characteristic of the surface-to-surface formulation frequently leads to better convergence behavior than the node-to-surface formulation. Existing models which are converted from the node-to-surface formulation to the surface-to-surface formulation can show different behaviors owing to the fundamentally different way in which penetrations are computed. However, both formulations converge to the same behavior as the mesh is refined.

Some usage notes are given below:

- By default, the contact constraints are centered at the nodes rather than at the slave face. This reduces the probability that there will be overconstraint issues. The face-centered approach is still available, but will likely be removed in a future version.
- The direct method of enforcing the contact constraint is available. This allows a hard contact condition to be enforced exactly. The penalty method is still the default, and is recommended for performance and robustness reasons unless a strict adherence to the hard contact constraint is essential.
- If a slave surface slides off, on, or around a corner of the master surface, contact force distributions vary in a smoother way, thus improving convergence behavior.

Some advantages and disadvantages of the finite sliding, surface-to-surface formulation relative to the node-to-surface formulation are listed in Table 7.6. The examples that follow demonstrate several important implications to be aware of when using the surface-to-surface formulation.

The snagging example is a typical example to illustrate surface-to-surface discretization and consists of a finite sliding between two deformable bodies: it serves to demonstrate how the inherent smoothing characteristic of the surface-to-surface

Table 7.6 Advantages and disadvantages of the surface-to-surface formulation

Advantages	Disadvantages
Relaxation of surface restrictions for instance with shell thickness taken into account, disconnected master surfaces can be used	Models converted from node-to-surface formulation can show different behaviors if the mesh is not sufficiently refined
Increased accuracy of contact stresses assuming an adequate surface representation	Increased memory and CPU cost
Reduced snagging of slave surface nodes, enhanced convergence rate	Sometimes requires asymmetric solver in cases where node to surface does not

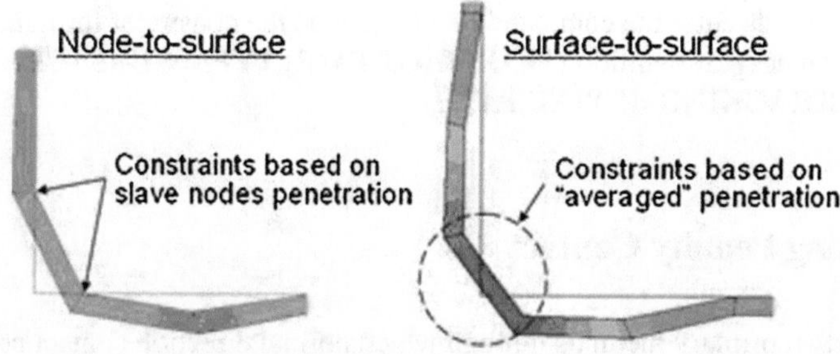

Fig. 7.41 Average surface facet penetration

formulation can help to avoid snagging. Representative configurations of the bodies as they slide relative to each other are depicted in Fig. 7.13. The node-to-surface image on the left side of the figure shows slave nodes snagging as they round the corner, whereas the surface-to-surface image on the right side of the figure shows a relatively smooth transition. The surface-to-surface formulation can also help to avoid hourglassing when reduced integration elements are used.

The average surface facet penetration example shown in Fig. 7.41 consists of a thin sheet that is wrapped around a sharp corner and serves to demonstrate the effect of enforcing contact constraints based on average surface facet penetration as opposed to slave node penetration.

Representative configurations of the sheet as it slides around the corner are depicted in Fig. 7.41. From inspection, it should be clear how the fundamental difference in penetration calculations can lead to dramatically different behaviors when an existing model that uses the node-to-surface formulation is converted for use of the surface-to-surface formulation.

The initial rigid body mode is also a typical example to illustrate surface-to-surface discretization and consists of a rigid cylinder in contact with a deformable body and demonstrates a complication that can arise because of the manner in which surface-to-surface penetrations are calculated. Using the node-to-surface formulation, the model depicted in Fig. 7.25 will start out properly without numerical difficulty because a slave node is in contact with the surface of the rigid body and a vertical constraint will be applied at the outset. Using finite sliding, surface to surface the average penetration is computed everywhere as negative (i.e., open) and the model will not run because of a rigid body mode in the vertical direction. Situations such as these are particularly common and when they arise one possible solution is to use contact stabilization. Another possible solution is to use displacement control instead of force control to move the rigid body into contact. The node-to-surface formulation remains the default. In order to activate the surface-to-surface formulation use the options listed below:

- with Abaqus CAE in Interaction module select the formulation, tracking, and other settings from the **Edit Interaction** dialog.

- with keywords enter the command line to include the constraint formulation type on the contact pair definition ***CONTACT PAIR, TYPE=SURFACE TO SURFACE, TRACKING=[STATE|PATH]**.

7.8 Using Penalty Contact

There are two primary methods through which normal direction contact constraints can be enforced in Abaqus standard with the traditional direct Lagrange multiplier method and a penalty-based method. A fundamental difference between the two methods is that the Lagrange multiplier method exactly enforces the contact constraint by adding degrees of freedom to the problem, while the penalty method approximately enforces the contact constraint through the use of springs without adding degrees of freedom. The penalty method is depicted schematically in Fig. 7.42. The upper surface is the slave, and the lower surface is the master. While the overclosure has been exaggerated, it is clear that the spring of stiffness k_p resists the penetration of the slave node into the master surface.

A large class of problems exists where the extra accuracy that is possible with the Lagrange multiplier method is not consistent with the approximations that are made (i.e., coarse meshes). In these problems, adequately capturing load transfer through the contacting interface is frequently more important than precise enforcement of the zero-penetration condition. The penalty method is attractive in such applications because it is usually possible to trade off a small amount of penetration for improved convergence rates.

The penalty method implementation in Abaqus attempts to choose a reasonable penalty stiffness based on the underlying element stiffness. If the default penalty stiffness is not suitable, options to scale the penalty stiffness are available. It is also possible to prescribe the penalty stiffness directly. If the scaled or user-prescribed penalty stiffness becomes excessively large, Abaqus automatically invokes special logic that minimizes the possibility of ill-conditioning. Advantages and disadvantages of each method are listed in Table 7.7.

Both linear and a nonlinear penalty stiffness definitions are available. With this approach, the penalty stiffness has constant initial and final values; these values serve as bounds for an intermediate overclosure regime in which the stiffness varies

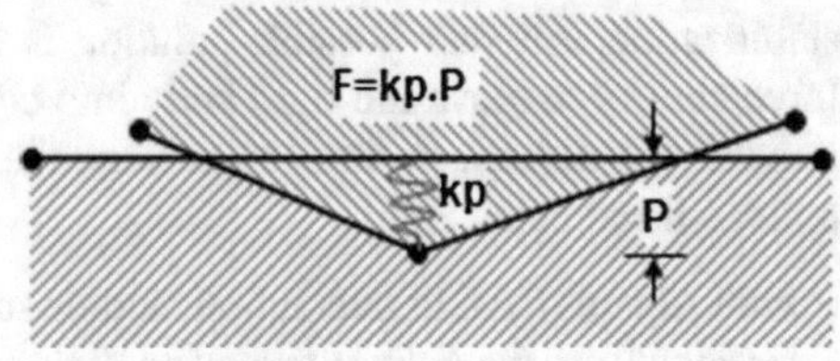

Fig. 7.42 Schematic diagram of the penalty method

Table 7.7 Direct Lagrange multiplier versus the penalty method

Direct Lagrange multiplier contact		Penalty contact	
Advantages	Disadvantages	Advantages	Disadvantages
Make an exact contact constraint enforcement with zero penetration	Larger system of equations	Number of equations does not increase	Approximate constraint enforcement (finite amount of penetration)
Easy to recover contact forces	Difficult to treat overconstraints	Easier to treat overconstraints	Difficult to choose proper penalty stiffness
No need to define contact stiffness	Sensitive to chattering		

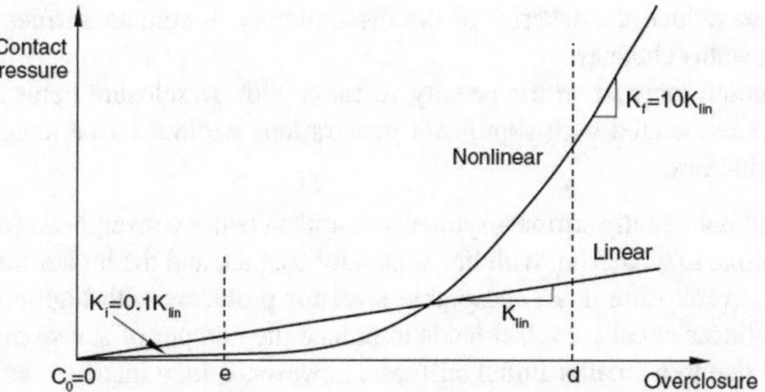

Fig. 7.43 Linear and nonlinear penalty stiffness behavior. Figure used by permission ©Dassault Systemes Simulia Corp

quadratically. A schematic comparison of the pressure–overclosure relationships for the linear and nonlinear penalty methods is given in Fig. 7.43.

The various parameters used for defining the nonlinear pressure–overclosure relationship are given below:

- K_{lin} is the linear stiffness used for linear penalty contact. The default value is ten times the representative underlying element stiffness.
- C_0 is the clearance at which contact pressure is zero. The default value is zero.
- K_i is the initial stiffness. The default value is one-tenth of the linear penalty stiffness.
- K_f is the final stiffness. The default value is ten times the linear penalty stiffness.
- d is the upper quadratic limit. The default value is 3% of the characteristic length computed by Abaqus standard to represent a typical facet size.
- e is the lower quadratic limit. The default value is 1% of the characteristic length computed by Abaqus standard to represent a typical facet size.
- $e_r = e/d$ is the lower quadratic limit ratio. From the default values of parameters d and e, the default value for e_r is 0.3333.

Table 7.8 Hertz contact results

	Penetration	Peak stress
Direct Lagrange	0	1.201E5
Linear penalty, default stiffness	4.482E-6	1.183E5
Scaled linear penalty, $S_f = 0.01$	2.492E-4	6.334E4

The default values for these parameters are based on the characteristics of the underlying elements of the slave surface. User control for changing the default values is provided. The nonlinear penalty method has the following characteristics:

- A relatively low penalty stiffness is used while the contact pressure is small. This serves to reduce the severity of the discontinuity in contact stiffness when the contact status changes.
- The smooth increase in the penalty stiffness with overclosure helps avoid inaccuracies associated with significant penetrations without introducing additional discontinuities.

The low initial penalty stiffness typically results in better convergence for problems that are prone to chattering with linear penalty contact, and the higher final stiffness keeps the overclosure at an acceptable level for problems with high contact pressure. Nonlinear penalty contact tends to reduce the number of severe discontinuity iterations due to a smaller initial stiffness; however, it may increase the number of equilibrium iterations due to the nonlinear pressure–overclosure behavior. Hence, it cannot be guaranteed that nonlinear penalty contact will result in a reduction of the total iteration count compared to linear penalty contact.

As discussed above, Abaqus attempts to choose reasonable penalty stiffness values based on the underlying element stiffness. Experience has shown that, for stiff or blocky problems, the default penalty stiffnesses chosen by Abaqus produces results that are comparable in accuracy to those produced using the direct Lagrange multiplier method but usually at less expense in terms of memory and CPU time. Experience has also shown that for bending-dominated problems the default linear penalty stiffness can often be scaled back without any significant loss of accuracy. Furthermore, scaling back the penalty stiffness for bending-dominated problems has been seen to sometimes dramatically increase the convergence rate. These experiences are demonstrated by the following examples.

The Hertz contact consists of two elastic cylinders in contact as depicted in Fig. 7.44. The node-to-surface formulation with matching meshes was used. Cases were run using the direct Lagrange multiplier method, the linear penalty method with default stiffness, and the linear penalty method with default stiffness scaled back by two orders of magnitude. Results from all three cases are presented in Table 7.8.

As expected, the direct Lagrange multiplier case produces zero penetration, while the cases that use the penalty method produce finite penetrations. In this example, scaling down the penalty stiffness by a factor of 100 results in a 55x increase in the

Fig. 7.44 Hertz contact

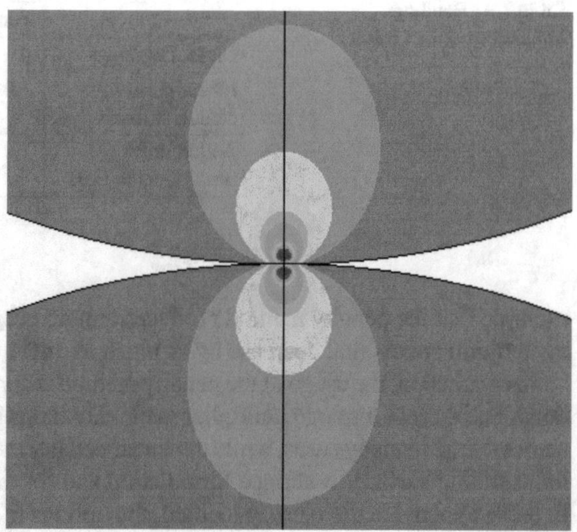

Fig. 7.45 Bending-dominated contact

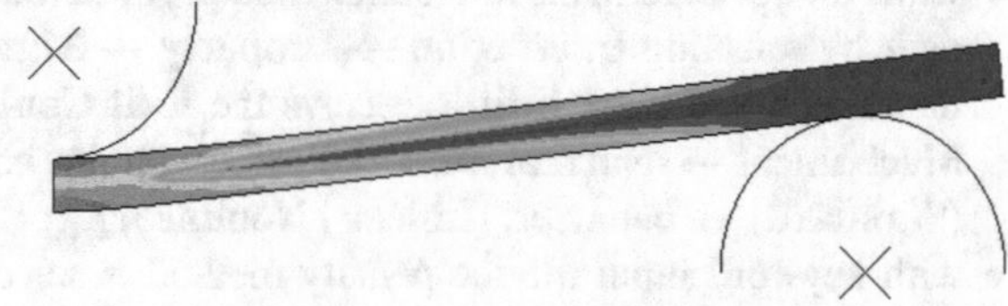

penetration. It is evident that for this example the default penalty stiffness predicts a peak stress that differs by only 1.5% from the peak stress that is computed using the direct Lagrange multiplier method. If the penalty stiffness is scaled back by a factor of 100, then the predicted peak stress decreases considerably and differs by 47% from the peak stress that is computed using the direct Lagrange multiplier method. The significant decrease in peak stress is due to the combination of displacement-controlled loading and the compliance at the contact interface with the penalty method.

The bending-dominated contact example consists of a three-point bending test of an elastic–plastic beam as depicted in Fig. 7.45. The node-to-surface formulation and half symmetry have been used. Cases were run using the direct Lagrange multiplier method, the linear penalty method with default stiffness, and the linear penalty method with default stiffness scaled back by two orders of magnitude. Results from the three cases are presented in Table 7.9.

As expected, the direct Lagrange multiplier case again produces zero penetration while the cases for which the penalty method is used produce finite penetrations. However, in this example, it can be seen that both cases that use the penalty method predict a peak stress that is practically identical to the peak stress computed using the direct Lagrange multiplier method. The behavior that is seen in this example, where a relatively small penalty stiffness produces quite accurate stress results, generalizes to a very large class of bending-dominated problems. It can also be seen that in this

Table 7.9 Bending-dominated contact results

	Penetration	Stress	Iterations
Direct Lagrange	0	2.416E4	130
Linear penalty, default stiffness	1.004E-7	2.416E4	117
Scaled linear penalty, $S_f = 0.01$	1.015E-5	2.416E4	112

example that the penalty method produces a more economical solution as measured by iteration counts that decrease by as much as 14%.

In conclusion, the usage of the penalty method is applicable to all contact formulations. The direct Lagrange multiplier method remains the default constraint enforcement method in many cases, while the linear penalty method is used by default for the finite sliding surface-to-surface formulation and for three-dimensional self-contact. In order to activate the penalty method, the options listed below can be employed.

- with Abaqus CAE in the Interaction module, open the **Interaction Property Manager** by selecting **Interaction** → **Property** → **Manager**. Select the appropriate interaction and click **Edit** to receive the **Edit Contact Property** dialog. Select **Mechanical** → **Normal Behavior** → constraint enforcement method: **Penalty (Standard)** → behavior: **[Linear | Nonlinear]**
- with keyword input file the penalty method is selected with the following keyword options ***SURFACE BEHAVIOR, PENALTY=[LINEAR | NONLINEAR]**. The data lines can be used to modify the default settings for either the linear or nonlinear approaches.

7.9 Using Augmented Lagrangian Contact

Three methods of contact constraint enforcement are available in Abaqus standard: the direct Lagrange multiplier method, the augmented Lagrangian method, and the penalty method. This section will discuss the differences between the direct and augmented Lagrangian methods.

Within the context of the classical hard contact problem, the direct and augmented Lagrangian methods differ as follows:

- Direct method:
  - Enforces the contact constraint exactly, so that no penetration of the slave nodes into the master surface is allowed.
- Augmented Lagrangian method:
  - Enforces the contact constraint approximately using a penalty method. The penalty stiffness is scalable.

– Nonzero penetration of the slave nodes into the master surface is allowed, and penetration tolerance is adjustable.
– Once solution convergence is obtained, it augments the contact pressure if a slave node penetrates the master surface by more than the relative penetration tolerance of 0.1% (default setting) of the characteristic interface length. Iterations continue again until convergence. The solution is accepted when the penetration tolerance is accepted.
– Can sometimes be more expensive than the Lagrange multiplier approach as a result of the augmentation scheme and additional iterations.

In general, the approximate nature of the augmented Lagrangian contact constraint can simplify the resolution of difficult contact problems and sometimes allow a solution to be found when the exact, Lagrange multiplier constraint is too restrictive. With the ability to scale the penalty stiffness and the penetration tolerance, the contact constraint can be relaxed to facilitate convergence; however, this must be done with caution and the results must be carefully checked for excessive penetration of the contact surfaces.

More specific situations where this approach can help include:

1. Very different mesh densities on the contact pair surfaces
   Nonuniform contact pressure distributions are more likely to occur when very different mesh densities are used on the two deformable surfaces making up a contact pair. If the Lagrange multiplier method is used, the nonuniformity can be particularly pronounced and oscillations and spikes in the contact pressure may occur when both surfaces are modeled with second-order elements (including modified, second-order tetrahedral elements). Smoother contact pressures may be obtained for surfaces modeled with second-order tetrahedral elements by using the augmented Lagrangian approach.
2. Overconstraint problems
   An overconstraint occurs when a contact constraint on the displacements, temperatures, electrical potentials, or pore fluid pressure at a slave node conflicts with a prescribed boundary condition or other kinematic constraints on that degree of freedom at the node. Specified boundary conditions on the master surface nodes typically do not cause overconstraints. Moreover, specified boundary conditions on slave nodes may create an overconstraint.
   Overconstraints can only be avoided by changing the contact definition or the boundary conditions. Overconstraint problems may also be alleviated by using the augmented Lagrangian contact constraint enforcement method. While this may help in certain difficult situations, it is generally preferable to remove the source of the overconstraint.

The augmented Lagrangian method cannot be used with a softened pressure–overclosure relationship.

7.10 Using Stiffness-Based Contact Stabilization

Automatic contact stabilization can be used to prevent rigid body motions in static problems before contact closure and friction restrains such motions. This capability is intended for use in cases where it is clear that contact will be established, but the exact positioning of multiple bodies is difficult during modeling.

Like any modeling capability, automatic contact stabilization must be used carefully to avoid unexpected results or convergence problems. It is not meant to simulate rigid body dynamics or to resolve contact chattering problems.

With this technique, rigid body motions are suppressed by applying viscous damping to the model. The damping coefficient is calculated automatically at each slave node based on the underlying element stiffness and the step time. The stabilization only works for the step in which it is specified.

By default, the damping coefficient is

- Ramped down linearly to zero over the step
  - If the contact is not fully established by the end of the step, or if the contact state changes during the step such that rigid body motions are allowed, convergence difficulties may arise. The manner in which the damping is ramped down can be changed in Abaqus CAE by selecting the Interaction Module and proceeding to **Contact Controls Editor**, then **Stabilization** → **Automatic Stabilization** → Fraction of damping at end of step and enter the value.
  - or in the input file by using the command line ***CONTACT CONTROLS, STABILIZE**, to ramp-down factor value. Using a value of 1 will cause the damping to remain constant over the step.
- Applied to all contact pairs equally in the normal and tangential directions
  The tangential component of the damping will introduce tangential forces in the model even if the contact is frictionless. Subsequently, care must be taken when interpreting the results. As noted above, by default, the damping is ramped to zero over the step. When using the default behavior, the tangential forces will be ramped to zero by the end of the step as well; however, during the step the forces are active and participate in the solution.
  - The damping in the tangential directions can be scaled or removed in Abaqus CAE by selecting Interaction Module and proceeding to **Contact Controls Editor**, then **Stabilization** → **Automatic Stabilization** → Tangent fraction and enter the value.
  - or in the input file by using the command line ***CONTACT CONTROLS, STABILIZE, TANGENT FRACTION=value**.

  In Abaqus CAE, the damping can be selectively applied to individual contact pairs by creating individual contact control definitions and associating them with the contact pairs. In the input file, the master and slave surfaces to be stabilized can be specified as ***CONTACT CONTROLS, STABILIZE, MASTER=master surface, SLAVE=slave surface**.

- Active only when the distance between the contact surfaces is smaller than a characteristic surface dimension
  By default, the opening distance over which the damping is applied is equal to the characteristic slave surface facet dimension. In the case of a node-based surface, a characteristic element length obtained for the whole model is used.
  - The distance over which the damping is applied can be changed in Abaqus CAE by selecting Interaction Module and proceeding to **Contact Controls Editor** then **Stabilization** → **Automatic Stabilization** → **Clearance** at which damping becomes zero → Specify the value.
  - or in the input file by using the command line ***CONTACT CONTROLS, STABILIZE, , damping range value**.
- Not guaranteed to be an optimal value
  While the default damping coefficient is typically appropriate, results must be checked to ensure sure that the damping is not distorting the solution. Output variable **ALLSD** measures the energy dissipated by viscous damping; the ratio of this quantity to the elastic strain energy or other appropriate general energy measure should be small. Note that if you are also using automatic stabilization for the entire model (***STATIC, STABILIZE**), **ALLSD** will include contributions from both sources of stabilization. Using both sources of damping simultaneously is discouraged as this will probably over damp the model; warnings will be issued in the analysis data (.dat) file when both are activated in the same step.
  Additionally, a comparison of the contact damping stresses with the output variable **CDSTRESS** to the true contact stresses **CSTRESS** after contact is established will allow for a more detailed understanding of the damping effect.
  If the stiffness of the underlying slave material is relatively high and the loading amplitude being applied to the slave body is small, the damping forces may balance the load and prevent contact from actually occurring. In this case, the level of damping should be reduced.
  - If the default damping coefficient is not appropriate, it can be scaled or specified directly in Abaqus CAE by selecting Interaction Module and proceeding to **Contact Controls Editor**, then **Stabilization** → **Automatic Stabilization, Factor** and enter the factor or the stabilization coefficient as the damping coefficient.
  - or in the input file by using the command line ***CONTACT CONTROLS, STABILIZE=factor** or ***CONTACT CONTROLS, STABILIZE** using the damping coefficient.

7.11 Modeling Contact with Second-Order Tetrahedral Elements

This section will discuss the best way to model contact surfaces using second-order tetrahedral elements in Abaqus standard. The analyst should consider the following guidelines when modeling contact with second-order tetrahedral elements:

1. If the finite sliding, surface-to-surface contact formulation with default settings is used, contact surfaces on C3D10 elements generally provide better accuracy than contact surfaces on C3D10M elements—without degraded contact robustness. Note that the penalty constraint enforcement method with node-centered constraint positioning is the default behavior.
2. C3D10 is preferable when the small sliding, surface-to-surface contact formulation is being used.
3. The finite sliding, surface-to-surface default formulation is not used as having C3D10 elements underlay the slave surface can degrade convergence and cause contact stress noise.

   - It can be difficult to obtain converged solutions when C3D10 elements are used for the slave surfaces if hard contact constraints are strictly enforced. In these cases, contact constraints associated with nodes at the corners of C3D10 elements are likely to chatter as a result of uneven force distributions for these elements. If the analysis does converge, the contact pressures are likely to be noisy even though the underlying element stresses are probably quite accurate.
   - Convergence with slave surfaces based on C3D10 elements is improved if a softened pressure–overclosure behavior is specified (note, however, that slave surfaces based on C3D10M elements are preferred with finite sliding and surface-to-surface default formulation is not used):
     – ***SURFACE BEHAVIOR, PENALTY**
     – ***SURFACE BEHAVIOR, AUGMENTED LAGRANGE**
     – ***SURFACE BEHAVIOR, PRESSURE-OVERCLOSURE = [EXPONENTIAL | LINEAR | TABULAR]**

     For these cases, Abaqus uses supplementary constraints to improve the distribution of contact forces and reduce contact pressure noise. While the supplementary constraints remove one source of convergence degradation (uneven contact force distribution), they cause another potential source of convergence degradation in that the total number of constraints exceeds the number of slave nodes.

   Note that if the finite sliding, surface-to-surface default formulation is used with C3D10 elements, but the formulation parameters (such as penalty stiffness) are changed, the effect on convergence will be problem dependent. For example, in some cases raising the penalty stiffness may harm the convergence behavior.
4. C3D10 elements can more accurately represent surface curvature than C3D10M elements. In many cases, this factor is predominant in the choice of element type underlying the master surface.

5. For a matched mesh across a contact interface, having different element types underlying the master and slave surfaces will cause noise in the contact stress solution; the exception is flat surfaces using the surface-to-surface formulation. For example, if a matched mesh is used and other factors imply that C3D10M elements should underlay the slave surface, then choosing C3D10M elements (rather than C3D10) for the master surface will provide a better contact stress solution.
6. The penalty method is recommended over strictly enforced hard contact, regardless of which contact formulation is used. In general, the penalty method increases the robustness and performance with insignificant degradation in accuracy.
7. If the underlying materials are incompressible or nearly incompressible, as is the case with many hyperelastic or elastic–plastic materials, C3D10 and C3D10M elements may have volumetric locking problems (with C3D10 being somewhat more susceptible). C3D10H and C3D10MH elements can be used to avoid these problems, at some added expense.
8. For problems with contact stress singularities or highly localized contact stress concentrations relative to the mesh refinement, C3D10M elements are more likely to smooth over a localized effect. For example, C3D10M elements tend to under predict the maximum contact stress for a highly localized contact stress concentration whereas, with the same mesh, C3D10 elements may tend to overpredict the maximum contact stress.
9. If the problem posed has a stress singularity near the edge of the contact region (where a corner of one surface meets a smooth part of another surface); with a fairly coarse mesh C3D10M elements may predict that the maximum contact stress is a bit high at these locations but the C3D10 elements with the same mesh will tend to show significantly higher peaks. With either element type, the peak value will continue to increase with more and more mesh refinement.

References

1. Dienwiebel M, Verhoeven GS, Pradeep N, Frenken JW, Heimberg JA, Zandbergen HW (2004) Superlubricity of graphite. Phys Rev Lett 92:26101
2. Cobb F (2008) Structural engineer's pocket book, 2nd edn: British standards edition. CRC Press, Boca Raton

Part III
A Toolbox to Do the Job

Some troubleshooting tasks involve multiforms and require a specific procedure to first identify the nature of the troubleshooting and then follow a certain procedure to fix the numerical issues or difficulties. The following chapters will provide a toolkit of the main troubleshooting procedures commonly met in structural analyses in order to fix an error or warning message. Numerical criteria will be discussed regarding some specific cases. Finally, some methods of carrying out control checks will be given in the case of users needing help to perform a specific task in order to minimize troubleshooting from sub-modeling, a restart, or shell techniques analyses.

Chapter 8
Troubleshooting in Job Diagnostics

8.1 Guidelines with Abaqus Standard

This section describes the main steps to be taken when an Abaqus standard analysis has not completed. How can users determine and correct the cause of such a failure in an analysis job? The content of this section will provide guidance to help solve many of the common errors that users may encounter when trying to do an Abaqus standard analysis.

1. Have you read the relevant documentation?
   Abaqus provides online documentation with extensive searching capabilities. Reading the documentation on your type of analysis before spending a lot of time working on it can save you time in the end. Most difficulties in using Abaqus can be solved by following the instructions in the manuals carefully.

   - Suggested order of reading
     - If the analyst is a new user, read "Getting Started with Abaqus" or "Getting Started with Abaqus/Standard: Keywords Edition." Abaqus uses these manuals to train new support engineers, so they are a reliable source and users can be confident that they are the best introduction to using Abaqus/Standard.
     - The Abaqus lecture notes on a particular topic provide a short and simple introduction to many common analysis issues. A list of available lecture notes is available on the Abaqus Home Page. Many of the lecture notes have associated workshops that can help the user to understand how to use Abaqus.
     - The Abaqus/Standard User's Manual (versions 6.3 and earlier) and the Abaqus Analysis User's Manual (versions 6.4-1 and later) provide details on using Abaqus/Standard. Use the online documentation to search for a specific topic.
     - The Example Problems, Benchmarks, and Verification manuals contain descriptions of many test cases that are useful to examine. The problems in the Example Problems and Benchmarks manuals are easy to understand, yet

R. J. Boulbes, *Troubleshooting Finite-Element Modeling with Abaqus*,
https://doi.org/10.1007/978-3-030-26740-7_8

realistic. The user can almost always find a problem in these manuals with features similar to the analysis performed.
- Use the online documentation to search for the particular topic in which users are interested. All the input files for these problems are also available, either from the online documentation in the Abaqus Input Files manual or from the **findkeyword** option. The user can fetch a particular input file using the Abaqus/Fetch utility.

2. Does your analysis simulation terminate prematurely?

- Does the analysis simulation terminate during the data-check phase of the analysis?
  If this is the case, the user will see *****ERROR** messages in the printed output (.dat) file. The printed output file gives detailed warnings and error messages that can help the user to find the modeling mistakes. Any errors that appear in the printed output file were encountered in the **data-check** phase and must be corrected before the analysis can start.
- Does the analysis simulation terminate during the analysis phase before the simulation has completed?
  Any warnings and errors encountered in the analysis phase are written to the message (.msg) file. The user must look into the message file to find the reasons why the analysis stopped prematurely.
  Abaqus viewer can be used to visualize job diagnostic information including detailed information about convergence behavior, numerical problems, and contact state changes.
  There are many reasons why an analysis might stop before it is complete. Frequently, the cause is a modeling error: either the physics is not modeled correctly or the approximation to the correct physics is poor. Abaqus probably cannot solve poorly modeled problems! The user probably does not want the answer to poorly modeled problems either!
  Assuming the physics has been modeled reasonably, the following questions can help you to determine why the analysis did not complete. Do not suppress or restrict the printout to the message (.msg) file, and do not restrict the frequency of the residual printout to the message file. To activate diagnostic printing:
  - Go to Step Module then **Output → Diagnostic Print... → [Contact | Plasticity | Residual]**
  - With keyword command lines enter ***PRINT, FREQUENCY=1, [CONTACT | PLASTICITY | RESIDUAL] = YES**

  In addition, remove all the solution controls and contact controls that can reduce the accuracy of the solution. These options are normally used to get around a convergence problem; they will obscure the causes of the problem in the analysis. The user may have to use them but should do so only as a last resort after considering all other possibilities.
  - Are too many increments needed when all increments converge?
    An insufficient number of increments is specified in the step definition to reach the total time for the step. The user needs to check why this has happened; the message file will have the clues.

– Were there too many cutbacks in the last time increment that was attempted? Abaqus standard is unable to converge during a particular time increment. In the message file, the user will see the printout indicating that Abaqus standard is making a number of attempts to solve the increment. Each subsequent attempt uses a smaller time increment until the time increment is less than the minimum time increment allowed, at which point the analysis stops. Look into the message file to find the cause. Abaqus standard may not be able to determine the correct contact state or may not be able to achieve equilibrium. It is also important to note when the analysis terminates is it in the first increment of a step or in a later increment?

3. Does the analysis complete but the results look suspicious.
The analyst is responsible for checking that the results obtained make sense. If the analysis completes, it is not a guarantee that the intended results have been obtained. If the results do look suspicious, consider the suggestions inside the next section to help determine the problem.

8.2 Job with Abaqus Standard Completes, But the Results Look Suspicious

Analysts must check that the results they obtain make sense. Always check the deformed mesh plot. Does the deformation look sensible?[1] If it does not, this may indicate a problem. It is quite easy to detect problems like hourglassing from deformed mesh plots: there will be a regular mode of deformation over a region of elements. Since it is unlikely that the deformation depends on the element size, verification is always suspicious. Check the hourglassing effect as a suggestion.

The following items are suggested for consideration when reviewing results:

1. Contact-related issues:

- Is there excessive penetration of the slave surface by the master surface.
  – Refine the underlying mesh of the slave surface. The slave surface should be more refined than the master surface.
  – Make sure the slave surface has the softer underlying material (lower Youngs modulus) if the mesh densities of the master and slave surfaces are roughly equal. Again, the stiffness is not just dependent on material properties; it also depends on the shape and the amount of restraint.
  – If small sliding is invoked, make sure the slave nodes initially, see the master surface.
- Strange looking contact output
  Make sure the slave nodes initially, see the master surface. Use ***PRINT,**

[1]If the user is using Abaqus/Viewer, they should make sure the deformed plot scale factor is set equal to one. Sometimes by default, it is set very high or very low. If it is very high, it will show elements being deformed or more distorted than they actually are, and if it is very low, the deformation may not be visible at all.

CONTACT=YES to get a detailed printout of the contact state as Abaqus is iterating. In this printout, you will see messages indicating that a slave node cannot see a master surface.
Does the problem truly involve small sliding (the slave node should slide less than about one-element length from the initial point of contact)
– The local contact plane may no longer accurately represent the surface geometry if excessive sliding occurs; use finite sliding in this case. See "Contact formulations in Abaqus standard" in Sect. 38.1.1 of Abaqus Analysis User's Guide (6.13).

2. Are geometrically nonlinear effects important. Were they considered in the analysis? See the section, "General and linear perturbation procedures" in Sect. 6.1.3 of the Abaqus Analysis User's Guide (6.13).
3. Abaqus standard performs a small strain, small rotation analysis unless the analyst used ***STEP, NLGEOM**; using the NLGEOM parameter makes Abaqus standard perform a large displacement rotation analysis. For most elements, large strains are also used, but some shells, beams, and special-purpose elements use small strain assumptions. See "Defining an analysis" in Sect. 6.1.2 of the Abaqus Analysis User's Guide (6.13).
4. Are the loading and boundary conditions reasonable. Check the regions on which loads and boundary conditions are applied in Abaqus viewer whenever possible. For further information, see "Conventions" in Sect. 1.2.2 of the Abaqus Analysis User's Guide (6.13).
5. Are the material properties correct and reasonable. See "Output" in Sect. 4.1.1 of the Abaqus Analysis User's Guide (6.13).
6. If users are using a new material model, do some simple one-element tests with this material to make sure that it behaves in the way the user expected. Often this simple check will show errors in the specification of the material model. In the Abaqus Verification Manual, there are many input files that test material models. Modifying one of these input files to check the material defined by the user, makes it easy to get the necessary results.
7. Problems when using a user subroutine:

- If a user subroutine is involved, was it verified by a simple test with one-element model.
  - For the user subroutine **UMAT**, always run uniaxial tension and pure shear tests (planar tension is the same as pure shear for incompressible materials). Make sure that the **UMAT** is checked for strain ranges greater than the user expects. Since the user does not know what the deformation will be, it is always possible that the material will deform by more than the user expects.
  - Make sure the argument list for the user subroutine is as documented in the manual. Check (by using a printout) that the variables passed into the user subroutine are what the user expects. If they are not, find out why; the user probably has a coding error. On some machines, incorrect argument lists can also cause link errors.

- For more information see "User subroutines: overview" in Sect. 18.1.1 of the Abaqus Analysis User's Guide (6.13).

8. When using ***CONTOUR INTEGRAL** for fracture mechanics, the contour values between node sets along the crack line show unexpectedly rapid variation. See "Fracture mechanics: overview" in Sect. 11.4.1 of the Abaqus Analysis User's Guide (6.13).

   - Check that the results for stress and strain look reasonable.
   - Check that the node sets defining the crack tip have been defined correctly. Print out all the node sets for the contours to the data file. Check the nodes in the node sets for the first contour. If there are more than 16 node sets defining the crack line, make sure there are only 16 node sets per line.
   - Check that the mesh is defined correctly. Indeed, the contouring algorithm around the crack area in Abaqus has limitations on how the mesh can be created.

9. Make sure that mesh mapped is adequately refined; it is the user's responsibility to make sure that refining the mesh will not change the results significantly. Good places to start refining the mesh are those where the mesh becomes severely distorted. In principle, an analyst should always carry out a mesh refinement study; in practice, this may not be possible mostly because of the project time consumption. If for some reason the adequacy of the mesh cannot be checked, the user should bear in mind that this important check on the results has not been done. Experience may allow the user to make the judgment.
10. Overly stiff behavior in a mesh made of the Abaqus fully integrated first-order elements can result from shear and volumetric locking effects. Mesh refinement or use of reduced integration or incompatible mode elements can help with these problems.

11 If the material definitions have field variable dependence and an initial stress state is defined (using ***INITIAL CONDITIONS, TYPE=STRESS**), it can be confusing to understand what happens if the values of the field variables (used in the material definitions) are not initialized using ***INITIAL CONDITIONS, TYPE=FIELD** and/or if the range of field variables used in the material definition table does not include the intended initial values of those field variables.

   Abaqus standard converts the initial stresses to initial elastic strains before it starts the analysis. To do this, the material stiffness for each element must be calculated. The values of the field variables used in this calculation are the values specified using ***INITIAL CONDITIONS, TYPE=FIELD** (if they are not specified, zero is used). Two errors are commonly made:

   - The user forgets to specify the initial values of the field variables but specifies nonzero values using ***FIELD** in a step in which the field variables remain constant (such as a ***GEOSTATIC** step). Abaqus uses values of zero for the field variables to calculate the initial elastic strains, whereas nonzero values of the field variables are used to evaluate the material properties during the step.

- For more information, see "Material data definition" in Sect. 21.1.2 of the Abaqus Analysis User's Guide (6.13).

The result is that the model deforms in an unexpected way or, in some cases, does not even converge. The corrections are as follows:

- Use ***INITIAL CONDITIONS, TYPE=FIELD** to define the initial values of the field variables; and/or
- Extend the range of the material definition data to include initial values of the field variables.

The user subroutine **USDFLD** modifies the field variables only at material integration points, and then only for the current iteration. The calculation of the field variable in **USDFLD** must be repeated in every iteration.
There are options in Abaqus standard that are not material properties but which can also have field variable dependence. An example is ***FILM PROPERTY**. The field variables for this type of option can be specified only by using ***INITIAL CONDITIONS**; ***FIELD**; or ***FIELD, USER**.

8.3 Model a Structure Undergoing a Global Instability

A nonlinear static problem can be unstable as a result of global buckling or material softening. If the load versus displacement response of the model seems to be reaching a load maximum and there is the possibility of global instability or negative stiffness, two approaches to solving the problem can be used: static or dynamic analysis.

- If the structure is reaching a buckling load in a static analysis, for instance, if there is a snap through then perform a Riks analysis. Otherwise, use Newton–Raphson or Modified Newton–Raphson solver.
The Riks method assumes that the global instability can be controlled by modification of the applied loads. This means the loss in stability cannot be too severe; that is, there cannot be a sharp bifurcation in the load versus displacement curve. Therefore, structures such as flat sheets, cylinders, and spheres that have a sudden significant loss of stiffness after buckling must have some imperfection built into the original geometry.
This can be done by using the ***IMPERFECTION** option to modify the original geometry by adding imperfections. The best approach is to use experimentally determined imperfections; however, since these measurements may not be available, the ***IMPERFECTION** option can use combinations of the eigenmodes from a previous buckling analysis as the imperfections added to the original geometry.
If the Riks method fails to converge near a limit or bifurcation point (buckling load), the problem may be that the loss in stiffness is too severe. Instability problems that exhibit a sharp transition often require a limit on the maximum incremental arc length to get past the transition point or need to have larger imperfections built into the geometry.

- If a dynamic analysis is desired, Abaqus explicit should be considered as the most robust approach, particularly in the presence of material failure, extreme deformation, or rapid changes in contact state. If the loss in stability is not too severe, or only the load maximum is to be computed rather than a fully collapsed configuration, then a dynamic analysis in Abaqus standard may be completed with less run time. Choose the APPLICATION parameter in ***DYNAMIC** to control the amount of numerical damping that is applied to the integration operator. If a dynamic analysis is used in Abaqus explicit, the structure will vibrate once it has passed the instability and you must decide how to damp the vibrations if a quasi-static solution is required.

Global instabilities can also be stabilized in a static analysis with viscous forces. Although not intended as a primary solution technique for global instabilities, automatic stabilization can be used in the static, coupled temperature–displacement, soils, and quasi-static procedures. Automatic stabilization will add viscous damping to the structure, which may allow the solution to go beyond the instability point.

Discrete dashpots can also be used to stabilize a problem of this type.

With either technique, the energy dissipated by the artificial viscous forces (output variable **ALLVD** for discrete dashpots or **ALLSD** for automatic stabilization) should remain small compared to the total internal energy (output variable **ALLIE**) in the problem. The nodal viscous forces should also be small when compared with typical forces in the problem (use the nodal output variable **VF**).

8.4 Correct Convergence Difficulties Caused by Local Instabilities

Some sources of instability include local yield, snap-through, surface wrinkling, and localized material failure. When local instabilities develop, it may not be possible to obtain a static solution; a dynamic method must be used or artificial damping must be introduced when using a static method. Local instabilities are indicated by sudden convergence difficulty with very small time increments. Unlike the case of global instabilities, the Riks method will not help in these cases.

The corrective action is normally the addition of a dissipation mechanism. Use automatic stabilization with the static, coupled temperature–displacement, soils, or quasi-static procedures; check that the viscous forces (use output variable **VF**) introduced are small compared to any other typical forces and that the energy dissipated by the automatic stabilization (use output variable **ALLSD**) is small relative to the internal energy **ALLIE**.

Alternatively, use discrete dashpots; apply them to all nodes in all directions; adjust the damping coefficient as necessary to ensure that dashpot forces are small relative to typical forces in the structure and that the viscous dissipation energy **ALLVD** is small relative to the internal energy **ALLIE**.

If the local instability is due to a loss of contact, contact damping can stabilize the contact region while the contact changes.

If the region of the model experiencing local instability is also simultaneously experiencing a large global deformation or rigid body motion, the automatic stabilization will not discriminate between the two scales of motion. Velocity proportional forces are applied to any motion; care must be taken to ensure that stabilization does not significantly alter the results. In some situations, it may be helpful to place dashpots between the locally unstable region and some other part of the structure so that the dashpots will react to relative rather than total velocities.

Discrete dashpot elements can also serve to prevent rigid body motion.

8.5 Correcting Errors During the Data-Check Phase of an Analysis

The data-check phase of an analysis checks whether the model is set up correctly. Abaqus cannot make sure that the physics of the model is correct; it will check as much as possible that the model makes sense. Details on the data-check execution procedure can be found in "Abaqus/Standard, Abaqus/Explicit and Abaqus/CFD execution" Sect. 3.2.2 of the Abaqus Analysis User's Guide (6.13).

When trying to correct data-check errors, consider the following points:

1. Check the geometry, boundary conditions, loadings, surface definitions, and material properties to make sure they are correct and reasonable.
   - The printout to the data (.dat) file in the data-check phase is switched off by default. It is wise to switch this printout on, at least until the user is sure to have the model set up correctly. If the user does not have the model set up correctly, it is useful to check how Abaqus interpreted the input data by looking at the data file. Switch on this printout by using
     – Go to Job Module then **Job** → **Create...** → **Continue...** → **General** → **Preprocessor printout** and request printout of the input data, model definition data and history data.
     – With keyword command line enter ***PREPRINT, MODEL=YES, HISTORY=YES**
     Material data and section property data are most easily checked in the (.dat) file. There is an environment file setting, **printed_output=ON**, that can be used to make this printout the default. Note, however, that this can sometimes cause the memory estimates for the problem to become erroneously large.
   - If the user has contact defined in the model, use the following in order to check the initial state of contact (open or closed and by how much):
     – Go to Job Module then **Job** → **Create...** → **Continue...** → **General** → **Preprocessor printout** and request printout of the contact constraint data.
     – With keyword command line enter ***PREPRINT, CONTACT=YES**

- Unless the user has a serious error in the input file, Abaqus will write information to the output database (.odb) file and the restart (.res) file (if the user requested restart output). The analyst can use Abaqus viewer to read the output database file to check the model graphically.
  Look at the model to make sure that the geometry is correct, the mesh is reasonable, the boundary conditions are applied correctly, the loads are in the correct places, and the surfaces are defined correctly. If the user has contact defined in the problem, it is quite easy to check the contact pairs in Abaqus viewer. In addition, check the normals to contact surfaces; if they are pointing in the wrong direction, the contact algorithm will try to apply this incorrect definition and probably cause severe distortion in the mesh.

2. Make sure the model is defined according to a valid and consistent set of units. Abaqus does not use any built-in units.
   A common error is to use the incorrect mass units for density in dynamic analyses, especially when using British units. In the inside cover of every printed Abaqus manual and at the beginning of the online manuals, there are tables that can assist the user in making unit conversions; alternatively Table. 3.1 can be used as an example. Using units that make numbers very large or very small can cause roundoff problems during the analysis in rare cases.
3. If the user gets the error message **THERE IS NOT ENOUGH MEMORY ALLOCATED TO PROCESS THE INPUT DATA**, to understand memory usage in Abaqus standard then read:
   - "Using the Abaqus environment settings" in Sect. 3.3.1 of the Abaqus Analysis User's Guide (6.13).
   - "Memory and disk management parameters" in Sect. 4.1.1 of the Abaqus Installation and Licensing Guide (6.13).
4. Particular kinds of analyses can use large amounts of memory in the data-check phase as well as the analysis phase, even if the number of degrees of freedom does not seem extremely large. These analyses include the following:
   - Automatic computation offers a reasonable sliding distance for three-dimensional finite sliding problems.
   - Cavity radiation problems with cavities that are finely meshed. The user may have to mesh the cavity more coarsely than the surrounding mesh. This will involve tying the heat transfer elements defining the cavity to the finer surrounding mesh using tie constraints. Performance enhancements have been made in Abaqus to alleviate this problem between memory usage and cavity radiation analysis.
   - Problems in which equations or multi-point constraints[2] are used to constrain a large number of nodes to a single node. If possible, use some other technique to tie the nodes together as described in Sect. A.2.

[2]See in Abaqus/CAE User's Guide v6.14 inside Sect. 15.15.6 Defining MPC constraints.

- Substructures with large numbers of retained degrees of freedom. If possible, reduce the number of retained degrees of freedom.

5. Other types of common errors can be found in input data file with improper syntax to define parameter in command line, for example, if a line has only one data item on it, this data item must be followed by a comma. For more information, see "Input syntax rules" in Sect. 1.2.1 of the Abaqus Analysis User's Guide (6.13).
6. If troubleshooting is due to an old version of the Abaqus model, the user can easily convert an input file from an older Abaqus version by using the **abaqus upgrade** utility. For more information see
   - "Fixed format conversion utility" in Sect. 3.2.25 of the Abaqus Analysis User's Guide (6.13).
   - "Input file and output database upgrade utility" in Sect. 3.2.17 of the Abaqus Analysis User's Guide (6.13).

When the user has finished the data-check phase, the data file should be reviewed for any warning messages. Make sure the user has a good reason to ignore any warnings before continuing. The warnings are there to help the user to create good models. Often they indicate serious errors. Sometimes they can be ignored. The user should understand the implications of ignoring any warning messages.

8.6 Analysis Ends Prematurely, Even Though All the Increments Have Converged

Even if all the increments eventually converge, a sufficient number of increments must be specified in the step definition to reach the total time for the step. To find out why the analysis ran out of increments, check the message (.msg) file; it will have the clues.

Abaqus standard may have had trouble achieving convergence at some point in the analysis (cutbacks in the time increment will occur). It is possible that the occurrence of too many cutbacks will mean that simulation cannot complete properly for no other reason than the fact that the number of time increments permitted is insufficient. If this is the case:

- restart the analysis and define a new step to complete the unfinished part of the step. If loads or boundary conditions use nondefault amplitude curves, make sure the amplitude curves are referred to correctly on restart;
  - In Abaqus CAE, go to Job Module then **Job → Create → Restart**
  - With keyword command line enter the following line ***RESTART, END STEP, READ, STEP =..., INC =...**
- or increase the number of allowable increments in the step definition, and rerun the analysis.

If the cutbacks in the time increment reduce the time increment to such a small value that the analysis will not complete in a reasonable amount of time, the cause may be similar to the situation in which an increment does not converge

8.7 Debugging Divergence with Too Many Cutbacks in the Last Attempted Increment

Possible sources of the problem include

- Failure to determine the contact state, which is identified by termination after too many severe discontinuity iterations (SDIs). These difficulties are generally caused by problems with the contact definitions. Consider the following:

  1. If the termination occurs in the first increment, possible causes include excessive over-closures.
  2. If the termination occurs after the first increment, possible causes are contact chattering (the slave surface contact state alternates between open and closed repeatedly so that the maximum number of SDIs is reached). Look at the causes of contact chatter for additional information.

- Failure to achieve equilibrium, which is identified by force/moment residuals and displacement/rotation corrections not getting smaller; a termination error indicates divergence. Consider the following items in diagnosing the problem:

  1. Contact-related issues.
  2. Excessive element distortion warnings, which normally occurs only after the analysis simulation has partially completed, unless the initial model was incorrect.
  3. Hourglassing with a deformed mesh that goes into a regular pattern that is clearly not physical; this is normally seen easily in deformed mesh plots.
  4. Excessive yielding, which is generally associated with a message that the current strain increment exceeds the strain to first yield by a factor of more than 50 times.
  5. Large elastic strains when using an elastic material model in a geometrically nonlinear analysis.
     Large elastic strains should be modeled with the hyperelastic or hyperfoam material models; elastic material models are intended for elastic strains that remain small ($\leq 5\%$). Elastic material models can be used with other material models, such as plasticity models, to simulate structures that undergo arbitrarily large strains (when using a plasticity model, the large strains are inelastic).
  6. Is the material response incompressible or nearly incompressible (frequently elastic–plastic materials are nearly incompressible.)

7. Possible unconstrained rigid body motion indicated by the presence of NUMERICAL SINGULARITY warnings and very large displacement corrections.
8. Over-constraints, which are frequently indicated by zero pivot messages in the message (.msg) file.
9. Local instabilities such as wrinkling or material localization developing during the analysis.
10. The analysis ends in a core dump; the analysis does not complete and output files end in the middle.
11. If the analysis seems to be approaching a load maximum and negative eigenvalue messages appear during the iterations of the last increment, the analysis may be reaching a global instability.
12. Very small displacement correction, but the tolerance on the residual force is not satisfied.

    This is probably due to numerical precision issues such as all the coordinates in the model being very large compared to the size of the model or using units in which typical quantities such as mass, force, or energy are very small or large. If the displacement corrections are truly small relative to the displacement increment, the increment has probably converged. The tolerance on the force residual can be relaxed in this case.
13. Follower loads (including distributed pressures) in analyses including nonlinear geometric effects.

8.8 Using Follower Loads in Nonlinear Analyses

This section describes what steps to take when a model with large displacement in Abaqus standard analysis uses follower forces (or distributed pressures) and is having convergence difficulties. The cause could be that the follower loads in large displacement analyses (when analysts are using ***STEP, NLGEOM**) introduce unsymmetric load stiffness terms in the model's stiffness matrix. Abaqus standard symmetrizes the load stiffness matrix unless you use the UNSYMM parameter in the ***STEP** option (or if the unsymmetric solver is activated automatically); generally, this will give a solution, provided that the deformation is not too large. However, when the deformation is large, using the symmetrized stiffness matrix will cause slower convergence during an increment and possibly no convergence.

Pressure loads on a surface are follower loads when the edges of the surface are not constrained (in the direction tangential to the surface at the edges). Users frequently do not realize this and forget to use the unsymmetric solver in finite strain analyses.

If the analyst uses the unsymmetric solver, it should be realized that one pass through the equation solver will take about four times as long as when the symmetric solver is used. In cases where convergence is not too slow, using the unsymmetric equation solver may not be worthwhile.

8.9 Understanding Negative Eigenvalue Messages

Negative eigenvalue messages are generated during the solution process when the system matrix is being decomposed. The messages can be issued for a variety of reasons, some associated with the physics of the model and others associated with numerical issues. An example of a message issued is

```
***WARNING: THE SYSTEM MATRIX HAS 16 NEGATIVE
EIGENVALUES.

IN AN EIGENVALUE EXTRACTION STEP THE NUMBER OF
NEGATIVE EIGENVALUES IS THIS MAY BE USED TO
CHECK THAT EIGENVALUES HAVE NOT BEEN MISSED.

NOTE: THE LANCZOS EIGENSOLVER APPLIES AN INTERNAL
SHIFT WHICH WILL RESULT IN NEGATIVE EIGENVALUES.

IN A DIRECT-SOLUTION STEADY-STATE DYNAMIC ANALYSIS,
NEGATIVE EIGENVALUES ARE EXPECTED. A STATIC ANALYSIS
CAN BE USED TO VERIFY THAT THE SYSTEM IS STABLE.

IN OTHER CASES, NEGATIVE EIGENVALUES MEAN THAT THE
SYSTEM MATRIX IS NOT POSITIVE DEFINITE:

FOR EXAMPLE, A BIFURCATION (BUCKLING) LOAD MAY HAVE
BEEN EXCEEDED.

NEGATIVE EIGENVALUES MAY ALSO OCCUR IF QUADRATIC
ELEMENTS ARE USED TO DEFINE CONTACT SURFACES.
```

Physically, negative eigenvalue messages are often associated with a loss of stiffness or solution uniqueness in the form of either a material instability or the application of loading beyond a bifurcation point (possibly caused by a modeling error). During the iteration process, the stiffness matrix can then be assembled in a state which is far from equilibrium, which can cause the warnings to be issued.

Numerically, negative eigenvalues can be associated with modeling techniques that make use of Lagrange multipliers to enforce constraints or local numerical instabilities that result in the loss of stiffness for a particular degree of freedom. Most negative eigenvalue warnings associated with Lagrange multipliers are suppressed; the exceptions are when quadratic three-dimensional elements are used to define contact surfaces or when hybrid elements are used in a geometrically nonlinear simulation and undergo large deformations.

Mathematically, the appearance of a negative eigenvalue means that the system matrix is not positive definite. If the basic statement of the finite-element problem is written as in Eq. 8.1 with the vector of nodal load F, then the stiffness matrix K and the nodal displacement vector x.

$$\mathbf{F} = (K)\,\mathbf{x} \tag{8.1}$$

Then a positive-definite system matrix K will be nonsingular and satisfy as shown in Eq. 8.2,

$$\mathbf{x}^T\,(K)\,\mathbf{x} > 0 \tag{8.2}$$

for all nonzero **x**. Thus, when the system matrix is positive definite, any displacement that the model experiences will produce positive strain energy.

In addition to the causes shown in the warning message, some situations in which negative eigenvalue messages can appear include

- Buckling analyses in which the prebuckling response is not stiff and linear elastic. In this case, the negative eigenvalues often point to spurious modes. Remember that the formulation of the buckling problem is predicated on the response of the structure being stiff and linear elastic prior to buckling.
- Unstable material response:
  - A hyperelastic material becoming unstable at high values of strain.
  - The onset of perfect plasticity.
  - Cracking of concrete or other material failure that causes softening of the material response.
- The use of anisotropic elasticity with shear moduli that are unrealistically very much lower than the direct moduli. In this case, ill-conditioning may occur triggering negative eigenvalues during shearing deformation.
- A nonpositive-definite shell section stiffness defined in a **UGENS** routine.
- The use of a pretension node that is not controlled by using the ***BOUNDARY** option and lack of kinematic constraint of the components of the structure. In this case, the structure could fall apart due to the presence of rigid body modes. The warning messages that result may include one related to negative eigenvalues.
- Some applications of hydrostatic fluid elements.
- Rigid body modes due to errors in modeling.

Negative eigenvalue warnings will sometimes be accompanied by other warnings, addressing such things as excessive element distortion or magnitude of the current strain increment. In cases where the analysis will not converge, resolution of the non-convergence will often eliminate the negative eigenvalue warnings as well.

For analyses that do converge, carefully check the results if the warnings appear in converged iterations. A common cause of negative eigenvalue warnings is the assembly of the stiffness matrix about a nonequilibrium state. In these instances, the warnings will normally disappear with continued iteration, and, if there are no warnings in any iterations that have converged, warnings that appear in non-converged iterations may safely be neglected. If the warnings appear in converged iterations, however, the solution must be checked to make sure it is physically realistic and acceptable. It may be the case that a solution satisfying the tolerance for convergence has been found for the model while it is in a nonequilibrium state.

8.10 Divergence with Numerical Singularity Warnings

These warnings indicate that so many digits are lost during linear equation solution that the results are not reliable. Most likely the model is experiencing an unconstrained rigid body motion, which is most commonly caused by a lack of constraints against such motion in a static stress analysis. Unconstrained rigid body motion results in a singular tangent stiffness matrix (i.e., one that cannot be inverted); this is indicated by the presence of NUMERICAL SINGULARITY warnings in the message (.msg) file and very large displacement corrections.

Frequently the reason for the rigid body motions is that contact has not been established and therefore cannot prevent rigid body motion.

The corrective action is to constrain the model properly (using boundary conditions or soft springs). In contact problems where contact and friction will prevent rigid body motion once contact is established use

- ***CONTACT STABILIZATION** to apply viscous damping stabilization to the normal and tangential directions in a general contact domain.
- ***CONTACT CONTROLS, STABILIZE** to apply viscous damping stabilization to the normal and tangential directions of a contact pair.

The ***STATIC, STABILIZE** option, discrete dashpots, or soft springs can be used as well, but it is **better** to avoid the problem by defining the correct boundary conditions or by establishing contact using displacement boundary conditions before applying loads. We recommend that this option not be used in combination with contact-specific stabilization as this will over damp the model.

Care must be taken when using any type of viscous stabilization. Check the viscous forces (output variable **VF**) and compare them with the expected nodal forces to make sure that the viscous forces do not dominate the solution. If necessary, follow the stabilized step with another step in which stabilization is not used or with a step in which a much smaller amount of damping is used. This will allow the model to re-equilibrate without the viscous forces.

With any of the aforementioned stabilization techniques, the energy dissipated by the artificial viscous forces (output variables **ALLSD**, **ALLVD**, **ALLCCSD**[3]) should remain small compared to the total internal energy (output variable **ALLIE**) in the problem.

---

[3]The sum of ALLCCSDN and ALLCCSDT. ALLCCSDN is the contact constraint stabilization dissipation in the normal direction, and ALLCCSDT is the contact constraint stabilization dissipation in the tangential direction.

8.11 Zero Pivot Warnings in the Message File

Warnings in the message (.msg) file about ZERO PIVOTs[4] are typically indicative of a model constraint. These occur during linear equation solution when there is a force term but no corresponding stiffness. Common causes are unconstrained rigid body modes and over-constrained degrees of freedom. Zero pivot warnings will be accompanied by NUMERICAL SINGULARITY warnings if unconstrained rigid body modes are present. Constraint problems are generally caused by a conflicting constraint at a node. For example, a slave node in contact with a master surface while also be constrained in the direction of contact by an MPC or a boundary condition.

In addition to the presence of ZERO PIVOT warnings in the message file, constraints are indicated by the presence of very large forces (orders of magnitude larger than a typical applied force), followed by very easy convergence. Sometimes the ZERO PIVOT warnings will not appear. A warning message will be given in most cases if the change in the force between iterations is an unreasonable amount.

Frequently, the constraint is due to a boundary condition being applied to a node on a slave surface that is in contact with a rigid body. It can also be because a slave node is part of an equation, MPC or tied contact pair and, through these options, is constrained. Constraints can also occur if displacements tangential to the contact surface are held fixed and the Lagrange or the rough friction model is used.

Abaqus standard checks for constraints caused by combinations of constraints applied to the same degrees of freedom (for example, a boundary condition and contact pair intersecting with tie constraints). For certain types of constraints detected during the batch model preprocessing or the analysis, Abaqus standard will automatically resolve the constraints.

If the automatic resolution of the constraints does not resolve the problem, the user will have to manually remove the superfluous constraint. In some contact cases, this can be done by using a penalty or softened contact; however, it is usually better to correct the modeling to eliminate the constraint.

ZERO PIVOT warnings can appear as soon as the analysis enters the solver, or they may appear later during the iteration process. Be cautious of accepting results for

[4]The pivot or pivot element is the element of a matrix, or an array, which is selected first by an algorithm (e.g., Gaussian elimination, simplex algorithm, etc.), to perform certain calculations. In the case of matrix algorithms, a pivot entry is usually required to be at least distinct from zero, and often distant from it; in this case, finding this element is called pivoting. Pivot refers to pivoting technique operations on the stiffness matrix used in linear algebra [1], which may be followed by an interchange of rows or columns to bring the pivot to a fixed position and allow the algorithm to proceed successfully, and possibly to reduce roundoff error. It is often used for verifying row echelon form.

Pivoting might be thought of as swapping or sorting rows or columns in a matrix, and thus it can be represented as multiplication by permutation matrices. However, algorithms rarely move the matrix elements because this would cost too much time; instead, they just keep track of the permutations.

Overall, pivoting adds more operations to the computational cost of an algorithm. These additional operations are sometimes necessary for the algorithm to work at all. At other times, these additional operations are worthwhile because they add numerical stability to the final result.

converged solutions in which ZERO PIVOTS were encountered during the iteration in which the increment converged. The results are generally (but not always) not useful.

A final consideration concerns the effects of zero pivots in Abaqus explicit. The Abaqus standard solvers are capable of finding a direct solution to the static equilibrium equation in a finite-element problem with some operation on the stiffness matrix. Indeed the direct linear equation solver (by default) operates by rearranging the system equations and then solving them via Gaussian elimination. This method is particularly suited to solving sparse matrix problems such as often occur when the individual element equations are combined into the global stiffness matrix. The second solver is the iterative linear equation solver which works in an iterative fashion, applying a Krylov preconditioner to the system equations, followed by solution via LU factorization. The iterative solver is generally not as efficient as the direct solver, and due to its iterative nature it can suffer from divergence. It is, therefore, generally only selected when there are a large number of degrees of freedom, (e.g., $>10^6$) and the elements are highly interconnected such as in blocky structures. Both solvers can be used with a technique such as Newton's method to enable them to solve nonlinear problems, by treating them as a combination of linear subproblems. Abaqus explicit does not solve the assembled stiffness matrix problem, it is not possible to perform these operation checks, and hence zero pivot warnings cannot be reported. This does not mean, however, that the problems associated with zero pivots (over/under-constraining, ill-conditioning, etc.) do not occur and have an influencing effect on an analysis performed in Abaqus explicit. The user should, therefore, be careful when generating the model to ensure it is properly constrained and should monitor the results to ensure that localized increases in nodal force are not a result of constraint and that rigid body motions are not resulting from under-constraints.

8.12 Convergence Difficulty in the First Increment of a Contact Analysis

One or more of the contact pairs in the analysis may have an excessive initial closure. The closure may be a feature of the model or maybe the result of a modeling error. In either case, an excessive closure may cause the analysis to terminate prematurely. In this section, some common sources of excessive initial closures are discussed.

1. In the case of a contact closure present in the model, or as with an interference fit analysis described in Sect. 7.1.5.8, the closure may be too large to resolve in one increment. To help with the diagnosis, information should be printed about the closures during the data-check and solution phases:

    - In Abaqus CAE go to Job Module then in Job Manager select **Edit** → **General** → **Preprocessor printout** → **Print contact constraint data**
      In Step Module select **Output** → **Diagnostic Print...** → **Contact**

- With keyword command line enter ***PREPRINT, CONTACT= YES** and ***PRINT, CONTACT= YES**

The solution, in this case, is to use the ***CONTACT INTERFERENCE** option to remove the closure. Sometimes it is necessary to resolve the closure over several increments. Adjusting the solution controls to increase the maximum number of severe discontinuity iterations allowed in an increment may also be necessary to overcome this difficulty.

Sometimes small closures (relative to element dimensions) can be removed by adjusting the slave nodes in the contact pair:

- In Abaqus CAE go to the Interaction Module then select Interaction Manager to **Edit → Slave node adjustment**
- With keyword command line enter ***CONTACT PAIR, ADJUST=[Node set label | Adjustment value].**

2. The master surface normals may be pointing away from the slave surface.
   Use Abaqus viewer to check the normals on the master surface. The solution in this case is to redefine the master surface so that the normals point towards the slave surface, assuming that the surfaces are open.
3. The model may be permitting a rigid body motion (*very* large closures detected sometimes of the order of 10^{15}), typically due to loading a body with forces when there are insufficient boundary conditions or insufficient active contact constraints to remove rigid body motion. The user will, in most cases, also see messages about NUMERICAL SINGULARITIES in the message (.msg) file. Print closure information during the solution (as described in Item 1) to diagnose the problem. This case can be resolved with different approaches:
   - Use boundary conditions to move the bodies until they are just in contact (do this in a dummy step). In the next step, remove these boundary conditions and replace them with forces that maintain the contact (this is the recommended technique).
   - Add soft springs to the model in the directions of the rigid body motion. The degrees of freedom requiring springs can be seen from the DOFs associated with the numerical singularity messages. The stiffness of the springs must be small enough that the forces in the springs are negligible compared to typical forces in the problem.
   - Use contact stabilization with ***CONTACT CONTROLS, APPROACH** or ***CONTACT CONTROLS, STABILIZE** to stabilize the motion (using damping effects) when, at some point in the analysis, contact is meant to prevent rigid body motion. If this is not done carefully, the viscous forces can dominate the solution. Print out the forces due to viscous effects applied with these options using the **VF** output variable. Additionally, output variable **ALLSD** measures the energy dissipated by viscous damping; the ratio of this quantity to the elastic strain energy or other appropriate general energy measures should be small.

If the analysis converges in an iteration during which NUMERICAL SINGULARITIES appear, check the solution carefully. Abaqus standard attempts to fix the solution, but sometimes this fix will not give a correct solution. The user should remove the cause of the NUMERICAL SINGULARITIES instead. If numerical singularities only occur in iterations prior to the iteration which converges, the solution should be correct; however, the user should always check that this is so.

4. Contact chattering: the contact state changes from one iteration to the next so that the maximum number of SDIs (severe discontinuity iterations) is reached.

8.13 Explicit Stable Time Increments When Using the Marlow Model with Noisy Test Data

A model solved with Abaqus explicit analysis uses a hyperelastic material implemented with uniaxial test data coming from different hyperelastic tests data table or measurement data. A significantly lower stable time increment occurs when the Marlow model is used than when the analyst uses other hyperelastic material models.

The reason is because material test data often contain scatter or noise. When fitted to hyperelastic models that are characterized by specific mathematical forms, the data are naturally smoothed, provided that the number of data points exceeds the number of unknown material coefficients.

The Marlow model, however, is quite different in that there is no specific equation to which the data are fitted; rather, the data are fitted exactly to a continuous curve, scatter included. If the material data contain significant scatter, it can create problems for numerical procedures, particularly those dependent on the material modulus, which is the instantaneous slope of the stress–strain curve. While data irregularity normally has only a modest impact on current values in the stress–strain relationship, the impact on the stress–strain slope can be quite large.

One such slope-dependent procedure is the selection of the stable time increment in Abaqus explicit, which is given approximately by Eq. 8.3

$$\Delta t \approx L_e \sqrt{\frac{\rho}{E}} \tag{8.3}$$

where L_e is a characteristic element length, ρ is mass density, and E is the current effective modulus. As can be seen, the stable time increment is inversely proportional to the material modulus, and if it is artificially high it will reduce the stable time increment.

To avoid such difficulties, Abaqus offers a smoothing option for all material test data, including a smoothing order, n:

- In Abaqus CAE go to the Property module then in material editor select **Mechanical** → **Elasticity** → **Hyperelastic**: **Input source**: **Test data** and **Test Data**

→ **Uniaxial Test Data**. In the **Test Data Editor** dialog box, toggle on **Apply Smoothing** and select a value for n.

- With keyword command line enter ***UNIAXIAL TEST DATA, SMOOTH=n**

It is strongly recommended that this option be employed, with **n** greater than or equal to two, when using the Marlow model in order to avoid artificially low stable time increments.

8.14 Cause of an Analysis Ending in a Core Dump

- If the analyst is using a user subroutine, verify the coding to determine whether the cause of the core-dump stems from the user subroutine. Abaqus is not responsible for user subroutines that are not coded correctly. Experience indicates that problems with analyses using user subroutines that end in core dumps are very frequently due to the user not making sure that the user subroutine is robust or that the argument list of the user subroutine corresponds to what is documented in the manual. Users should make sure that the dimensions of arrays are correct and that users are not writing to positions in an array that are out of bounds.
- If the user is not using a user subroutine, check the log (.log) file and the message (.msg) file for the job. A message in one of these files may indicate the cause. It may be the case that the analysis is running out of disk space in the Abaqus scratch directory or the directory to which the output files are being written. If the core dump is occurring on a UNIX machine when a file written by Abaqus is close to 2 GB, this may indicate that the operating system environment variable controlling the file size has not been set to allow unlimited file sizes. If this is the case, change this limit. On some UNIX machines, the command will correct this problem. If not, speak to the system administrator.

```
limit filesize unlimited
```

- If the above cases are not the cause, before contacting Abaqus support get a copy of the system and environment information from **abaqus info=env** and **abaqus info=sys** with the MSDOS command line. Send this information to the online Abaqus support together with the log (.log) file and the input (.inp) file.

8.15 Debugging User Subroutines and Post Processing Programs

Abaqus provides a debugging utility to help troubleshoot issues with user subroutines. The utility is designed to work with the following debuggers:

- Microsoft Visual Studio .NET 2003 (msdev)
- Microsoft Visual Studio 2005 (devenv)
- Windows Debugger (windbg)
- Etnus Totalview (totalview)
- GNU Debugger (gdb)
- DBX Debugger (dbx)
- GNU DDD: The Data Display Debugger (ddd)
- Intel Debugger (idb)
- Workshop Debugger (cvd)
- HP-UX Debugger (wdb)

The Data Display Debugger works only as a graphical front end for the Intel debugger.

1. Verify that the debugger that you intend to use is functional AND accessible from the command line. For example, if using Totalview debugger, the command **totalview** (without the path) should start the debugger.
2. Review the debug utility usage syntax by running the debug command. Please note that this utility is maintained for internal use; therefore, not all debugging options may be available.

```
d:\temp>abq6141 debug help=
Error: must use "-" syntax.  For example:

USAGE:
abq debug <-executable> [<debugger>] -job c1 ...
abq debug cae [ -exe ker|gui ] ...
abq debug [ -exe <<path to exe>>  [ -core <<path
 to core>> ]]

OPTIONS:
-db debugger     -- Debugger to launch
-exe executable -- Executable to debug
-stop function  -- Function to stop at
-wait            -- Wait for license (if unavailable)
-dbscript <file> -- Load custom debugger script at
 startup
-dbargs "..."    -- Pass arguments to the debugger

EXAMPLES -- SOLVERS
abq debug -pre         dbx  -job c1  ...
abq debug -standard  cvd  -job c1  ...
abq debug -standard  gdb  -job c1  -stop step_
abq debug -explicit        -job x1
abq debug -explicit        -job x1  -double
abq debug -explicit        -job x1  -recover
```

```
EXAMPLES -- CAE
abq debug cae -exe ker ...
abq debug cae -exe gui ...

EXAMPLES -- TEST PROGRAMS
abq debug test cowT_String

Default debugger on this platform is: devenv

Recognized executables:
pre,package,standard,explicit,select,state,
Calculator,Extrapolator,ker,gui,cse

Recognized debuggers:
devenv,msdev,msdev10,windbg,totalview,tvbeta,
tvold,gdb,gdb64,dbx,cvd,wdb,ddd,idb
```

3. Copy the attached **debug.env** code to a local directory **abaqus_v6.env** file to add the appropriate compiler debug option to the Abaqus compile and link parameters. Compiler performance options will also be removed.

Listing 8.1 debug.env

```
import os

def prepDebug(var, dbgOption):
    import types
    varOptions = globals().get(var)
    if varOptions:
        # Add debug option
        if type(varOptions) == types.StringType:
            varOptions = varOptions.split()
        varOptions.insert(6, dbgOption)
        # Remove compiler performance options
        if var[:4] == 'comp':
            optOptions = ['/O', '-O', '-xO',   \
                          '-fast', '-depend', \
                          '-vpara','/Qx', '/Qax']
            for option in varOptions[:]:
                for opt in optOptions:
                    if len(option) >= len(opt) and \
                       option[:len(opt)] == opt:
                        varOptions.remove(option)
    return varOptions
```

```
if os.name == 'nt':
    compile_fortran = prepDebug('compile_fortran', \
    '/debug')
    compile_cpp = prepDebug('compile_cpp', '/Z7')
    link_sl = prepDebug('link_sl', '/DEBUG')
    link_exe = prepDebug('link_exe', '/DEBUG')
else:
    compile_fortran = prepDebug('compile_fortran', \
    '-g')
    compile_cpp = prepDebug('compile_cpp', '-g')

del prepDebug
```

4. Run the job with the user subroutine according to the examples above:

```
abaqus j job user user debug standard
```

5. When the debugger starts, open the source of the user subroutine, add watch-points, breakpoints, and so on, and start the run. Some debuggers like Code Viser Debugger (CVD) will not work by following the method described above as they do not recognize the way in which Abaqus spawns a process. However, there is an indirect way of getting around this problem. First, launch the CVD debugger. Insert an artificial breakpoint in the user subroutine which can be used to hang the process. This can be achieved in two ways. One is to prompt the user for a value in the user subroutine. Alternatively, the user can have an infinite loop in the user subroutine by inserting something like

```
do while(JFLAG .ne. 999)
     JFLAG = 1
end do
```

Run the Abaqus job the usual way:

```
abaqus -j job -user user
```

Attach the analysis process (the process spawned by the Abaqus job) to CVD. The user may see more than one process. Attach the one that shows the CPU time increasing. Once the user attached to the process CVD will show the subroutine and the program counter will hang in the artificially inserted breakpoint. If the user is using the infinite loop as a breakpoint, then the value of JFLAG can be changed. In CVD, the analyst can use **View** → **Variable Browser** to set it to 999. Set a breakpoint in the subroutine and continue debugging.

Debugging postprocessing routines

1. Copy the attached **debug.env** code to a local directory **abaqus_v6.env** file to add the appropriate compiler debug option to the Abaqus compile and link parameters. Compiler performance options will also be removed.
2. Build the program using the Abaqus/Make procedure.

```
abaqus make -j prog
```

3. In some circumstances, the user may need to execute the processing program without the driver. In this case, the appropriate library search path must be set up. Run the attached Python script to determine the appropriate environment setting that the user will need to apply.

```
abaqus python getLibPath.py
```

Listing 8.2 getLibPath.py

```
# Extracts the Abaqus library search path
import driverUtils, os

libName = os.environ['ABA_LIBRARY_PATHNAME']
abaPath = os.environ['ABA_PATH']
subdir = os.path.join('exec','lbr')
try:
    libs = driverUtils.getBundleLibs(driverUtils.\
           getPlatform(), os.environ)
except:
    libs = []
libs += driverUtils.locateAllDirectories(abaPath, \
        subdir)
libs += driverUtils.locateAllDirectories(abaPath, \
        'External')

libPath = os.pathsep.join(libs)
if os.name == 'nt':
    print "\nset %s=%s\n" % (libName, libPath)
else:
    print "\nIf C-SHELL use:\n"
    print "    setenv %s %s\n" % (libName, libPath)
    print "\nIf BASH use:\n"
    print "    export %s=%s\n" % (libName, libPath)
```

4. Launch the debugger and load the executable. The debugger may report a warning that the source code for main was not found. The user can just ignore this message. Open the code in the debugger, set breakpoints, and start the run.
5. A classic debugging protocol consists of adding the combination of **pause** and **print** instructions into the code and following the pause by pause execution of the program in the command or debugger window.

8.16 No Free Memory Available on Linux at the End of an Analysis

When Abaqus is running a large analysis, the job's processes consume a significant amount of memory, while also performing a large amount of I/O operations, reading and writing to several large files.

The process' memory usage can be seen by running Linux utilities such as top. However, due to the high number of I/O operations, a significant amount of memory will be utilized by the operating system. The operating system will allocate currently unused memory to disk cache buffers in order to improve I/O operation performance. This is a normal occurrence of all modern operating systems. Due to I/O-intensive applications such as Abaqus, almost all available memory may be utilized by the operating system in this manner. When user processes request additional memory, the OS will immediately flush some of the cache buffers as necessary to meet the application request. User application memory has priority over cache buffer memory.

Operating system cache buffer memory is not automatically freed at the completion of an Abaqus analysis or any other user process/application. The OS will free this memory according to the kernel algorithms in place, but the general rule is that the oldest last-access data will be removed first, when user applications request memory that is not available.

When the Abaqus analysis completes, the Linux free and top utilities may report almost no free memory available, which may cause worry about possible memory leaks and poor performance. This should not be a matter of concern, because the memory allocated in the cache buffers will be automatically released by the Linux Operating System if a new application requests additional physical memory. The OS will often perform this quick enough such that the user application performance will not be affected. The ability of modern operating systems to perform the above-described memory management is of significant benefit to the overall performance of all user applications.

If the user prefers to have all applications flush the OS cache buffers before the next application begins its work, the memory allocator utility given below can be used. This small program is designed to request and use all the physical memory of the machine. In doing so, the operating system will be forced to release any unused cache buffers in order to satisfy the program's memory requests. When the program exits, both the processes and cache buffer memory will be freed. This process is designed to be performed on machines that are idle. The performance of any other application running in the background may be impacted. The user needs to proceed as follows:

1. To perform the task manually, the user will call the Python **memAllocator.py** script, shown below:

Listing 8.3 memAllocator.py

```
import uti, os, sys
try:
    from numpy.oldnumeric import zeros
except ImportError:
    from Numeric import zeros
if uti.getVersion()[2] in ['8', '9', '1']:
    memSize=uti.MemoryInfo().GetPhysicalMemory()[:-2]
else:
    #Physical memory size must be determined manually
    memSize = '4096'
if len(sys.argv) > 1:
    memSize = sys.argv[1]
blSize = 200
a = []
print "Flushing OS cache buffers."
if not os.name == 'nt':
    os.system('free')
for i in range(0, int(memSize)/blSize):
    print "Allocating block: ", i
    try:
        a.append(zeros((blSize * 1000000,1), 'c'))
    except:
        print "Reached memory limit."
        break

if not os.name == 'nt':
    os.system('free')
```

2. Then run the program specifying the memory size of the machine in Megabytes. For example, a machine with 32GB of RAM:

```
abaqus python memAllocator.py 32000
```

The memory size value is an optional parameter. The default value is the amount of physical memory when using Abaqus Version 6.8 and later, or 4096MB when using Abaqus Version 6.7 and earlier.

3. To run this program at the completion of each Abaqus analysis, the user must append the **pythonfree code onJobCompletion()** code to the **abaqus_version/site** or the **$HOME abaqus_v6.env** file.

Listing 8.4 pythonfree.py

```
# onJobstartup routine to allocate all the memory and
# free it. This causes all I/O buffers to be flushed
def onJobStartup():
```

```
import uti, os
# The default is flush for every job.  To turn
# this off, change the following line 'ALL' to ''
memSize=os.environ.get('ABA_FLUSH_BUFFERS','ALL')
if memSize:
    try:
        from numpy.oldnumeric import zeros
    except ImportError:
        from Numeric import zeros
    if memSize == 'ALL':
        memSize=uti.MemoryInfo()
                   .GetPhysicalMemory()[:-2]
    blSize = 200
    a = []
    print "Flushing OS cache buffers."
    if not os.name == 'nt':
        os.system('free')
    for i in range(1, int(memSize)/blSize):
        print "Allocating block: ", i
        try:
            a.append(zeros((blSize*1000000,1),'c'))
        except:
            print "Reached memory limit."
            break
    del a
    if not os.name == 'nt':
        os.system('free')
```

Listing 8.5 onJobCompletion.py

```
def onJobCompletion():
  import os, driverUtils, uti
  if uti.getVersion()[2] in ['8', '9', '1']:
    memSize=uti.MemoryInfo().GetPhysicalMemory()[:-2]
  else:
    #Physical memory size must be determined manually
    memSize = '4096'
  parent = os.path.join(os.environ['ABA_HOME'], '..')
  platform = driverUtils.getPlatform().capitalize()
  malloc = 'memAllocator' + platform + '.exe'
  found = 0
  for location in [os.environ['HOME'],savedir,parent]:
      prog = os.path.join(location, malloc)
      if os.path.isfile(prog):
          found = 1
          break
```

```
if found:
    if os.uname()[0] == 'Linux': os.system('free')
    print "Flushing␣OS␣cache␣buffers"
    os.system(prog + '␣' + memSize)
    if os.uname()[0] == 'Linux': os.system('free')
```

Reference

1. Stroud KA (2003) Advanced Engineering Mathematics. Palgrave Macmillan, Basingstoke

Chapter 9
Numerical Acceptance Criteria

9.1 Generalities

The principal analysis convergence control parameters will be presented here to understand the numerical criteria used by the different solvers to compute the solution and how to modify such parameter values if necessary to optimize the accuracy of output with computational time or improve the solution convergence. A good understanding of parameter effects and assigned values can help the user to figure out how to properly operate on the numerical solver criteria, in a range to keep the accuracy of the computed solution as quickly as possible.

9.1.1 *Commonly Used Control Parameters*

Solution control parameters can be used to control nonlinear equation solution accuracy, time increment adjustment with FSI stabilization, and mesh distortion in Abaqus CFD, Abaqus standard or Abaqus explicit co-simulation. These solution control parameters need not be changed for most analyses. In difficult cases, however, the solution procedure may not converge with the default controls or may use an excessive number of increments and iterations. In other words, if the troubleshooting is not due to modeling errors then the user would have to consider some change in control parameters. This section presents a brief synopsis of the more important solution control parameters, together with a description of the circumstances in which they can be used effectively.

The most significant solution control parameters for field equation tolerances are R_n^α, C_n^α, $\tilde{q}_0^\alpha$ and $\tilde{q}_u^\alpha$. They may have to be modified in cases where the residuals are

R. J. Boulbes, *Troubleshooting Finite-Element Modeling with Abaqus*,
https://doi.org/10.1007/978-3-030-26740-7_9

large relative to the fluxes[1] or in cases where the incremental solution is essentially zero.

- In Abaqus CAE go to Step module then **Other** → **General Solution Controls** → **Edit** and toggle on Specify: Field Equations: Apply to all applicable fields or Specify individual fields: field.
- With keyword command line enter ***CONTROLS, PARAMETERS=FIELD, FIELD=field**.

The residual control R_n^α is the convergence criterion for the ratio of the largest residual to the corresponding average flux norm $\tilde{q}^\alpha$ for convergence. The default value is 5×10^{-3}, which is rather strict by engineering standards but in all but exceptional cases will guarantee an accurate solution to complex nonlinear problems. The value for this ratio can be increased to a larger number if some accuracy can be sacrificed for computational speed.

The solution correction control C_n^α is the convergence criterion for the ratio of the largest solution correction to the largest corresponding incremental solution value. The default value is $C_n^\alpha = 10^{-2}$. In addition to sufficiently small residuals, Abaqus standard requires that the largest correction to the solution value be small in comparison to the largest corresponding incremental solution value. Some analyses may not require such accuracy, thus permitting this ratio to be increased. To avoid testing the magnitude of the solution correction, the user can set C_n^α equal to 1.0.

The average flux $\tilde{q}^\alpha$ is the value used by Abaqus standard for checking residuals. The default value is the time-averaged flux calculated by Abaqus standard, as defined in: Convergence criteria for nonlinear problems. The user may, however, define a constant value $\tilde{q}_u^\alpha$ for the average flux, in which case $\tilde{q}^\alpha = \tilde{q}_u^\alpha$ throughout the step. Analyst may wish to use absolute tolerances for the residual checks. The absolute tolerance value is then equal to the product of the average flux $\tilde{q}_u^\alpha$ and the ratio R_n^α.

The initial time-averaged flux $\tilde{q}_0^\alpha$ is the initial value of the time-averaged flux for the current step. The default value is the time-averaged flux from the previous step or 10^{-2} if this is Step 1. Redefining $\tilde{q}_0^\alpha$ is sometimes helpful when a coupled problem is analyzed and some of the fields in the problem are not active in the first step; for example, if a static step is carried out before a fully coupled thermal–stress step. Redefinition of $\tilde{q}_0^\alpha$ can also be useful if the first step is essentially a null step; for example, in a contact problem before any contact occurs, the initial fluxes (forces) generated are zero. In such cases, $\tilde{q}_0^\alpha$ should be given as a typical flux magnitude that will occur when the field α first becomes active. The initial value of $\tilde{q}^\alpha$ is retained until an iteration is completed for which $\bar{q}^\alpha > \varepsilon^\alpha.\tilde{q}^\alpha$ at which time we redefine $\tilde{q}^\alpha = \bar{q}^\alpha$. This new $\tilde{q}^\alpha$ can become less than $\tilde{q}_0^\alpha$. If the user specifies the average flux $\tilde{q}_u^\alpha$ directly, the value given for $\tilde{q}_0^\alpha$ is ignored.

[1]In this section the word flux means the variable whose discretized equilibrium is being sought and for which the equilibrium equations may be nonlinear: force, moment, heat flux, concentration volumetric flux, or pore liquid volumetric flux. The word field refers to the basic variables of the system, such as the components of the displacement in a continuum stress analysis or temperature in a heat transfer analysis. The superscript refers to one such type of equation.

Table 9.1 Example of customized solution control parameters

Field equation	R_n^α	C_n^α	$\tilde{q}_0^\alpha$	$\tilde{q}_u^\alpha$	ε^α
Displacement	0.01	1.0	10.0	–	10^{-4}
Rotation	0.02	2.0	20.0	2×10^3	–

The controls in effect for an analysis are listed in the data (.dat) and message (.msg) files. Non-default controls are marked by ***. For example, specifying the following controls (Table 9.1):
would result in the following output:

```
 CONVERGENCE TOLERANCE PARAMETERS FOR FORCE
*** CRIT. FOR RESIDUAL FORCE FOR A NONLINEAR PROBLEM          1.000E-02
*** CRITERION FOR DISP. CORRECTION IN A NONLINEAR PROBLEM          1.00
*** INITIAL VALUE OF TIME AVERAGE FORCE                            10.0
    AVERAGE FORCE IS TIME AVERAGE FORCE
    ALT. CRIT. FOR RESIDUAL FORCE FOR A NONLINEAR PROBLEM     2.000E-02
*** CRIT. FOR ZERO FORCE RELATIVE TO TIME AVRG. FORCE         1.000E-04
    CRIT. FOR DISP. CORRECTION WHEN THERE IS ZERO FLUX        1.000E-03
    CRIT. FOR RESIDUAL FORCE WHEN THERE IS ZERO FLUX          1.000E-08
    FIELD CONVERSION RATIO                                         1.00

 CONVERGENCE TOLERANCE PARAMETERS FOR MOMENT
*** CRIT. FOR RESIDUAL MOMENT FOR A NONLINEAR PROBLEM       2.000E-02
*** CRIT. FOR ROTATION CORRECTION IN A NONLINEAR PROBLEM         2.00
*** INITIAL VALUE OF TIME AVERAGE MOMENT                         20.0
*** USER DEFINED VALUE OF AVERAGE MOMENT NORM               2.000E+03
    ALT. CRIT. FOR RESID. MOMENT FOR A NONLINEAR PROBLEM    2.000E-02
    CRIT. FOR ZERO MOMENT RELATIVE TO TIME AVRG. MOMENT     1.000E-05
    CRIT. FOR ROTATION CORRECTION WHEN ZERO FLUX            1.000E-03
    CRIT. FOR RESIDUAL MOMENT WHEN ZERO FLUX                1.000E-08
    FIELD CONVERSION RATIO                                       1.00
```

9.1.2 Controlling the Time Incrementation Scheme

Solution control parameters can be used to alter both the convergence control algorithm and the time incrementation scheme. The time incrementation parameters are the most significant since they have a direct effect on convergence. They may have to be modified if convergence is (initially) non-monotonic or if convergence is non-quadratic. Non-monotonic convergence may occur if various nonlinearities interact; for example, the combination of friction, nonlinear material behavior, and geometric nonlinearity may lead to non-monotonically decreasing residuals. Non-quadratic convergence will occur if the Jacobian is not exact, which may occur for complex material models. It may also occur if the Jacobian is nonsymmetric but the symmetric equation solver is used. In that case, the unsymmetric equation solver should be specified for the step.

- In Abaqus CAE go to Step module then **Other → General Solution Controls → Edit** and toggle on Specify: Time Incrementation.
- With keyword command line enter ***CONTROLS, PARAMETERS=TIME INCREMENTATION**.

The equilibrium iteration for a residual check I_0 is the number of equilibrium iterations after which it is checked that the residuals are not increasing in two consecutive iterations. The default value is $I_0 = 4$. If the initial convergence is non-monotonic, it may be necessary to increase this value.

The equilibrium iteration for a logarithmic rate of convergence check I_R is the number of equilibrium iterations after which the logarithmic rate of convergence check begins. The default value is $I_R = 8$. In cases where convergence is non-quadratic and this cannot be corrected by using the unsymmetric equation solver for the step, the logarithmic convergence check should be eliminated by setting this parameter to a high value.

To avoid premature cutbacks in difficult analyses sometimes it is useful to increase both I_0 and I_R. For example, in a difficult analysis involving both friction and the concrete material model, it may be helpful to set $I_0 = 8$ and $I_R = 10$ to avoid premature cutbacks of the time increment. These two parameters can be raised to more appropriate values for severely discontinuous problems by increasing them individually.

The user can automatically set the parameters described above to the values $I_0 = 8$ and $I_R = 10$. In this case, any values that you specified previously for I_0 and I_R are overridden. However, if I_0 and I_R are specified multiple times in a step with different solution control settings, the last definition will be used.

In order to improve the solution convergence in a problem that involves a high coefficient of friction, it might sometimes help in an analysis to set a high coefficient of friction with the time incrementation parameters by using the unsymmetric equation solver. The controls in effect for an analysis are listed in the data (.dat) and message (.msg) files. Non-default controls are marked by ***. For example, specifying the time incrementation parameters $I_0 = 7$ and $I_R = 10$ would result in the following output:

```
TIME INCREMENTATION CONTROL PARAMETERS:
*** FIRST EQUIL. ITERATION FOR CONSECUTIVE DIVERGENCE CHECK          7
*** EQUIL. ITER. AT WHICH LOG. CONVERGENCE RATE CHECK BEGINS        10
    EQUIL. ITER. AFTER WHICH ALTERNATE RESIDUAL IS USED              9
    MAXIMUM EQUILIBRIUM ITERATIONS ALLOWED                          16
    EQUIL. ITERATION COUNT FOR CUT-BACK IN NEXT INCREMENT           10
    MAX EQUIL. ITERS IN TWO INCREMENTS FOR TIME INC. INCREASE        4
    MAXIMUM ITERATIONS FOR SEVERE DISCONTINUITIES                   12
    MAXIMUM CUT-BACKS ALLOWED IN AN INCREMENT                        5
    MAX DISCON. ITERS IN TWO INCS FOR TIME INC. INCREASE             6
    CUT-BACK FACTOR AFTER DIVERGENCE                             0.250
    CUT-BACK FACTOR FOR TOO SLOW CONVERGENCE                     0.500
    CUT-BACK FACTOR AFTER TOO MANY EQUILIBRIUM ITERATIONS        0.750
```

9.1.3 Activate the Line Search Algorithm

In strongly nonlinear problems, the Newton algorithms used in Abaqus standard may sometimes diverge during equilibrium iteration. The line search algorithm used for problems as discussed in Sect. 4.9 will detect these situations automatically, apply a scale factor to the computed solution correction, and help prevent divergence. The line search algorithm is particularly useful when the quasi-Newton method is used. By default, the line search algorithm is enabled only during steps where the quasi-Newton method is used. Set the maximum number of line search iterations N^{ls} to a reasonable value (such as 5) to activate the line search procedure or to zero to forcibly deactivate the line search.

- In Abaqus CAE go to Step module then **Other** $\rightarrow$ **General Solution Controls** $\rightarrow$ **Edit** and toggle on Specify: Line Search Control set with N^{ls} value.
- With keyword command line enter ***CONTROLS, PARAMETERS= LINE SEARCH,** N^{ls}.

9.1.4 Controlling the Solution Accuracy in Direct Cyclic Analysis

Solution control parameters can be used in direct cyclic analysis to specify when to impose the periodicity conditions and to set tolerances for stabilized state and plastic ratcheting detections.

- In Abaqus CAE go to the Step module: **Other** $\rightarrow$ **General Solution Controls** $\rightarrow$ **Edit** and toggle on Specify, Direct Cyclic: I_{PI}, CR_n^α, CU_n^α, CR_0^α, CU_0^α.
- With keyword command line enter ***CONTROLS, TYPE=DIRECT CYCLIC** with I_{PI}, CR_n^α, CU_n^α, CR_0^α, CU_0^α values.

To impose the periodicity condition, the user can specify the iteration number at which the periodicity condition is first imposed, I_{PI}. The default value is $I_{PI} = 1$, in which case the periodicity condition is imposed for all iterations from the beginning of an analysis. This solution control parameter rarely needs to be reset from its default value.

To define the tolerances for stabilized state and plastic ratcheting detections, the user can specify the stabilized state detection criteria CR_n^α and CU_n^α. CR_n^α is the maximum allowable ratio of the largest residual coefficient on any terms in the Fourier series to the corresponding average flux norm and CU_n^α is the maximum allowable ratio of the largest correction to the displacement coefficient on any terms in the Fourier series to the largest displacement coefficient. The default values are $CR_n^\alpha = 5 \times 10^{-3}$ and $CU_n^\alpha = 5 \times 10^{-3}$.

The solution converges to a stabilized state if both these criteria are satisfied. If plastic ratcheting occurs, the shape of the stress versus strain curves remains unchanged but the mean value of the plastic strain over a cycle continues to shift from one iteration to the next. In that case it is desirable to use separate tolerances for the constant term in the Fourier series to detect the plastic ratcheting.

The user can also specify the plastic ratcheting detection criteria CR_0^α and CU_0^α. CR_0^α is the maximum allowable ratio of the largest residual coefficient on the constant term in the Fourier series to the corresponding average flux norm and CU_0^α is the maximum allowable ratio of the largest correction to the displacement coefficient on the constant term in the Fourier series to the largest displacement coefficient. The default values are $CR_0^\alpha = 5 \times 10^{-3}$ and $CU_0^\alpha = 5 \times 10^{-3}$. Plastic ratcheting is expected if the residual coefficients and the corrections to the displacement coefficients on any of the periodic terms are within the tolerances set by CR_n^α and CU_n^α respectively. But the maximum residual coefficient on the constant term and the maximum correction to the displacement coefficient on the constant term exceed the tolerances set by CR_0^α and CU_0^α respectively.

The controls in effect for an analysis are listed in the data (.dat) and message (.msg) files. Non-default controls are marked by **.

For example, specifying the following controls with $I_{PI} = 5, CR_n^\alpha = 10^{-4}, CU_n^\alpha = 10^{-4}, CR_0^\alpha = 10^{-4}$ and $CU_0^\alpha = 10^{-4}$ would result in the following output:

```
STABILIZED STATE AND PLASTIC RATCHETTING DETECTION
PARAMETERS FOR FORCE
** CRIT. FOR RESI. COEFF. ON ANY FOURIER TERMS              1.0E-04
** CRIT. FOR CORR. TO DISP. COEFF. ON ANY FOURIER TERMS     1.0E-04
** CRIT. FOR RESI. COEFF. ON CONSTANT FOURIER TERM          1.0E-04
** CRIT. FOR CORR. TO DISP. COEFF. ON CONST. FOURIER TERM 1.0E-04

PERIODICITY CONDITION CONTROL PARAMETER:
** ITERATION NUMBER AT WHICH PERIODICITY CONDITION
** STARTS TO IMPOSE                                               5
```

9.1.5 Controlling the Solution Accuracy and Mesh Quality in a Deforming Mesh Analysis with Abaqus CFD

Solution control parameters can be used to control the mesh motion and to maintain the mesh quality in deforming mesh problems involving moving boundaries or deforming geometries. They can also be used to control FSI stabilization when performing Abaqus CFD to Abaqus standard or to Abaqus explicit co-simulation.

9.1.5.1 Controlling Mesh Smoothing and FSI Stabilization

When the implicit algorithm (default) for mesh smoothing is used, the user can specify the number of iterations before performing a convergence check, the maximum number of iterations, the FSI penalty scale factor, the solid/fluid density ratio, the linear convergence criterion, and the stiffness scale factor to control mesh motion and FSI stabilization.

The implicit algorithm uses the matrix-free iteration method to solve the pseudo-elastic problem. The number of iterations and the linear convergence criterion control the accuracy when solving the linear elasticity equations during the ALE process for FSI or deforming mesh problems. Reducing the number of iterations or relaxing the linear convergence criterion can help reduce the computational time. Similarly, increasing the number of iterations or the linear convergence criterion can help to ensure that the mesh quality remains good. The stiffness scale factor can be used to scale the elastic stiffness. Decreasing the elastic stiffness produces an ALE mesh with more local deformation.

When the explicit algorithm for mesh smoothing is used, the analyst can specify the minimum number of mesh smoothing increments, the maximum number of mesh smoothing increments, the FSI penalty scale factor, the solid/fluid density ratio, and the stiffness scale factor to control the mesh motion and FSI stabilization.

The minimum and maximum numbers of mesh smoothing increments control the number of mesh smoothing steps taken during the ALE process for FSI or deforming mesh problems. Reducing the minimum and maximum numbers of mesh smoothing increments can help reduce computational time. Similarly, increasing the minimum and maximum numbers of smoothing increments helps to ensure that the mesh quality remains good and avoids potential element collapse during the evolution of a deforming mesh problem.

The FSI penalty scale factor is used to control FSI stabilization and has a default value of 1.0. Increasing this parameter in increments of 0.1 may be necessary for extremely flexible structures in high-density fluids when the structural accelerations are high.

The solid/fluid density ratio is also used to control FSI stabilization. By default, the solid/fluid density ratio is ignored if its value is not specified. When multiple solid–fluid interfaces are present, the user should choose the smallest solid/fluid density ratio.

- Controlling FSI stabilization in an Abaqus CFD to Abaqus standard or to Abaqus explicit co-simulation is not supported in Abaqus CAE.
- Use one of the following options to control the mesh smoothing or FSI stabilization:
  - ***CONTROLS, TYPE=FSI, MESH SMOOTHING=IMPLICIT**, number of iterations before convergence check, maximum number of iterations, FSI penalty scale factor, solid/fluid density ratio, stiffness scale factor, linear convergence criterion.
  - ***CONTROLS, TYPE=FSI, MESH SMOOTHING=EXPLICIT**, minimum number of mesh smoothing increments, maximum number of mesh smoothing increments, FSI penalty scale factor, solid/fluid density ratio, stiffness scale factor.

9.1.5.2 Controlling Mesh Distortion

Similarly to the distortion control used in Abaqus explicit, Abaqus CFD offers distortion control to prevent elements from inverting or distorting excessively in fluid mesh movement when the explicit mesh smoothing algorithm is used. By default, distortion control is turned off during the co-simulation and ignored if the implicit mesh smoothing algorithm is used.

- Controlling mesh distortion in an Abaqus CFD to Abaqus standard or to Abaqus explicit co-simulation is not supported in Abaqus CAE.
- Use the following option to deactivate distortion control (default) when the implicit mesh smoothing algorithm is used with the command line ***CONTROLS, TYPE=FSI, MESH SMOOTHING=EXPLICIT, DISTORTION CONTROL=OFF**.
- Use the following option to activate distortion control with the command line ***CONTROLS, TYPE=FSI, MESH SMOOTHING=EXPLICIT, DISTORTION CONTROL=ON**.

9.1.6 Convergence Criteria for Nonlinear Problems

The information in this section is provided for users who may wish to adjust the convergence criteria for the solution of nonlinear systems. In most cases these criteria need not be adjusted.

Where possible, Abaqus standard uses Newton's method to solve nonlinear problems. In some cases, it uses an exact implementation of Newton's method, in the sense that the Jacobian of the system is defined exactly, and quadratic convergence is obtained when the estimate of the solution is within the radius of convergence of the algorithm. In other cases, the Jacobian is approximated so that the iterative method is not an exact Newton method. For example, there are some non-isotropic material and surface interface models such as nonassociated with flow plasticity models or Coulomb friction which create a nonsymmetric Jacobian matrix, but the user may choose to approximate this matrix by its symmetric part.

Many problems exhibit discontinuous behavior. A common example is contact, at a particular point on a surface, the contact constraint is either present or absent. Another (usually less severe) example is strain reversal in plasticity at a point where the material is yielding.

The analyst can choose to use the quasi-Newton technique for a particular step instead of the standard Newton method for solving nonlinear equations. The quasi-Newton technique can save substantial computational cost in some cases by reducing the number of times the Jacobian matrix is factorized.

Generally, it is most successful when the system is large and many iterations are needed per increment or when the stiffness matrix is not changing much from iteration to iteration, such as in a dynamic analysis using implicit time integration or in a small displacement analysis with local plasticity. It can be used only for symmetric systems of equations. Therefore, it cannot be used when the unsymmetric solver is specified for a step, nor can it be used for procedures that always produce an unsymmetric system of equations, such as Fully coupled thermal–stress analysis and Abaqus Aqua analysis. In addition, it cannot be used for a static Riks procedure. The quasi-Newton method works well in combination with the line search method. Line searches help to prevent divergence of equilibrium iterations resulting from the inexact Jacobian produced by the quasi-Newton method. The line search method is activated by default for steps that use the quasi-Newton method. The user can override this action by specifying line search controls.

The user can specify the number of quasi-Newton iterations allowed before the kernel matrix is reformed. The default number of iterations is eight. Additional matrix reformations may occur automatically during the iteration process depending on the convergence behavior. Since quadratic convergence is not expected during quasi-Newton iterations, the logarithmic rate of convergence check is not applied during the time incrementation. Furthermore, the iteration count used in the time incrementation is a weighted sum of quasi-Newton iterations, with the weight factor depending on whether or not a kernel matrix has been reformed.

- In Abaqus CAE go to the Step module then in step editor go to **Other: Solution technique: Quasi-Newton** and enter the number of iterations allowed before the kernel matrix is reformed: *n*.
- With keyword command line enter ***SOLUTION TECHNIQUE, TYPE= QUASI-NEWTON, REFORM KERNEL=n**.

Alternatively, the analyst can choose to use the separated technique instead of the standard Newton method for solving nonlinear equations for fully coupled thermal–stress and coupled thermal–electrical procedures. The separated technique approximates the Jacobian by eliminating inter-field coupling terms and can save substantial computational cost in cases where there is relatively weak coupling between the fields.

- In Abaqus CAE go to Step module then in step editor go to **Other: Solution technique: Separated**.
- With keyword command line enter ***SOLUTION TECHNIQUE, TYPE= SEPARATED**.

9.1.6.1 Controlling the Accuracy of the Solution

The default solution control parameters defined in Abaqus standard are designed to provide reasonably optimal solutions to complex problems involving combinations of nonlinearities as well as efficient solutions to simpler nonlinear cases. However, the most important consideration in the choice of the control parameters is that any solution accepted as converged is a close approximation to the exact solution of the nonlinear equations. In this context, close approximation is interpreted rather strictly by engineering standards when the default value is used, as described below.

The user can reset many solution control parameters related to the tolerances used for field equations. If the user defines less strict convergence criteria, results may be accepted as converged when they are not sufficiently close to the exact solution of the system. Use caution when resetting solution control parameters. Lack of convergence is often due to modeling issues, which should be resolved before changing the accuracy controls.

The user can select the type of equation for which the solution control parameters are being defined; for example, the user can redefine the default controls for the displacement field and warping degree of freedom equilibrium equations only. By default, the solution control parameters will apply to all active fields in the model.

- In Abaqus CAE go to the Step module then **Other** → **General Solution Controls** → **Edit** and toggle on **Specify: Field Equations:** Apply to all applicable fields or **Specify individual fields:** field.
- With keyword command line enter:
  ***CONTROLS, PARAMETERS=FIELD, FIELD=field**
  $R_n^\alpha, C_n^\alpha, \tilde{q}_0^\alpha, \tilde{q}_u^\alpha, R_P^\alpha, \varepsilon^\alpha, C_\varepsilon^\alpha, R_l^\alpha$
  $C_f, \varepsilon_l^\alpha, \varepsilon_d^\alpha$.

Each field α that is active in the problem is tested for convergence of the field equations. The following measures are used in deciding whether an increment has converged:

- r_{max}^α the largest residual in the balance equation for field α.
- Δu_{max}^α the largest change in a nodal variable of type α in the increment.
- c_{max}^α the largest correction to any nodal variable of type α provided by the current Newton iteration.
- e^j the largest error in a constraint of type j.
- $\bar{q}^\alpha\,(t)$ the instantaneous magnitude of the flux for field α at time t, averaged over the entire model (spatial average flux). This average is by default defined by the fluxes that the elements apply to their nodes and any externally defined fluxes according to Eq. 9.1

$$\bar{q}^\alpha\,(t) = \frac{1}{\sum\limits_{e=1}^{E}\sum\limits_{n_e=1}^{N_e} N_{n_e}^\alpha + N_{ef}^\alpha}\left[\sum_{e=1}^{E}\sum_{n_e=1}^{N_e}\sum_{i=1}^{N_{ne}^\alpha}|q|_{i,n_e}^\alpha + \sum_{i=1}^{N_{ef}^\alpha}|q|_i^{\alpha,ef}\right] \tag{9.1}$$

where E is the number of elements in the model, N_e is the number of nodes in element e, $N_{n_e}^\alpha$ is the number of degrees of freedom of type α at node n_e of element e, $|q|_{i,n_e}^\alpha$ is the magnitude of the total flux component that element e applies at its ith degree of freedom of type α at its n_e^{th} node at time t, N_{ef}^α is the number of external fluxes for field α (depending on element type, loading type, and number of loads applied to an element), and $|q|_i^{\alpha,ef}$ is the magnitude of the ith external flux for field α.

- $\tilde{q}^{\alpha}(t)$ is an overall time-averaged value of the typical flux for field α so far during this step including the current increment. Normally $\tilde{q}^{\alpha}(t)$ is defined as $\bar{q}^{\alpha}$ averaged over all the increments in the step in which $\bar{q}^{\alpha}$ is nonzero. The $\bar{q}^{\alpha}$ for the current increment is recalculated after every iteration of the current increment, according to Eq. 9.2

$$\tilde{q}^{\alpha}(t) = \frac{1}{N_t}\sum_{i=1}^{N_t}\bar{q}^{\alpha}(t|_i) \tag{9.2}$$

where N_t is the total number of increments so far in the step, including the current increment, in which $\bar{q}^{\alpha}(t|_i) > \varepsilon^{\alpha}\tilde{q}^{\alpha}(t|_i)$. Here $\bar{q}^{\alpha}(t|_i)$ is the value of $\bar{q}^{\alpha}$ at increment i and ε^{α} is a small number. The default for ε^{α} is 10^{-5}, but in rare cases the user can change this default. Alternatively, the user can define a value for the average flux in the step $\tilde{q}_u^{\alpha}$. In this case $\tilde{q}^{\alpha}(t) = \tilde{q}_u^{\alpha}$ throughout the step. At the start of the step $\tilde{q}^{\alpha}$, is normally the value from the previous step (except for Step 1, when $\tilde{q}^{\alpha} = 10^{-2}$ by default). Alternatively, the user can define an initial value for the time-averaged flux $\tilde{q}_0^{\alpha}$. $\tilde{q}^{\alpha}$ retains its initial value until an iteration is completed for which $\bar{q}^{\alpha} > \varepsilon^{\alpha}\tilde{q}^{\alpha}$, at which time it redefines $\tilde{q}^{\alpha} = \bar{q}^{\alpha}$. If $\tilde{q}_u^{\alpha}$ is defined, the value defined for $\tilde{q}_0^{\alpha}$ is ignored.
- $\tilde{q}_{max}^{\alpha}$ the time-averaged value of the largest flux corresponding to the field α during this step, excluding the current increment.
- q_{max}^{α} the largest flux corresponding to the field α during the current iteration.

The time-averaged value of the flux $\tilde{q}^{\alpha}(t)$ is computed from the spatial average of the flux $\bar{q}^{\alpha}(t)$ at various instants in time. In some situations where only a small part of the model is active (the fluxes over the rest of the model are zero or very small), the spatial average of a flux over the entire model can be very small when compared to the spatial average over the active part of the model. Over a period of time, this can result in a small time-averaged value of the flux and in turn may lead to a convergence criterion that is very strict by engineering standards. To avoid such an excessively strict convergence criterion, Abaqus standard uses an algorithm to determine the active parts of a model at any given instant.

During an iteration, any flux $|q_i^{\alpha}(t)| < \varepsilon_l^{\alpha}\tilde{q}_{max}^{\alpha}$ is treated as inactive, and the corresponding degree of freedom is also marked inactive. $\tilde{q}_{max}^{\alpha}$ is the time-averaged value of the largest flux in the model during the current step. The default value of ε_l^{α} is 10^{-5}; the user can redefine this parameter.

At the end of an iteration, the largest flux in the model during the current iteration q_{max}^{α} is compared with the time-averaged value of the largest flux $\tilde{q}_{max}^{\alpha}$. If $q_{max}^{\alpha} \geq 0.1\tilde{q}_{max}^{\alpha}$ the spatial average is computed over only the active parts of the model; if $q_{max}^{\alpha} < 0.1\tilde{q}_{max}^{\alpha}$ all inactive parts of the model are reclassified as active and the spatial average is computed over the entire model. The appropriate spatial average of the flux obtained in this manner is then used to compute the time-averaged flux $\tilde{q}_{max}^{\alpha}(t)$ that is used in the convergence criterion. Setting $\varepsilon_l^{\alpha} = 0$ forces the spatial averages of a flux to be always computed over the entire model.

Regarding the residuals term, most nonlinear engineering calculations will be sufficiently accurate if the error in the residuals is less than 0.5%. Therefore Abaqus standard normally uses $r^{\alpha}_{max} \leq R^{\alpha}_{n}\tilde{q}^{\alpha}$ as the residual check, where the user can define R^{α}_{n} (it is 0.005 by default). If this inequality is satisfied, convergence is accepted if the largest correction to the solution c^{α}_{max}, is also small compared to the largest incremental change in the corresponding solution variable Δu^{α}_{max}, such as $c^{\alpha}_{max} \leq C^{\alpha}_{n}\,\Delta u^{\alpha}_{max}$ or if the magnitude of the largest correction to the solution that would occur with one more iteration, estimated as shown in Eq. 9.3 satisfies the same criterion $c^{\alpha}_{est} \leq C^{\alpha}_{n}\,\Delta u^{\alpha}_{max}$, the user can define C^{α}_{n} the default value is 10^{-2}.

$$c^{\alpha}_{est} = \frac{\left(r^{\alpha}_{max}\right)^{i}}{min\left(\left(r^{\alpha}_{max}\right)^{i-1}, \left(r^{\alpha}_{max}\right)^{i-2}\right)} c^{\alpha}_{max} \tag{9.3}$$

In some cases there may be zero flux in the equations of type α anywhere in the model during some increments. Zero flux is defined as $\bar{q}^{\alpha} \leq \varepsilon^{\alpha}\tilde{q}^{\alpha}$ here ε^{α} has a default value of 10^{-5} and the solution for field α is accepted if $r^{\alpha}_{max} \leq \varepsilon^{\alpha}\tilde{q}^{\alpha}$. If not c^{α}_{max} is compared to Δu^{α}_{max} and the convergence for field α is accepted when $c^{\alpha}_{max} \leq C^{\alpha}_{\varepsilon}\,\Delta u_{max}$. The default value of C^{α}_{ε} is 10^{-3}; the user can redefine this parameter.

Cases may arise where more than one field is active in the model yet there is negligible response in some of the fields in some increments. If some type of physical conversion factor f^{α}_{β}, exists between active fields α and β, $\tilde{q}^{\alpha}$ in the above paragraph can be replaced by $f^{\alpha}_{\beta}C_{f}\tilde{q}^{\beta}$ for those particular increments where $\tilde{q}^{\alpha}$ is deemed too small, such as $\bar{q}^{\alpha} \leq \tilde{q}^{\alpha} < f^{\alpha}_{\beta}C_{f}\tilde{q}^{\beta}$ to be used realistically as part of the convergence criteria for field α. An example of f^{α}_{β} is a characteristic length to convert between the relation with force and moment. Here f^{α}_{β} is a factor calculated by Abaqus standard based on the problem definition and the fields involved and C_{f} is a field conversion ratio that the user can define. The default value for C_{f} is 1.0. Currently, this concept is used only for converting between the fields associated with forces and moments when f^{α}_{β} represents a characteristic element length.

Linear cases do not require more than one equilibrium iteration per increment. If $r^{\alpha}_{max} \leq R^{\alpha}_{l}\tilde{q}^{\alpha}$ for all α, the increment is considered to be linear. The user can define R^{α}_{l} it is intended to be very small. The default value of R^{α}_{l} is 10^{-8}. Any case that passes such a stringent comparison of the largest residual with the average flux magnitude in each field is considered linear and does not require further iteration. If this requirement is satisfied at some iteration after the first, the solution is accepted without any check on the size of the correction to the solution.

In some cases, quadratic convergence of the iterations is not possible because the Jacobian of the Newton scheme is approximated. If after iterations the convergence rate is only linear, Abaqus standard uses a looser tolerance, $r^{\alpha}_{max} \leq R^{\alpha}_{p}\tilde{q}^{\alpha}$, as the residual check. This tolerance modification is not applied when the quasi-Newton method is used, since it is normal for this method to require a larger number of iterations to converge. The user can define R^{α}_{p} (2×10^{-2} by default) and/or I_{P} (9 by default). Convergence also requires that $c^{\alpha}_{max} \leq C^{\alpha}_{n}\,\Delta u^{\alpha}_{max}$ and iteration continues until both criteria are satisfied for all active fields or the increment is abandoned. When the active field is the displacement, the convergence criterion requiring the largest displacement correction to be small relative to the maximum displacement increment $c^{\alpha}_{max} \leq C^{\alpha}_{n} \Delta u^{\alpha}_{max}$ is ignored when the maximum displacement increment itself is very small, as defined by $\Delta u^{\alpha}_{max} < \varepsilon^{\alpha}_{d} f^{\alpha}_{\beta}$, where f^{α}_{β} is the characteristic element length. The default value of ε^{α}_{d} is 10^{-8}; the user can redefine this parameter.

9.1.6.2 Controlling Iteration

Each increment of a nonlinear solution will usually be solved by multiple equilibrium iterations. The number of iterations may become excessive, in which case the increment size should be reduced and the increment attempted again. On the other hand, if successive increments are solved with a minimum number of iterations, the increment size may be increased. The user can specify a number of time incrementation control parameters.

- In Abaqus CAE go to the Step module then **Other** → **General Solution Controls** → **Edit** and toggle on Specify: Time Incrementation; click More to see additional data tables.
- With keyword command line enter:
  ***CONTROLS, PARAMETERS=TIME INCREMENTATION**
  $I_0, I_R, I_P, I_C, I_L, I_G, I_S, I_A, I_J, I_T, I^c_S, I^a_J, I^c_A$
  $D_f, D_C, D_B, D_A, D_S, D_H, D_D, W_G$
  $D_G, D_M, D^{dyn}_M, D^{diff}_M, D_L, D_E, D_R, D_F$
  D_T.

Abaqus standard may have trouble with the element calculations because of excessive distortion in large displacement problems or because of very large plastic strain increments. If this occurs and automatic time incrementation has been chosen, the increment will be attempted again with a time increment of D_H times the current time increment, where the user can define D_H. By default, $D_H = 0.25$. If fixed time stepping has been chosen, the analysis will terminate with an error message.

Sometimes the increment is too large for the solution to converge at all, the initial state is outside the radius of convergence of the Newton method. This condition can be detected by observing the behavior of the largest residuals r^{α}_{max}. These will not decrease from iteration to iteration throughout an iteration sequence that leads to convergence, but we assume that, if they fail to decrease over two consecutive iterations, the iterations should be abandoned. Thus, if $min\left(\left(r^{\alpha}_{max}\right)^{i}, \left(r^{\alpha}_{max}\right)^{i-1}\right) > \left(r^{\alpha}_{max}\right)^{i-2}$, where (i) is the iteration counter, the iterations are abandoned. This check is first made after I_0 iterations following a solution discontinuity. The user can define I_0; it must be at least 3. The default value of I_0 is 4. If fixed time stepping has been chosen, the analysis will terminate with an error message. With automatic time stepping, the increment is begun again, using a time increment of D_f times the previous attempt, where the user can define D_f. By default $D_f = 0.25$. This subdivision continues until a successful time increment is found or the minimum time increment allowed has failed, in which case the job ends with an error message. Using the line search algorithm with $N^{ls} = 4$ sometimes helps in such cases.

In case quadratic convergence cannot be obtained, the logarithmic rate of convergence equal to $ln\left(\left(r^{\alpha}_{max}\right)^{i}/\left(r^{\alpha}_{max}\right)^{i-1}\right)$ will often be maintained throughout the iteration process. This rate can be established during the early iterations. If convergence has not been achieved after I_R or more iterations following a solution discontinuity, if automatic time incrementation has been selected, and if the slowest convergence rate overall fields α suggests that more than I_C total iterations subsequent to the last solution discontinuity are expected to be required, the increment is begun again with a time increment of D_C times the one abandoned. If fixed time incrementation has been chosen, the iterations are continued but if convergence is not achieved within I_C iterations after the last solution discontinuity in the increment, the analysis will terminate with an error message. The user can define I_R, I_C and D_C (by default $I_R = 8$, $I_C = 16$ and $D_C = 0.5$).

Increasing or reducing the size of the time increment for efficiency is always a tricky question when automatic time incrementation is chosen, the effectiveness of the nonlinear equation solution is used in the selection of the next time increment. If no more than I_G iterations are required in two consecutive increments, the time increment may be increased by a factor of D_D. If an increment converges but takes more than I_L iterations, the next time increment is reduced to D_B times the current time increment. The user can define the values of I_G, I_L, D_D, and D_B. By default $I_G = 4$, $I_L = 10$, $D_D = 1.5$, and $D_B = 0.75$.

At each increment after the first increment of a nonlinear analysis step Abaqus standard estimates the solution to the increment by extrapolating the solution from the previous increment (or increments). By default, 100% linear extrapolation is used (1% for the Riks method). Extrapolation is abandoned if $\Delta t_i \leq D_E \Delta t_{i-1}$, where Δt_i is the proposed new time increment and Δt_{i-1} is the last successful time increment. The user can define the value of D_E; it is 0.1 by default.

9.1.6.3 Convergence of Strain Constraints in Hybrid Elements

Strain constraint convergence in hybrid elements is checked by comparing the largest error in each strain constraint e^j with an absolute tolerance for the corresponding error T^j. The magnitudes of these errors are reported in the message (.msg) file after each iteration as compatibility errors. For example, the volumetric compatibility error is a measure of the accuracy with which the incompressibility constraint is satisfied. Since nonlinearity in constraint equations is generally reflected in the field equations in the same problem, no attempt is made to estimate convergence rates in these constraint equations: we assume that the measures of convergence rate in the field equations are sufficient. The user can define the T^j (T^{vol}, T^{axial}, and T^{shear}). By default, all of the values of $T^j = 10^{-5}$.

- In Abaqus CAE go to the Step module then **Other → General Solution Controls → Edit** and toggle on **Specify: Constraint Equations**.
- With keyword command line enter ***CONTROLS, PARAMETERS= CONSTRAINTS** T^{vol}, T^{axial}, T^{shear}.

9.1.6.4 Severe Discontinuity Iterations

Abaqus standard distinguishes between regular, equilibrium iterations in which the solution varies smoothly and SDIs in which abrupt changes in stiffness occur. By default, Abaqus standard will continue to iterate until the severe discontinuities are sufficiently small (or no severe discontinuities occur) and the equilibrium (flux) tolerances are satisfied. Alternatively, Abaqus standard will continue to iterate until no severe discontinuities occur and the equilibrium (flux) tolerances are satisfied. This more traditional method can cause convergence difficulties if the contact conditions are only weakly determined and contact chattering occurs or if a large number of severe discontinuity iterations are required to settle the contact conditions. The user can define the contact and slip compatibility tolerance, the soft contact compatibility tolerance for low pressure, and the contact force error tolerance. Defining the contact force error tolerance is not supported in Abaqus CAE.

- In Abaqus CAE go to the Step module then **Other → General Solution Controls → Edit** and toggle on **Specify: Constraint Equations**.
- With keyword command line enter ***CONTROLS, PARAMETERS= CONSTRAINTS** ,,,T^{cont},T^{soft},,,T^{cfe}.

In implicit dynamic analysis, the average time of all contact changes in the increment is estimated and the time incrementation is interrupted to solve impact equations at that time. With augmented Lagrange or penalty constraint enforcement methods or with softened contact, no contact constraints are imposed when impact equations are solved. However, if the contact constraints are not satisfied within given tolerances, a severe discontinuity iteration is forced.

By default, Abaqus applies sophisticated criteria involving changes in penetration, changes in the residual force, and the number of severe discontinuities from one iteration to the next to determine whether iteration should be continued or terminated. Hence, it is in principle not necessary to limit the number of severe discontinuity iterations. This makes it possible to run contact problems that require large numbers of contact changes without having to change the control parameters. It is still possible to set a limit I_S^c for the maximum number of severe discontinuity iterations; by default, $I_S^c = 50$, which in practice should always be more than the actual number of iterations in an increment.

- In Abaqus CAE go to the Step module then **Other → General Solution Controls → Edit** and toggle on **Specify: Time Incrementation**; click **More** to see additional data tables.

- With keyword command line enter:
***CONTROLS, PARAMETERS=TIME INCREMENTATION**
,,,,,,I_S
,,,,D_S.

9.1.6.5 Improving the Efficiency of the Solution by Using the Line Search Algorithm

Abaqus standard provides the option of including a line search algorithm. The purpose of the line search is to improve the robustness of the Newton or quasi-Newton methods. By default, the line search is active only for steps that use the quasi-Newton method. During equilibrium iterations where residuals are large, the line search algorithm scales the correction to the solution by a line search scale factor s^{ls}. An iterative process is used to find the value of s^{ls} that minimizes the component of the residual vector in the direction of the correction vector; this component is called y^j, where (j) is the line search iteration number. Each line search iteration requires one pass through the Abaqus standard element loop but does not require any operations using the global stiffness matrix.

It is usually sufficient to determine s^{ls} only to modest accuracy. Several controls are used to limit this accuracy. A maximum of $j = N^{ls}$ line search iterations are performed. There is a limit on the allowable range of s^{ls}, such as $s^{ls}_{min} \leq s^{ls} \leq s^{ls}_{max}$.

The line search ceases when $|y^j| \leq f_s^{ls}|y^0|$, where y^0 is evaluated before the first equilibrium iteration. The residual reduction factor at which the line search ceases, f_s^{ls} typically set to a rather loose tolerance. The line search algorithm will also cease when the change in s^{ls} provided by a line search iteration is less than η^{ls} times s^{ls}.

The user can define the values of N^{ls}, s^{ls}_{max}, s^{ls}_{min}, f_s^{ls}, η^{ls}. By default $N^{ls} = 0$ with the Newton method, and $N^{ls} = 5$ with the quasi-Newton method. Set N^{ls} to a nonzero value to activate the line search algorithm or to zero to forcibly deactivate line search. Default values for the additional line search parameters are $s^{ls}_{max} = 1.0$, $s^{ls}_{min} = 0.0001$, $f_s^{ls} = 0.25$, and $\eta^{ls} = 0.10$. These defaults are chosen to achieve modest accuracy for the line search scale factor while minimizing the additional cost of line search iterations. More aggressive line searching can be beneficial in some simulations, especially when many nonlinear iterations and/or cutbacks are needed to resolve sharp discontinuities in the solution. In these cases, the user could try allowing more line search iterations ($N^{ls} = 10$) and requiring more accuracy in the line search scale factor ($\eta^{ls} = 0.01$). This may result in more line search iterations but fewer nonlinear iterations and cutbacks and an overall reduction in solution cost.

- In Abaqus CAE go to the Step module then **Other** → **General Solution Controls** → **Edit** and toggle on **Specify: Line Search Control**.
- With the keyword command line enter ***CONTROLS, PARAMETERS=LINE SEARCH** N^{ls}, s^{ls}_{max}, s^{ls}_{min}, f_s^{ls}, η^{ls}.

Table 9.2 Time integration accuracy measures for various procedures

Procedures	Accuracy measure S^J	Tolerance T^J
Implicit dynamics[a]	Half increment residual	Half increment residual tolerance
Transient heat transfer analysis[b]	Temperature increment $\Delta\theta$	$\Delta\theta_{max}$
Consolidation analysis[c]	Pore pressure increment Δu_ω	Δu_ω^{max}
Creep and viscoelastic material behavior[d]	$\left(\dot{\bar{\varepsilon}}^{cr}\vert_{t+\Delta t} - \dot{\bar{\varepsilon}}^{cr}\vert_t\right)\Delta t$	Creep tolerance

[a]See "Implicit dynamic analysis using direct integration", Sect. 6.3.2 of the Analysis User's Guide
[b]See "Uncoupled heat transfer analysis", Sect. 6.5.2 of the Analysis User's Guide
[c]See "Coupled pore fluid diffusion and stress analysis", Sect. 6.8.1 of the Analysis User's Guide
[d]See "Rate-dependent plasticity: creep and swelling", Sect. 23.2.4 of the Analysis User's Guide

9.1.7 *Time Integration Accuracy in Transient Problems*

Abaqus standard usually uses automatic time-stepping schemes for the solution of transient problems. Factors influencing the increment size for transient problems include convergence aspects related to the degree of geometric, material, contact nonlinearity and the ability of the time integration operator to accurately resolve variations in the accelerations, velocities, and displacements over an increment. This section discusses tolerance parameters and adjustments to the time increment size related to the latter aspect.

Table 9.2 lists the tolerance parameters available for specific analysis procedures. Descriptions of time integrators for the transient procedure types and, in the case of implicit dynamics, discussion of additional factors influencing the time increment size related to the accuracy of time integration are provided in the respective sections referenced in Table 9.2.

In any transient analysis where automatic time incrementation is used, some of these tolerances, T^J $(J = 1, 2, \ldots)$, will be active. Corresponding measures of the integration accuracy, S^J, will be calculated for each increment in the step. Abaqus standard will use these values to adjust the time incrementation using the criteria described in this section. The smallest time increment required by all criteria is used if more than one accuracy measure is active.

Reducing the time increment size is recommended if $S^J > T^J$ for any control (J), that is active in the step, the time increment Δt is too large to satisfy that time integration accuracy requirement. Therefore, the increment is begun again with a time increment of $D_A\left(\frac{T^J}{S^J}\right)\Delta t$, where the user can define the value of D_A. By default, $D_A = 0.85$.

- In Abaqus CAE go to the Step module then **Other** → **General Solution Controls** → **Edit** and toggle on **Specify: Time Incrementation**; click **More** to see additional data tables.

- With keyword command line enter:
  ***CONTROLS, PARAMETERS=TIME INCREMENTATION**
  first line of data
  ,,,D_A.

Increasing the time increment size if at the current time increment Δt, the equation $\Delta t\left(\frac{S^J}{\Delta t}\right)_i < W_G T^J$ is satisfied for all values of (J) in each of consecutive increments (i) and if no cutback has occurred within those increments because of nonlinearity, the next time increment will be increased to $min\left(D_G \Delta t_p, D_M \Delta t\right)$. The user can define the values of I_T, W_G, and D_G. By default, $I_T = 3$, $W_G = 0.75$, and $D_G = 0.8$. Δt_p is the proposed new time increment, which is defined as $\Delta t_p = \left(\frac{T^J}{S^J}\right) \Delta t$ for transient heat transfer and transient mass diffusion problems and defined as, for example, $\Delta t_p = I_T\left(\frac{T^J}{\sum_{i=1}^{I_T}(S^J/\Delta t)_i}\right)$ for other transient problems.

A limit D_M, is placed on the time increment increase factor. The default value of D_M depends on the type of analysis:

- $D_M = 1.25$ for dynamic analysis,
- $D_M = 2.0$ for diffusion dominated processes: creep, transient heat transfer, coupled temperature displacement, soils consolidation, and transient mass diffusion,
- $D_M = 1.5$ for all other cases.
- The user can redefine D_M for each analysis type.

If the problem is nonlinear, the time increment may be restricted by the rate of convergence of the nonlinear equations.

- In Abaqus CAE go to the Step module then **Other → General Solution Controls → Edit** and toggle on **Specify: Time Incrementation**; click **More** to see additional data tables.
- With keyword command line enter:
  ***CONTROLS, PARAMETERS=TIME INCREMENTATION**
  ,,,,,,,,,I_T
  ,,,,,,W_G
  D_G,D_M,D_M^{dyn},D_M^{diff}.

9.1.8 Avoid Small Changes to the Time Increment Size During Implicit Integration Procedures

In linear transient problems when Abaqus standard uses implicit integration, the system of equations must be reformed and decomposed whenever the time increment changes even though the stiffness matrix does not change. Therefore, to reduce the number of increments at which the system matrix changes, Abaqus standard makes use of the factor D_L, where $D_L = min\left(\frac{\Delta t_p}{D_M \Delta t}\right)$.

The definition of D_L results in the following inequality between the proposed and the current time increments $\Delta t_p \geq D_L D_M \Delta t$. Based on this inequality, the time increment is allowed to increase only when its value computed by the criteria described earlier in this section, or computed using the value of **PNEWDT** specified in certain user subroutines (**UMAT**, for example), is greater than or equal to $D_L D_M \Delta t$. The default value of D_L is 1.0, but the user can redefine it to be a smaller number. Reducing D_L to a value less than 1.0 allows the time increment to increase by a factor that is smaller than D_M, thereby forcing a time increment change, even if the change is small. Otherwise, the solution continues with the same Δt.

- In Abaqus CAE go to the Step module then **Other** → **General Solution Controls** → **Edit** and toggle on **Specify: Time Incrementation**; click **More** to see additional data tables.
- With keyword command line enter:
  ***CONTROLS, PARAMETERS=TIME INCREMENTATION**
  first line data
  second line data
  ,,,,D_L.

9.2 How Much Hourglass Energy Is Acceptable

Knowing how much hourglass energy can be acceptable is a question especially when the FEA model uses a first-order element with reduced integration and thus has hourglass control.

The formulation for reduced integration elements considers only the linearly varying part of the incremental displacement field in the element for the calculation of the increment of physical strain. The remaining part of the nodal incremental displacement field is the hourglass field and can be expressed in terms of hourglass modes. Excitation of these modes may lead to severe mesh distortion, with no stresses resisting the deformation. Hourglass control attempts to minimize this problem without introducing excessive constraints on the element's physical response.

Excessive use of hourglass control can cause an overly stiff response. For example, yielding may be delayed or prevented altogether in problems involving elastic–plastic material response.

The level of hourglass control in a model is evaluated by plotting the model energy histories. The artificial energy (output variable **ALLAE**) used to control hourglassing should be small relative to the internal energy (output variable **ALLIE**); a general rule of thumb is to limit **ALLAE** to be no more than 1% of **ALLIE**.

To reduce the amount of hourglass control energy in a model, modify the default settings of the hourglass formulation or refine the mesh. Mesh refinement is the recommended approach.

9.2.1 Enhanced Hourglass Control and Elastic Bending Moment

The enhanced hourglass control formulation is tuned to give accurate results for regularly shaped elements undergoing elastic bending. Where the latter conditions apply, a coarse mesh may give acceptable results despite the artificial energy being greater than a few percent. An independent check of the results should be made to determine whether they are acceptable.

9.2.2 Enhanced Hourglass Control and Plastic Bending Moment

When plasticity is involved (e.g., plastic bending), for either Abaqus standard or Abaqus explicit, the usual rule of thumb regarding the acceptable level of artificial energy should be followed, but additional precautions should be taken as outlined below.

The enhanced hourglass stiffness is determined from the current material data at the centroid of the element. Until the plasticity reaches the integration point closest to the surface, the moduli used in the enhanced hourglass formulation does not include the effect of the plasticity. Subsequently, the enhanced hourglass stiffness may be too large, and incorrect results (e.g., delayed yielding or excessive springback effect) may be obtained. The effect becomes more pronounced as the mesh is coarsened; care must be taken to ensure that enough elements are used through the thickness of the structure.

When plasticity is present, the stiffness-based hourglass control is less stiff than the enhanced hourglass control. The stiffness-based hourglass moduli use a significant knockdown factor on the elastic moduli, causing the element to be less stiff than in the enhanced control case. This may give better results with plastic bending; the user is urged to examine results thoroughly and exercise careful judgment.

9.2.3 Kelvin Viscoelastic Hourglass Control

A model using viscoelastic material properties can also be faced with the hourglass control problem, and therefore it is important to know how are the stiffness (K) and linear viscous coefficient (C) used in the Kelvin viscoelastic method of hourglass control have been determined [1].

The Kelvin viscoelastic method of suppressing hourglass modes is based on the use of a pure stiffness approach and a pure viscous approach. The two components can be used in isolation or in combination and are based on a linear stiffness K and a linear viscous coefficient C. The K and C terms are used to calculate the forces and moments that are conjugate to the hourglass deformation. Letting the function q be the hourglass mode magnitude, the forces and moments used to suppress the hourglass deformation are calculated as shown in Eq. 9.4,

$$Q = s\left[(1-\alpha)\,Kq\,(t) + \alpha C\frac{dq}{dt}\right] \tag{9.4}$$

The stiffness K and viscous coefficient C depend on the material properties. The stiffness is defined as shown in Eq. 9.5,

$$K = sB\frac{L^2}{V} \tag{9.5}$$

where s is the scaling factor specified on the dataline of the ***SECTION CONTROLS** keyword, B is the effective bulk modulus, L is the smallest element dimension and V is the element volume. The viscous coefficient is defined as shown in Eq. 9.6,

$$C = s\rho C_d V^{2/3} \tag{9.6}$$

where C_d is the dilatation wave speed and ρ the material mass density.

9.3 Errors Printed to the Message File for a Model with Hybrid Elements

The compatibility errors information printed to the message file when hybrid elements are used does not indicate that an error has occurred in the analysis. In the case of solid hybrid elements, volumetric strain is computed, both, from the displacement degrees of freedom and from the independent pressure degree of freedom. Abaqus checks that the difference between these two strain measures lies within a specific tolerance in order to ensure convergence. This is similar to the checks on force and moment equilibrium. In each iteration, information regarding the hybrid element compatibility is printed to the message file.

For example, the following is from the last iteration of an increment using hybrid elements:

```
COMPATIBILITY ERRORS:
TYPE        NUMBER EXCEEDING     MAXIMUM      IN ELEMENT
            TOLERANCE            ERROR
VOLUMETRIC  0                   -9.595E-10           49

INSTANCE: PART-1
```

In the case above, the maximum error (or difference) between the two strain values is well within the acceptable tolerance of 10^{-5}. It is safe to ignore messages like these. The situation may arise when force and moment equilibrium has been satisfied but the hybrid element compatibility check fails.

For example, this happens in step 2, increment 4, iteration 3 of the **bootseal.inp** example problem. In this analysis, the compatibility check is satisfied in iteration 4, so the analysis proceeds and again the message can be ignored.

There may be situations where Abaqus has difficulty satisfying the compatibility check and the analysis fails to converge. Usually, this means that inappropriate elements are being used for the analysis or that the material is completely incompressible. In these situations, it is generally necessary to change the model or the analysis to solve the problem.

Reference

1. Flanagan DP, Belytschko T (1981) A uniform strain hexahedron and quadrilateral with orthogonal hourglass control. Int J Numer Methods Eng 17:679–706

Chapter 10
Need Some Help?

10.1 Retrieving Files Referred to Examples in the Abaqus Documentation

Use the Abaqus fetch utility to retrieve example files referred to in the Abaqus documentation. For example, use the following command to fetch the input file boltpipeflange_3d_cyclsym.inp where **abaqus** is the command you use to run Abaqus.

```
abaqus fetch job=boltpipeflange_3d_cyclsym.inp
```

Similarly, the Abaqus fetch utility may be used to extract other documented input files, python script files (such as journal files and parametric study script files) user subroutine files, or post-processing programs from the compressed archive files provided with the release.

10.2 Using the Abaqus Verification, Benchmarks, and Example Problems Guides

If the user is searching for a demonstration of a particular feature or keyword. The applicable example from the Abaqus Example Problems Guide is quite advanced and combines several features. Does Abaqus include smaller, simpler examples that isolate certain features? Are there any comparisons of Abaqus solutions with analytical solutions available in manual handbooks or published work.

Included with every Abaqus installation there is an extensive amount of documentation designed to illustrate the features of the program. The Abaqus Verification, Benchmarks and Example Problems guides have been created to demonstrate the use of individual features as well as the application of Abaqus to real-world engineering simulations.

R. J. Boulbes, *Troubleshooting Finite-Element Modeling with Abaqus*,
https://doi.org/10.1007/978-3-030-26740-7_10

The Abaqus Verification, Benchmarks, and Example Problems Guides demonstrate the features of Abaqus in an ascending level of complexity. Taken together, the three manuals serve as a very detailed, complete collection of examples and illustrations that will allow users to quickly gain familiarity with the features and capabilities of Abaqus. The manuals are described below:

Abaqus Verification Guide: This volume contains more than 5000 basic test cases, providing verification of each individual program feature (procedures, output options, MPCs, loads, boundary conditions, material models, etc.) against exact calculations and other published results.
The problems contained in this manual are small, typically ranging from 1 to 20 elements in size. It is an ideal source of examples for individual features, and it may be useful to run these problems when learning to use a new capability. In addition, the supplied input data files provide good starting points to check the behavior of elements, materials, and so on.

Abaqus Benchmarks Guide: This volume contains over 200 benchmark problems and standard analyses used to evaluate the performance of Abaqus; the tests are multiple element tests of simple geometries or simplified versions of real problems. The NAFEMS[1] benchmark problems are included in this manual.

Abaqus Example Problems Guide: This volume contains more than 75 detailed examples designed to illustrate the approaches and decisions needed to perform meaningful linear and nonlinear analyses in Abaqus standard and Abaqus explicit. It is generally useful to look for relevant examples in this manual and to review them when embarking on a new class of problem. All the model examples are categorized into the following physics problems:

1. **Static Stress/Displacement Analyses**
   - 1.1. *Static and quasi-static stress analyses*
     - 1.1.1. Axisymmetric analysis of bolted pipe flange connections
     - 1.1.2. Elastic–plastic collapse of a thin walled elbow under in plane bending and internal pressure
     - 1.1.3. Parametric study of a linear elastic pipeline under in plane bending
     - 1.1.4. Indentation of an elastomeric foam specimen with a hemispherical punch
     - 1.1.5. Collapse of a concrete slab
     - 1.1.6. Jointed rock slope stability
     - 1.1.7. Notched beam under cyclic loading
     - 1.1.8. Uniaxial ratchetting under tension and compression
     - 1.1.9. Hydrostatic fluid elements: modeling an airspring
     - 1.1.10. Shell-to-solid sub-modeling and shell-to-solid coupling of a pipe joint

[1]NAFEMS is the International Association for the Engineering Modelling, Analysis and Simulation Community. It is a not for profit organization, which was established in 1983. https://www.nafems.org/.

- 1.1.11. Stress-free element reactivation
- 1.1.12. Transient loading of a viscoelastic bushing
- 1.1.13. Indentation of a thick plate
- 1.1.14. Damage and failure of a laminated composite plate
- 1.1.15. Analysis of an automotive boot seal
- 1.1.16. Pressure penetration analysis of an air duct kiss seal
- 1.1.17. Self-contact in rubber/foam components: jounce bumper
- 1.1.18. Self-contact in rubber/foam components: rubber gasket
- 1.1.19. Submodeling of a stacked sheet metal assembly
- 1.1.20. Axisymmetric analysis of a threaded connection
- 1.1.21. Direct cyclic analysis of a cylinder head under cyclic thermal–mechanical loadings.
- 1.1.22. Erosion of material (sand production) in an oil wellbore
- 1.1.23. Submodel stress analysis of pressure vessel closure hardware
- 1.1.24. Using a composite layup to model a yacht hull
- 1.1.25. Energy computations in a contact analysis

1.2. *Buckling and collapse analyses*

- 1.2.1. Snap through buckling analysis of circular arches
- 1.2.2. Laminated composite shells: buckling of a cylindrical panel with a circular hole
- 1.2.3. Buckling of a column with spot welds
- 1.2.4. Elastic–plastic K frame structure
- 1.2.5. Unstable static problem: reinforced plate under compressive loads
- 1.2.6. Buckling of an imperfection sensitive cylindrical shell

1.3. *Forming analyses*

- 1.3.1. Upsetting of a cylindrical billet: quasi-static analysis with mesh-to-mesh solution mapping (Abaqus standard) and adaptive meshing (Abaqus explicit)
- 1.3.2. Superplastic forming of a rectangular box
- 1.3.3. Stretching of a thin sheet with a hemispherical punch
- 1.3.4. Deep drawing of a cylindrical cup
- 1.3.5. Extrusion of a cylindrical metal bar with frictional heat generation
- 1.3.6. Rolling of thick plates
- 1.3.7. Axisymmetric forming of a circular cup
- 1.3.8. Cup/trough forming
- 1.3.9. Forging with sinusoidal dies
- 1.3.10. Forging with multiple complex dies
- 1.3.11. Flat rolling: transient and steady state
- 1.3.12. Section rolling
- 1.3.13. Ring rolling
- 1.3.14. Axisymmetric extrusion: transient and steady state
- 1.3.15. Two-step forming simulation

- 1.3.16. Upsetting of a cylindrical billet: coupled temperature–displacement and adiabatic analysis
- 1.3.17. Unstable static problem: thermal forming of a metal sheet
- 1.3.18. Inertia welding simulation using Abaqus standard and Abaqus CAE
- 1.3.19. Moldflow translation examples

1.4. *Fracture and damage*
- 1.4.1. A plate with a part through crack: elastic line spring modeling
- 1.4.2. Contour integrals for a conical crack in a linear elastic infinite half-space
- 1.4.3. Elastic–plastic line spring modeling of a finite-length cylinder with a part through axial flaw
- 1.4.4. Crack growth in a three-point bend specimen
- 1.4.5. Analysis of skin stiffener debonding under tension
- 1.4.6. Failure of blunt notched fiber–metal laminates
- 1.4.7. Debonding behavior of a double cantilever beam
- 1.4.8. Debonding behavior of a single-leg bending specimen
- 1.4.9. Postbuckling and growth of delaminations in composite panels

1.5. *Import analyses*
- 1.5.1. Springback of two-dimensional draw bending
- 1.5.2. Deep drawing of a square box

2. **Dynamic Stress/Displacement Analyses**

2.1. *Dynamic stress analyses*
- 2.1.1. Nonlinear dynamic analysis of a structure with local inelastic collapse
- 2.1.2. Detroit Edison pipe whip experiment
- 2.1.3. Rigid projectile impacting eroding plate
- 2.1.4. Eroding projectile impacting eroding plate
- 2.1.5. Tennis racket and ball
- 2.1.6. Pressurized fuel tank with variable shell thickness
- 2.1.7. Modeling of an automobile suspension
- 2.1.8. Explosive pipe closure
- 2.1.9. Knee bolster impact with general contact
- 2.1.10. Crimp forming with general contact
- 2.1.11. Collapse of a stack of blocks with general contact
- 2.1.12. Cask drop with foam impact limiter
- 2.1.13. Oblique impact of a copper rod
- 2.1.14. Water sloshing in a baffled tank
- 2.1.15. Seismic analysis of a concrete gravity dam
- 2.1.16. Progressive failure analysis of thin-wall aluminum extrusion under quasi-static and dynamic loads
- 2.1.17. Impact analysis of a pawl ratchet device
- 2.1.18. High-velocity impact of a ceramic target

2.2. *Mode-based dynamic analyses*

- 2.2.1. Analysis of a rotating fan using substructures and cyclic symmetry
- 2.2.2. Linear analysis of the Indian Point reactor feedwater line
- 2.2.3. Response spectra of a three-dimensional frame building
- 2.2.4. Brake squeal analysis
- 2.2.5. Dynamic analysis of antenna structure utilizing residual modes
- 2.2.6. Steady-state dynamic analysis of a vehicle body in white model

2.3. *Eulerian analyses*

- 2.3.1. Rivet forming
- 2.3.2. Impact of a water filled bottle

2.4. *Co-simulation analyses*

- 2.4.1. Dynamic impact of a scooter with a bump

3. **Tire and Vehicle Analyses**

3.1. *Tire analyses*

- 3.1.1. Symmetric results transfer for a static tire analysis
- 3.1.2. Steady-state rolling analysis of a tire
- 3.1.3. Subspace-based steady-state dynamic tire analysis
- 3.1.4. Steady-state dynamic analysis of a tire substructure
- 3.1.5. Coupled acoustic structural analysis of a tire filled with air
- 3.1.6. Import of a steady-state rolling tire
- 3.1.7. Analysis of a solid disc with Mullins effect and permanent set
- 3.1.8. Treadwear simulation using adaptive meshing in Abaqus standard
- 3.1.9. Dynamic analysis of an air-filled tire with rolling transport effects
- 3.1.10. Acoustics in a circular duct with flow

3.2. *Vehicle analyses*

- 3.2.1. Inertia relief in a pick up truck
- 3.2.2. Substructure analysis of a pick up truck model
- 3.2.3. Display body analysis of a pick up truck model
- 3.2.4. Continuum modeling of automotive spot welds

3.3. *Occupant safety analyses*

- 3.3.1. Seat belt analysis of a simplified crash dummy
- 3.3.2. Side curtain airbag impactor test

4. **Mechanism Analyses**

4.1. *Mechanism analyses*

- 4.1.1. Resolving overconstraints in a multi body mechanism model
- 4.1.2. Crank mechanism
- 4.1.3. Snubber arm mechanism
- 4.1.4. Flap mechanism
- 4.1.5. Tail skid mechanism
- 4.1.6. Cylinder cam mechanism
- 4.1.7. Driveshaft mechanism
- 4.1.8. Geneva mechanism

4.1.9. Trailing edge flap mechanism
4.1.10. Substructure analysis of a one piston engine model
4.1.11. Application of bushing connectors in the analysis of a three-point linkage
4.1.12. Gear assemblies

5. **Heat Transfer and Thermal–Stress Analyses**

5.1. *Heat transfer and thermal–stress analyses*
5.1.1. Thermal–stress analysis of a disc brake
5.1.2. A sequentially coupled thermal–mechanical analysis of a disc brake with a Eulerian approach
5.1.3. Exhaust manifold assemblage
5.1.4. Coolant manifold cover gasketed joint
5.1.5. Conductive, convective, and radiative heat transfer in an exhaust manifold
5.1.6. Thermal–stress analysis of a reactor pressure vessel bolted closure

6. **Fluid Dynamics and Fluid–Structure Interaction**

6.1. *Fluid dynamics and fluid–structure interaction*
6.2. *Conjugate heat transfer analysis of a component mounted electronic circuit board*

7. **Electromagnetic Analyses**

7.1. *Piezoelectric analyses*
7.1.1. Eigenvalue analysis of a piezoelectric transducer
7.1.2. Transient dynamic nonlinear response of a piezoelectric transducer
7.2. *Joule heating analyses*
7.2.1. Thermal–electrical modeling of an automotive fuse

8. **Mass Diffusion Analyses**

8.1. *Mass diffusion analyses*
8.1.1. Hydrogen diffusion in a vessel wall section
8.1.2. Diffusion toward an elastic crack tip

9. **Acoustic and Shock Analyses**

9.1. *Acoustic and shock analyses*
9.1.1. Fully and sequentially coupled acoustic structural analysis of a muffler
9.1.2. Coupled acoustic structural analysis of a speaker
9.1.3. Analysis of a speaker using Abaqus Dymola co-simulation
9.1.4. Response of a submerged cylinder to an underwater explosion shock wave
9.1.5. Convergence studies for shock analyses using shell elements

9.1.6. UNDEX analysis of a detailed submarine model
9.1.7. Coupled acoustic structural analysis of a pickup truck
9.1.8. Long duration response of a submerged cylinder to an underwater explosion
9.1.9. Deformation of a sandwich plate under CONWEP blast loading

10. **Soils Analyses**

10.1. *Soils analyses*
10.1.1. Plane strain consolidation
10.1.2. Calculation of phreatic surface in an earth dam
10.1.3. Axisymmetric simulation of an oil well
10.1.4. Analysis of a pipeline buried in soil
10.1.5. Hydraulically induced fracture in a well bore
10.1.6. Permafrost thawing pipeline interaction

11. **Structural Optimization Analyses**

11.1. *Topology optimization analyses*
11.1.1. Topology optimization of an automotive control arm
11.2. *Shape optimization analyses*
11.2.1. Shape optimization of a connecting rod
11.3. *Sizing optimization analyses*
11.3.1. Sizing optimization of a gear shift control holder
11.3.2. Sizing optimization of a car door

12. **Abaqus Aqua Analyses**

12.1. *Abaqus Aqua analyses*
12.1.1. Jack up foundation analyses
12.1.2. Riser dynamics

13. **Particle Methods Analyses**

13.1. *Discrete element method analyses*
13.1.1. Mixing of granular media in a drum mixer

14. **Design Sensitivity Analyses**

14.1. *Overview*
14.2. *Examples*
14.2.1. Design sensitivity analysis of a composite centrifuge
14.2.2. Design sensitivities for tire inflation, footprint, and natural frequency analysis
14.2.2. Design sensitivity analysis of a windshield wiper
14.2.4. Design sensitivity analysis of a rubber bushing

15. **Post-processing of Abaqus Results**

15.1. *Post-processing examples*
15.1.1. User post-processing of Abaqus results files: overview

15.1.2. Joining data from multiple results files and converting file format: FJOIN
15.1.3. Calculation of principal stresses and strains and their directions: FPRIN
15.1.4. Creation of a perturbed mesh from original coordinate data and eigenvectors: FPERT
15.1.5. Output radiation view factors and facet areas: FRAD
15.1.6. Creation of a data file to facilitate the post-processing of elbow element results: FELBOW
15.1.7. Translating Abaqus data to modal neutral file format for analysis in MSC.ADAMS

Each of the input files contained in the Abaqus Verification, Benchmarks, and Example Problems guides is included with the Abaqus release. The Abaqus Fetch utility is used to extract sample input files, user subroutine files, journal files, parametric study script files, or post-processing programs from the compressed archive files provided with the release. For example, to extract the file **t1-std.inp** from the archive and rename it is **mytest.inp**, type

```
abaqus fetch job=mytest input=t1-std
```

The archive of demonstration and example files provided with Abaqus can be searched using the Abaqus findkeyword utility. This utility will also query the archive for problems used in training seminars, benchmark timing problems, and those in the keyword versions of Getting Started with Abaqus standard; Getting Started with Abaqus explicit and the Getting Started with Abaqus; also Getting Started with Abaqus standard: Keywords Version, Getting Started with Abaqus explicit: Keywords Version tutorial guides.

Specify which keywords, parameters, and values are of interest; and then this utility will list the input files that contain those keywords, parameters, and values. Multiple keywords may be specified, which causes the utility to list those input files that contain all the specified keywords. For example, to query the archive for all input files that contain the ***INITIAL CONDITIONS, TYPE= STRESS** option and send the output to a file named **output.dat**, type

```
abaqus findkeyword job=output
```

When prompted with a * symbol, type

```
initial conditions, type=stress
```

For additional information, see Execution procedure for querying the keyword/problem database in the Abaqus Analysis User's Guide.

10.3 Excessive Memory Usage with Cavity Radiation Problems

The memory requirements for an Abaqus standard, heat transfer, cavity radiation analysis may be excessively large if the cavity includes shell elements with multiple temperature degrees of freedom (i.e., multiple section points) through the shell thickness. It is possible to work around this problem in the following manner:

- Over the original shell elements define a set of dummy shell elements with only one section point through the thickness.
- Assign very low conductivity, and very low specific heat properties to the dummy elements.
- Use the dummy shell elements for the cavity radiation definition. Tie ***TIE** the dummy shell elements to the original elements using the dummy elements as the master surface.

An outline of the keyword options that may be used to implement this workaround is shown below. Items in red need to be replaced by values appropriate for the model being analyzed. For example, **originalShellNodes** is a set of nodes in the original model used to define the elements used for the cavity definition.

```
*** Copy nodes and elements of shells that define the cavity ***
**
*NCOPY,CHANGE NUMBER=nodeNumbOffset,OLDSET=originalShellNodes,SHIFT,
NEWSET=<fontcolor="#FF0000">dummyShellNode
0,0,0
0,0,0,1,0,0,0
**
** Tip: *NSET,NSET=originalShellNodes,ELSET=originalShellElem
** may be used to create the node set 'originalShellNodes' from
** an element set containing the cavity elements.
**
*ELCOPY,ELEMENT SHIFT=elNumbOffset,OLDSET=originalShellElem,
SHIFT NODES=nodeNumbOffset,NEWSET=dummyShellElem
**
*** Define properties for the dummy shells - ***
*** must have only 1 section point ***
**
*SHELL SECTION,ELSET=dummyShellElem,MATERIAL=DummyMat
thickness, 1
**
*MATERIAL,NAME=DummyMat
*SPECIFIC HEAT
select a value MUCH smaller than model materials
*DENSITY
density
*CONDUCTIVITY
select a value MUCH smaller than model materials
**
*** dummy shell surface definition ***
**
*SURFACE,NAME=dummySurf,PROPERTY=propertyName
```

```
Same data as original but with new dummy elements
**
*** Tie the dummy shell to the original shell ***
*** IMPORTANT NOTE: ***
*** TO WORK AROUND THE PROBLEM THE DUMMY SHELL ***
*** MUST BE THE MASTER SURFACE ***
**
*TIE,NAME=Tie,TIED NSET=originalShellNodes
originalCavitySurface,dummySurf
**
*** Define the cavity on the dummy shells ***
*** (remove the original cavity) ***
**
*CAVITY DEFINITION,NAME=cavityName
dummySurf,
```

10.4 Perform a Sub-model Analysis

The user first obtains results for the entire model using a relatively coarse mesh and perhaps simplified geometry. This model is called the global model. Data from the output database that is generated by the global model are used to drive the sub-model. As a result, the output requests in the global model must include the driven variables. The user can also drive a sub-model using the global model results stored in a results file that was generated from the Abaqus execution procedure.

The main steps of the procedure are as follows:

1. Get a global solution using a coarse mesh.
2. Interpolate this solution onto the boundary of a locally refined mesh.
3. Get a more accurate solution in the local area of interest.

The assumptions are fundamental to determine what kinds of partitioning strategies will be more relevant to use the function of the design model. The Saint-Venant's principle[2] applies.

The boundary of a sub-model is sufficiently far from the region within the sub-model where the response changes, thus

- The global model solution defines the response on the sub-model boundary.
- Detailed modeling of the local region has a negligible effect on the global solution.

[2] The Saint-Venant's principle allows elasticians to replace complicated stress distributions or weak boundary conditions with ones that are easier to solve, as long as that boundary is geometrically short. Quite analogous to the electrostatics, where the electric field due to the ith moment of the load (with zeroth being the net charge, first the dipole, second the quadrupole) decays as $1/r^{i+2}$ over space, Saint-Venant's principle states that high order momentum of mechanical load (moment with order higher than torque) decays so fast that they never need to be considered for regions far from the short boundary. Therefore, the Saint-Venant's principle can be regarded as a statement on the asymptotic behavior of the Green's function by a point-load.

Warning: The user must ensure that the sub-modeling approach will provide physically meaningful results.

- There are no default protections when sub-modeling; it is a matter of the user's judgment that the sub-model is done correctly.
- Examine contour plots of important variables near the boundaries of the sub-modeled regions.
  - If the contour plots are created with the same contour range, the results will be valid if contour values coincide at the boundary of the sub-modeled region.

Driven variables: The nodal degrees of freedom on the boundary of a defined sub-model surface, whose values are calculated by interpolation of the global solution. Therefore, the only link is the transfer of values for the driven variables.

10.4.1 Implementation

The sub-modeling implementation in Abaqus is very general.

- The following types of sub-modeling are provided:
  - Solid to Solid.
  - Shell to Shell.
  - Membrane to Membrane.
  - Shell to Solid.
- The transfer of solution variables from the global model to the sub-model is based on the position of the sub-model boundary nodes.
- Submodel boundary nodes need not align with mesh lines in the global model.
- Element types that are different from those used in the global model can be used in the sub-model.
  - Second-order elements can be used in the local model and first-order elements can be used in the global model, or the opposite.
- Materials in the sub-model can be different from those in the global model.
  - Metal plasticity can be used in the local model, and linear elasticity can be used in the global model.
- Procedures in the sub-model can be different from those in the global model.
  - The static response of the local model is based on the dynamic response of the global model.
- The sub-modeling implementation accommodates both linear and nonlinear analyses.

- Both the sub-model and the global model can follow sequences of analysis procedures (multi-step analysis), and the sequences can be different for the local and global models.
- Submodeling can be repeated for any number of levels.

10.4.2 Loading Conditions

The surface-based sub-modeling technique is an alternative sub-modeling technique that uses the stress field to interpolate global model results onto the sub-model integration points on the driven element-based surface facets. To use surface-based submodeling, the user creates a sub-model load.

- All types of loading and prescribed boundary conditions can be applied to the sub-model.
- The user must apply loading and prescribed boundary conditions in the sub-model in a manner that is consistent with the loading of the global model. Otherwise, incorrect results will be obtained.
- Abaqus interpolates values only of the driven nodal variables onto the sub-model. Predefined fields, such as temperature in a stress analysis, must be provided for all nodes in the sub-model where they are required.
- Initial conditions should be consistent between the global model and the sub-model.

10.4.3 Sub-model Boundary Conditions

The most common sub-modeling technique is node-based sub-modeling, which uses a nodal results field (including displacement, temperature, or pressure degrees of freedom) to interpolate global model results onto the sub-model nodes. Node-based sub-modeling is also a more general technique. To use node-based sub-modeling, the user creates a sub-model boundary condition.

If the user applies a sub-model boundary condition to nodes that are constrained by either a displacement/rotation boundary condition or a connector displacement boundary condition on the global model in a previous step and the global model's boundary condition is fixed using the **Fixed at Current Position** method, Abaqus CAE disregards the sub-model boundary condition for those nodes and retains the specifications in the boundary condition on the global model instead. Abaqus CAE reports this replacement of boundary conditions in the data file for the analysis.

- Usually, all active degrees of freedom are specified at the driven nodes in the sub-model.
- Only the fundamental solution variables can be driven.
  - In either Solid-to-Solid or Shell-to-Shell sub-modeling: displacements, temperature, electric potential, pore pressure, and so on.

For example, the user cannot drive velocities or accelerations on the sub-model boundary because they represent the first or second derivative of the displacements, as the electric field is the gradient of the electric potential, and so on.
  - Abaqus selects the driven variables itself when a global shell is used to drive a local solid model.

- Submodel boundary conditions can be removed or reintroduced in the usual way.
- When using the keywords interface (.inp), only one ***BOUNDARY, SUBMODEL** option can be specified per step of the sub-model analysis.
- When using Abaqus CAE, multiple sub-model boundary conditions may be specified.
  - These are later combined automatically to satisfy the keywords requirement stated above.
- Multiple sub-model definitions should be included when different global sets must be used to drive nodes that occupy almost identical spatial positions.

10.4.4 Interpolation

The sub-model results from global model results computed are interpolated in accordance with the following points.

- Abaqus determines the values of the driven nodal variables throughout the step of the sub-model analysis.
  - Interpolation is done in both space and in time.
- The order of spatial interpolation of the driven variables is dictated by the order of elements used at the global level.
- Automatic time incrementation is applied independently in the global and sub-model analyses.
  - Independent time incrementation is accommodated by the temporal interpolation of the driven variables.
  - Linear temporal interpolation is used between the values read from the output database or results file.

10.4.5 Step-by-Step Procedure for a Sub-model

Please follow the procedure below to make a proper sub-model with Abaqus:

1. Open Abaqus and create the Job.cae file inside the current work directory.
2. Make the global model called xyz_global and submit the job as usual to get the xyz.odb

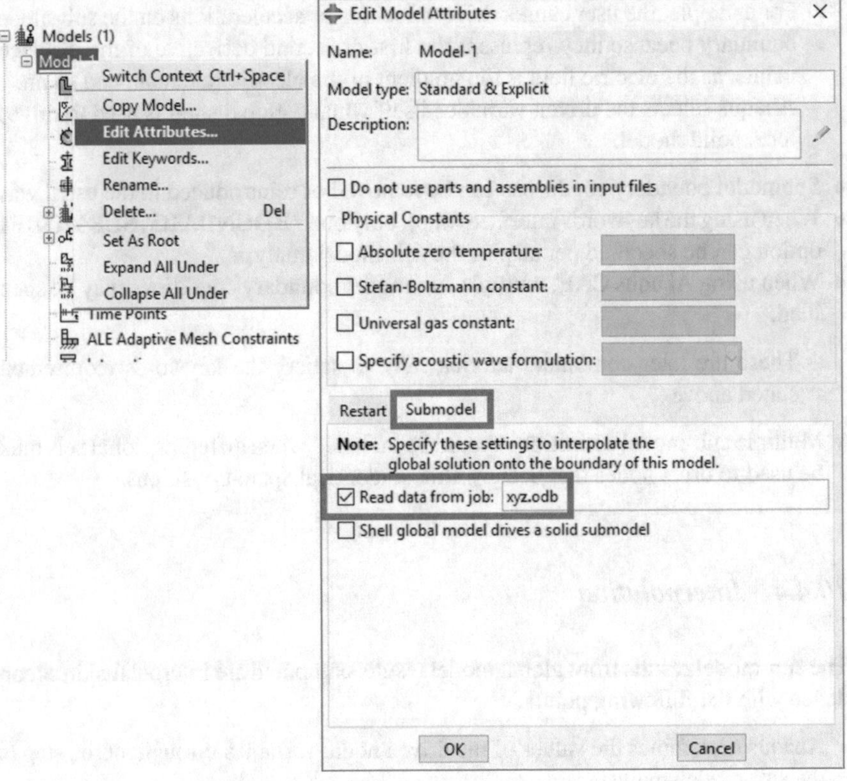

Fig. 10.1 Create sub-model and read data from global model

- Do **NOT** create a partition zone for sub-modeling in the global model.

3. Make a sub-model xyz_sub-model by copying the global model, then click right and go on **Edit Attributes…** to link the transfer of data from the global model xyz.odb as shown in Fig. 10.1.
4. Open and work with the model xyz_sub-model.
5. Set the partition strategies. The user will need to make the sub-model zone(s) at the concerned part of the assembly.
6. Delete **ALL** mapped mesh in order to remap **ONLY** the mesh inside the sub-model zone(s).
7. Suppress **ALL** existing boundary and loading conditions.
8. Sub-model boundary conditions in **BCs** for the analysis step the user needs:

- Driven nodes are defined through the sub-model boundary condition.
- The user may specify that only particular degrees of freedom can be driven at the boundary of the sub-model.
- The usage form of the boundary condition is used in sub-modeling analysis, except when a shell model is used to drive a solid sub-model.

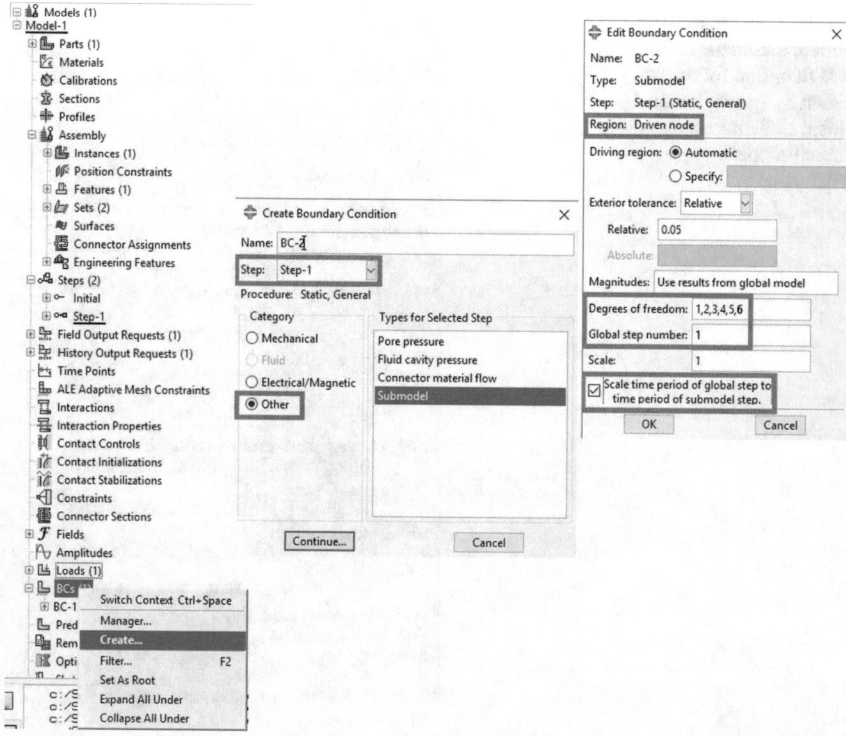

Fig. 10.2 Create sub-model boundary condition

- The selected analysis step is the step in the global solution that is to be mapped to this step of the sub-model analysis.
- The node set Driven Node contains the nodes that are to be driven in this step of the sub-model analysis.
- The degrees of freedom define either displacements or rotations, which will be interpolated onto the driven nodes thanks to the shape function of the kinematics selected with a finite-element type, and meshed on a structure model. For example, the Cartesian's notation regarding the kinematic of a finite element used for a beam element is $1 = U_x$, $2 = U_y$, $3 = U_z$, $4 = Rot_x$, $5 = Rot_y$, $6 = Rot_z$.
- The user must ensure that the step time in the sub-model analysis matches the step time in the global analysis, by ticking off the scale time period option as shown in Fig. 10.2. Otherwise, the interpolation with respect to the time increment will be incorrect.

9. Submit the job as usual.

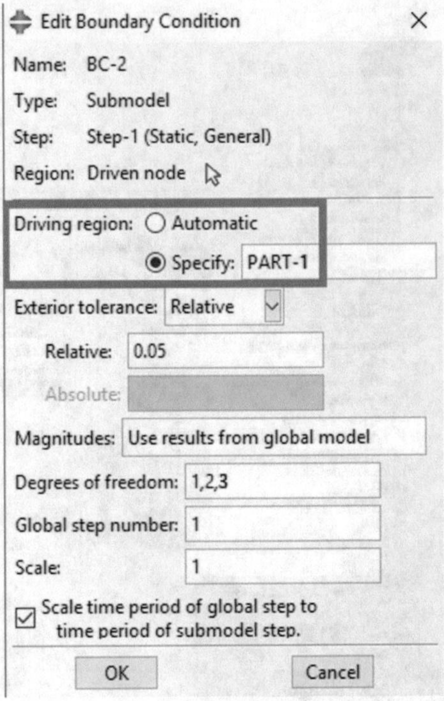

Fig. 10.3 In the **driving region**, automatic is the default option for driving region by specifying the global elements used to drive the sub-model

10.4.6 Setting Options

The most important settings for the driving region, the exterior tolerance and the linear perturbation are described in Fig. 10.3, Fig. 10.4, and Fig. 10.5, respectively.

As shown in Fig. 10.3, the driving region settings are described below:

- By default, the global model in the vicinity of the sub-model is searched for elements that contain driven nodes.
  - The sub-model is then driven by the response of these elements.
- To preclude certain elements from driving the sub-model, a global set can be defined to limit the search to this subset of the global model.
  - This is necessary if the model has closely spaced entities.

As shown in Fig. 10.4, the exterior tolerance settings are described as follows:

- The exterior surface is used to interpolate spatially in the global solution onto the sub-model nodes.
- By default, this tolerance is 0.05 times the average element size in the global model.
- The absolute exterior tolerance parameter can be used to specify the distance in the consistent model of units.

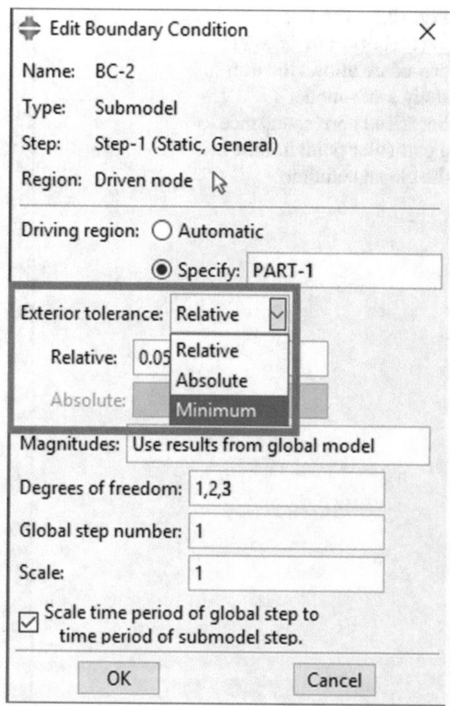

Fig. 10.4 The **exterior tolerance** parameter of the sub-model option defines how far a boundary node in the sub-model can lie outside the exterior surface of the global model

As shown in Fig. 10.5, the linear perturbation setting is described as follows:

- If the time period of the sub-model analysis is different from the time period of the global analysis, you can choose to scale the time period of the global step to match the time period of the sub-model step. For example, Abaqus determines the displacements of the global model at the time when 20% of the global step has been completed and applies those displacements at the time when 20% of the sub-model step has been calculated.

10.4.7 Shell to Solid

Starting with a global model combining Shell and Solid elements, it is possible to make a sub-model such that the Shell nodes in the global model will drive the Solid nodes in the sub-model.

- Sub-modeling also allows the user to use the results from a global shell model as the load for a more detailed, continuum element sub-model.
- This capability is useful to model details of shell joints, to analyze 3D crack problems, or to obtain more accurate solutions through the shell thickness.

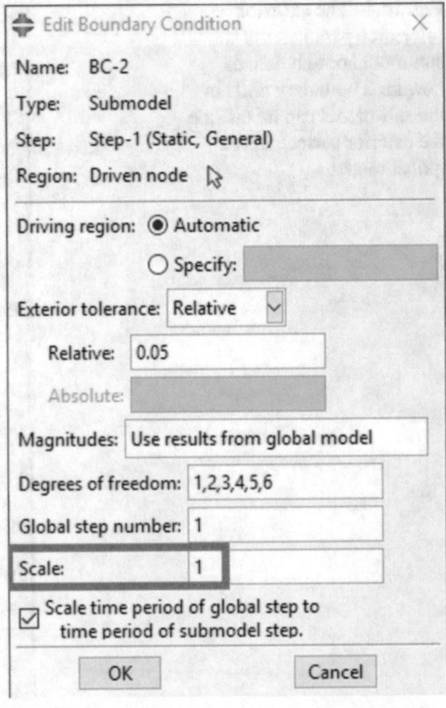

Fig. 10.5 The **linear perturbation** in the static procedure allows the user to study a sub-model's linearized correspondence to a particular point in time in the global solution

The edit model attribute in such a case is shown in Fig. 10.6 and the usage is shown in Fig. 10.7.

As shown in Fig. 10.7, the different usage parameters are described as follows:

- The **Shell thickness** is the maximum thickness of the shell elements in the sub-model region of the global model.
  - Thicker elements away from the sub-model region should not be considered.
  - If the shell section was offset in the global model, set this value to twice the shell thickness.
- The **Center zone size** is the size given in the length scale of the model of a zone around the global shell model's reference surface.

Driven variables: at the nodes are chosen automatically.

- Which variables are driven depends on the location of these nodes relative to the center zone around the shell's reference surface.
  - All three displacement components are driven at nodes within this center zone.
  - For driven nodes that are further away and outside this zone, only the displacement components parallel to the global shell model's reference surface are driven.

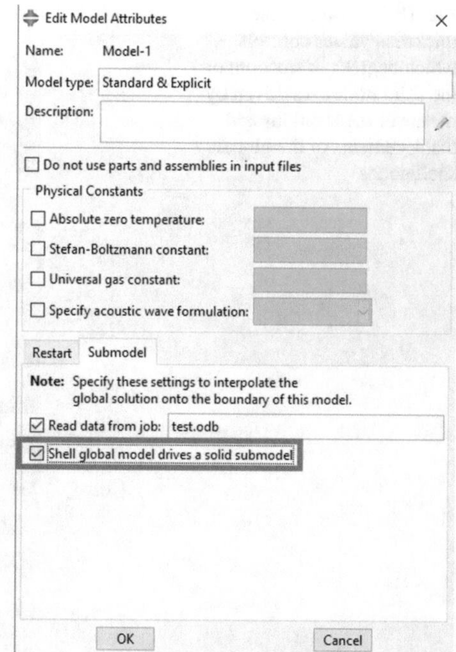

Fig. 10.6 Edit model attributes with Abaqus CAE, the model attributes must be set appropriately to define a shell-to-solid sub-modeling

– Driven nodes lying inside and outside the center zone can be visualized and differentiated from one another using Abaqus viewer by displaying the model boundary conditions.

10.4.8 Changing Procedures

Some changes in procedure regarding sub-modeling techniques are explained below:

- It is possible to treat general analysis steps as linear perturbation steps during sub-modeling, and vice versa.
- For example, the following three steps are applicable in

1. Static preload (general analysis step).
2. Natural frequency extraction including preload effects (perturbation step).
3. Analysis of 5 s of modal dynamic response (perturbation step).

10.4.9 Frequency Domain

The main options for sub-modeling techniques with transient analysis in the frequency domain are described below:

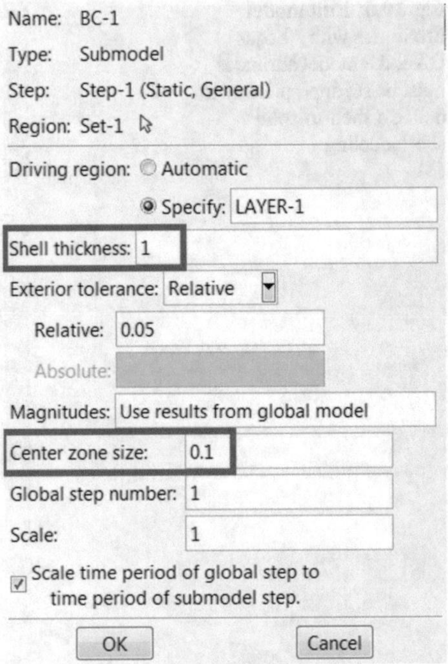

Fig. 10.7 **Usage** of these thickness values controls which degrees of freedom of the solid nodes are driven by means of the rotations and displacements of the global shell nodes

Name: BC-1
Type: Submodel
Step: Step-1 (Static, General)
Region: Set-1
Driving region: Automatic
Specify: LAYER-1
Shell thickness: 1
Exterior tolerance: Relative
Relative: 0.05
Absolute:
Magnitudes: Use results from global model
Center zone size: 0.1
Global step number: 1
Scale: 1
Scale time period of global step to time period of submodel step.
OK
Cancel

- Submodeling in the frequency domain is available only with the direct steady-state dynamics procedure. Other options do not allow driven boundary conditions.
- The frequency range of the sub-model must lie within the maximum and minimum frequencies calculated in the global model, or the results will not be accurate.
- Both the amplitude and the phase of the nodal displacements in the global model must be saved to drive the sub-model.
- Abaqus will interpolate the global solutions spatially and in the frequency domain.
- Results are most accurate when the frequencies requested in the sub-model match the frequencies at which the response was calculated in the global model, especially in the vicinity of the global model's natural frequencies.

10.4.10 Thermal and Stress Analysis

The main options when using sub-modeling techniques with coupled thermal–stress analysis are described below:

- The sub-modeling analysis technique can be used for either fully coupled or sequentially coupled thermal–stress analysis.
- For a fully coupled thermal–stress analysis, the procedure is identical to that for a static structural analysis:

1. Run a global coupled thermal–stress job.
2. Run a sub-model coupled thermal–stress job.
   - Use the global coupled thermal–stress job results to drive the sub-model boundary displacements, rotations and temperatures.

- For a sequential-coupled thermal–stress analysis, you can read the temperature as a field variable from the **Output data** base for either **similar or dissimilar** meshes. The steps for a sequential-coupled case are as follows:

1. Run a global heat transfer job (with a a mesh 1).
2. If necessary, run a sub-model heat transfer job (with a mesh 2) and use the global heat transfer job results to drive the sub-model boundary temperature.
3. Run a global thermal–stress job (with a mesh 3, which could be different from mesh 1), and then read the temperatures from the global heat transfer job.
4. Run a sub-model thermal–stress job (with a mesh 4, which may be different from mesh 2)
   - Use the global thermal–stress job results to drive the sub-model boundary displacements and rotations.
   - Read the temperatures from any of the preceding analyses.

10.4.11 Dynamic Analysis

The main options when using sub-modeling techniques in dynamic analysis are described below:

- The sub-modeling feature can be used in dynamic procedures using explicit integration (Abaqus explicit) and implicit direct integration (Abaqus standard)
  - Thus, we can drive an Abaqus standard sub-model with either an Abaqus standard or an Abaqus explicit global model.
  - Alternatively, we can drive an Abaqus explicit sub-model with either an Abaqus standard or an Abaqus explicit global model.
- In general, the same time scales should be used in the global model and sub-model. This is mostly true in problems in which inertia forces are significant.
- For quasi-static analyses, the time period of the global model and sub-model can be different. Scale the time variable of each driven node's amplitude function, in order to be correlated with the sub-model analysis step time.
- In dynamic sub-modeling, output from the global model should be at a high enough frequency to avoid aliasing problems (under sampling). The displacement results for the nodes that are used to drive the sub-model should be saved for each increment.

10.4.12 Limitations of Sub-modeling

There are some limitations regarding the elements, procedures, and Shell-to-solid configuration as briefly explained below:

10.4.12.1 Elements

- Elements that can be used at the global and sub-model levels include:
  - First- and second-order triangular and quadrilateral continuum elements.
  - First- and second-order tetrahedral, wedge, or brick continuum elements.
  - First- and second-order triangular and quadrilateral shell and membrane elements.
- The sub-model boundary nodes cannot lie in regions of the global model, where there are only 1D elements such as beams, trusses, links, or axisymmetric shells (no 1D interpolation).
- The sub-model boundary nodes cannot lie in regions of the global model, where there are only user elements, substructures, springs, dashpots, or other special elements (no basis for interpolation).
- The sub-model boundary nodes cannot lie in regions of the global model where there are only axisymmetric solid elements with nonlinear asymmetric deformation.
- Five degrees of freedom per node for shell elements (S4R5, S8R5, etc.) should be avoided at the global level; rotations are not saved.
  - Should these elements be necessary at the global level, driving the displacements for two rows of nodes along the sub-model boundary should effectively transfer the correct rotations to the sub-model.
  - They cannot be used in shell-to-solid sub-modeling.

10.4.12.2 Procedures

Sub-models cannot use any of the following analysis procedures:

- Coupled thermal–electrical analysis.
- Mode-based, linear dynamics analsysis:
  - ***MODAL DYNAMIC**
  - ***RANDOM RESPONSE**
  - ***RESPONSE SPECTRUM**
  - ***STEADY-STATE DYNAMICS**

10.4.12.3 Shell to Solid

Different kinematics of finite element like, for instance, a model using both Shell and Solid finite elements must consider the following indications:

- Shell-to-solid sub-modeling cannot be used with any other type of sub-modeling in the same model.
- Global models can contain both solid and shell elements. However, all driven nodes must lie within the shell elements in the global model.
- Shell elements with five degrees of freedom per node cannot be used.
- Temperature degrees of freedom cannot be driven.
- There is an approximation of the distribution of material away from the mid-surface of shells at corners.
  - Sub-models will not be driven correctly if the driven nodes lie within the shell thickness from a corner.
  - Include a corner as part of the sub-model, and drive it at nodes well away from the corner. Knowing that the corners are a source of stress concentration and high-stress gradients.

10.5 Perform a Restart Analysis

When the user runs an analysis, the analyst can write the model definition and state to the files required for restart. Scenarios for using the restart capability include the following:

Continuing an interrupted run: If an analysis is interrupted by a computer malfunction, the Abaqus restart analysis capability allows the analysis to complete as originally defined.
Continuing with additional steps: After viewing results from a successful analysis, the user may decide to append steps to the load history.
Changing an analysis: Sometimes, having viewed the results of the previous analysis, the user may want to restart the analysis from an intermediate point and change the remaining load history data in some manner. In addition, the user may want to add additional steps to the load history if the previous analysis completed successfully.

The motivations for performing a restart analysis are listed below:

- Continue analyses that stop at intermediate points.
  - The job may have stopped because:
    The maximum number of increments specified for the step was reached.
    There was not enough disk space or the machine failed.
    The job failed to converge.
  - The user may wish to continue the job after:

Examining results up to a specific point.
Modifying the history: procedure, loading, output, controls, and so on.

- Transfer results between Abaqus standard and Abaqus explicit.

The model used in the restart analysis must be the same as the model used for the original analysis up to the restart location.

- It must not modify or add any geometry, mesh, materials, or others that are already defined in the original analysis model.
- It must not modify any step, load, boundary condition, field, or interaction at/or before the restart location.
- New sets and amplitude curves may be defined in the restart analysis model.

Some modifications that may be performed when carrying out a restart analysis are, for example:

- Modification of the model attributes to define the restart data:
  - The job from which data will be read.
  - The step and increment/interval from which the analysis will begin.
  - If the increment/interval does not correspond to the end of the step, Abaqus can
    Try to finish the original step before trying any new steps.
    Terminate the original step.

When the user restarts an analysis, Abaqus creates a new output database file, **job-name.odb**, and a new results file **job-name.fil**; it writes output data to those files according to the criteria described below.

The Abaqus output database file **job-name.odb** contains results that can be used for post-processing in Abaqus CAE. By default, the output database file is not made continuous across restarts; Abaqus creates a new output database file each time a job is run. The user can combine the (X-Y) data extracted from multiple output database files in the Visualization module of Abaqus CAE. Alternatively, the user can also join field and history results from an original analysis and a restart analysis by running the Abaqus restart join execution procedure.

The Abaqus results file created in Abaqus standard and Abaqus explicit **job-name.fil** contains the user-specified results that can be used for post-processing in external post-processing packages. In Abaqus explicit, results are also written to the selected results file **job-name.sel**, which is then converted to the results file for post-processing.

Upon restart, Abaqus standard will copy the information from the old results file into the results file for the new job up to the restart point and begin writing the new results to the new file following that point. Abaqus explicit will copy the information from the old selected results file into the selected results file for the new job up to the restart point and begin writing the new results to the new file following that point.

If the old results file is not provided, Abaqus standard will continue the analysis, writing the results of the restart analysis only to the new results file. Therefore, the user will have segments of the analysis results in different files, which should be avoided in most cases since post-processing programs assume that the results are in a single continuous file. The user can merge such segmented results files, if necessary, by using the **abaqus append** execution procedure.

A restart analysis in Abaqus standard can use the restart files generated from the same or any previous maintenance delivery of the same general release. Restart is not compatible between general releases. In Abaqus explicit and Abaqus CFD, the original analysis and the restart analysis must use precisely the same release. A restart analysis in Abaqus and a recover analysis in Abaqus explicit must be run on a computer that is binary compatible with the computer used to generate the restart files.

10.5.1 Step-by-Step Procedure for a Restart

The same model performed will be reused as a restart model afterward.

1. To save computational time with at least two different load cases, the same analysis model will be used to get all model solutions from an initial model called Model-1 after analysis Step-1 as shown in Fig. 10.8.
2. Create a restart point, by setting the frequency equal to 1, and select the overlay option as shown in Fig. 10.9.

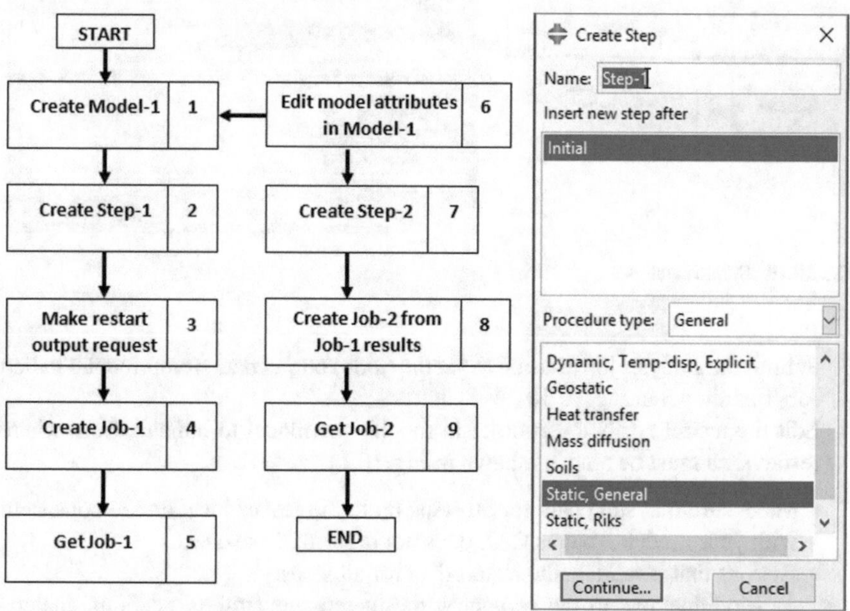

Fig. 10.8 Create the final analysis step before restarting the analysis

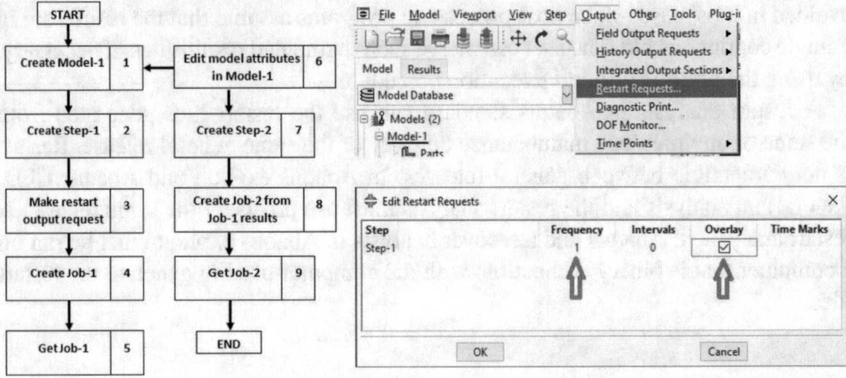

Fig. 10.9 Create the restart request

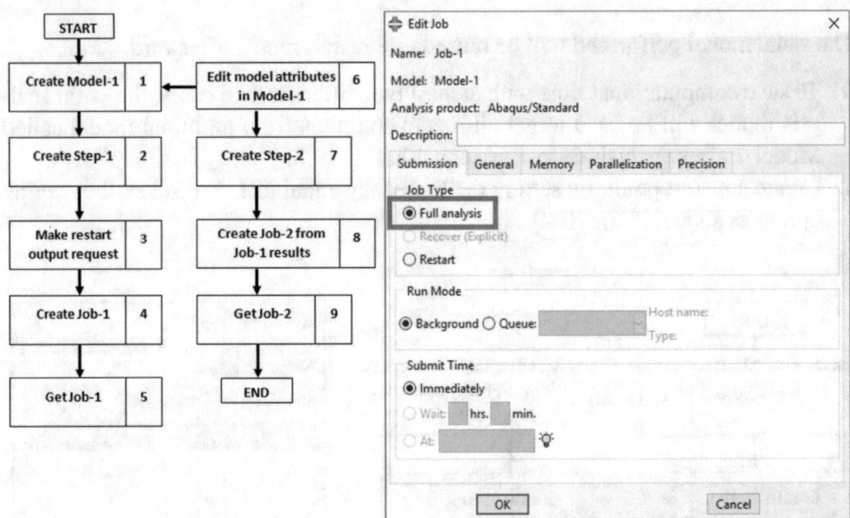

Fig. 10.10 Submit analysis

3. Submit the analysis job as usual to get the (.odb) output data from Model-1 called Job-1 as shown in Fig. 10.10.
4. Edit the model attributes in order to modify the model to indicate from where restart data must be read, as shown in Fig. 10.11.

   - Make sure that state data for all steps up to the restart location are consistent with your model. Abaqus CAE does not perform these checks yet.
   - Actions that invalidate the restart data for all steps:
     – Modifications to the geometry, mesh, sets, constraints, sections, materials, profiles, beam section profiles, skins, material orientations, and beam section orientations.

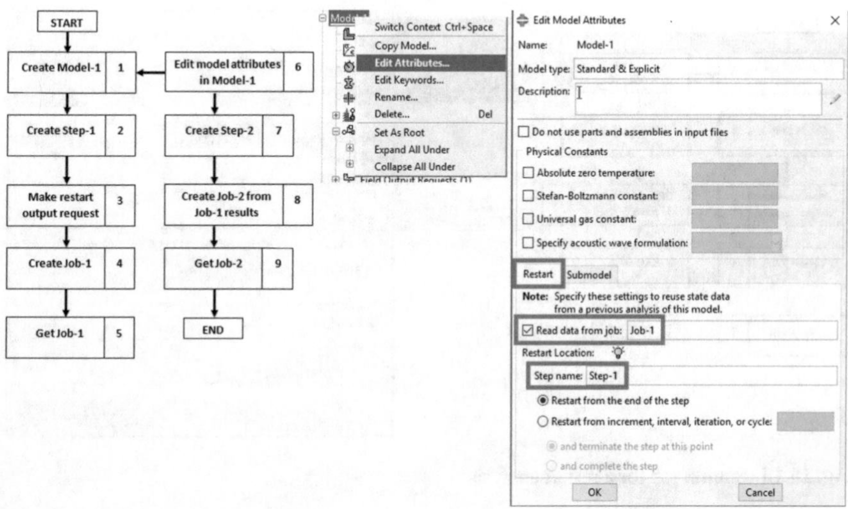

Fig. 10.11 Edit model for restart

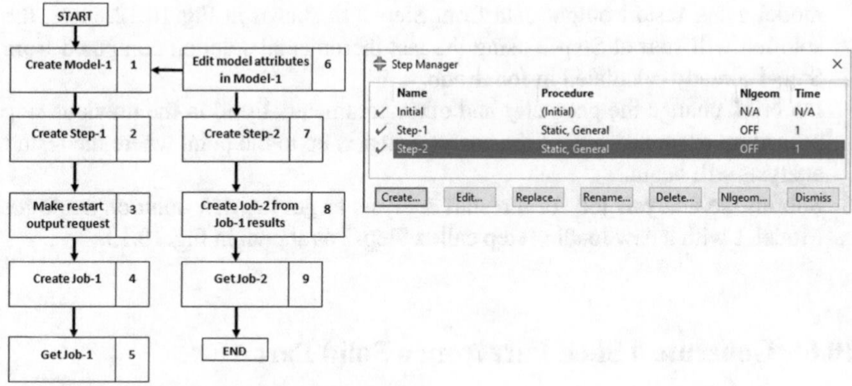

Fig. 10.12 Add new analysis step(s)

- Actions that invalidate restart data for a step and for subsequent steps:
  - Changes for propagating object in that step. Propagated objects include loads, fields, boundary conditions, and interactions.
  - Changes to the step object.
- Limitations of the restart features in Abaqus CAE:
  - Changes that are made with the Keywords Editor are not included in the job analysis.
  - New surfaces should not be defined (or used) in an Abaqus Explicit restart analysis.

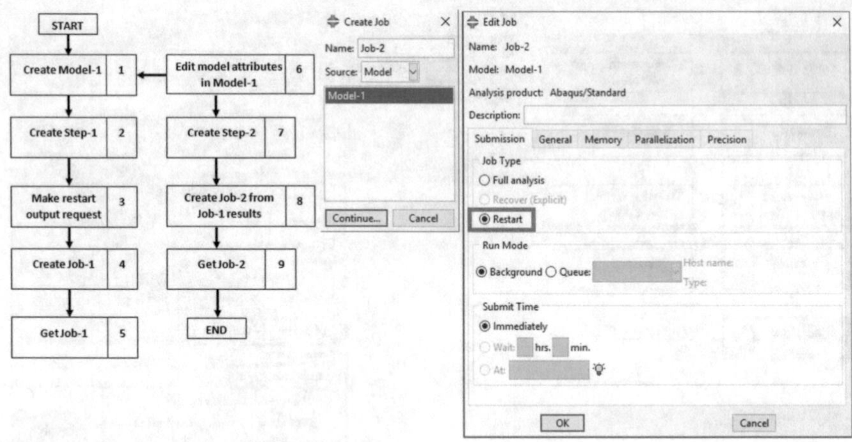

Fig. 10.13 Submit job for a restart analysis

5. Add the new analysis step(s) function of the loading condition to apply to the model using restart output data from Step-1 as shown in Fig. 10.12. Here, the solution will start at Step-2 using the last incremental solution computed from Step-1 already calculated in Job-1.odb.
   Do **NOT** change the geometry and other parameters listed in the previous step procedure or modify any of the analysis step(s) up to the point where the restart analysis will begin.
6. Submit the analysis job for a restart analysis to get the new solution made for Model-1 with a new loading step called Step-2 as shown in Fig. 10.13.

10.6 Generate a Shell Part from a Solid Part

The Abaqus mid-surface modeling feature offers tools to manually create a mid-surface representation of thin solids to make your shell model. The thickness of the parent solid geometry is associated with the shell mesh.

10.6.1 Benefits for Using Shell Structures

In general, shell structures can be used to investigate a slender structure or trusses design that is applicable to structural engineering or metal forming analysis. Solid elements have three degrees of freedom: u_x, u_y, and u_z (all displacements with no rotation). On the contrary, the shell elements have a different kinematics with five degrees of freedom all displacements plus two rotations θ_x and θ_y. As shown in

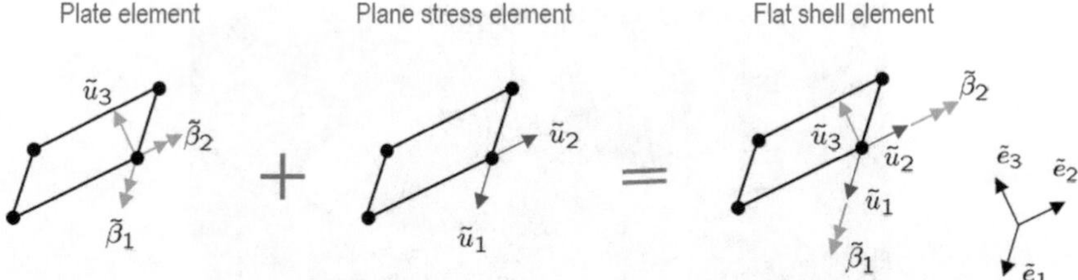

Fig. 10.14 Example of a flat shell element kinematics

Fig. 10.15 **Architecture and building**; the development of masonry domes and vaults in the Middle Ages made possible the construction of more spacious buildings. In more recent times, the availability of reinforced concrete has stimulated interest in the use of shells for roofing purposes

Fig. 10.14, the stresses of an element shell are the contribution of the summation of the membrane stresses (u_z, θ_x, θ_y) plus the plane stresses (u_x, u_y). Shell elements can offer huge time savings and minimize the risk of convergence problems since they allow thin features to be modeled with far fewer elements than are needed when solid elements are meshed. They are also easier to mesh and less prone to negative Jacobian errors which might occur when using extremely thin solid features.

10.6.2 Applications to Model Shell Structures

It is instructive to assemble a list of applications from a historical point of view as shown in Figs. 10.15, 10.16, 10.17, 10.18, 10.19, and 10.20, to take as a connecting theme the way in which the introduction of the thin shell as a structural form has made an important contribution to the development of several branches of engineering. The following is a brief list, which is by no means complete.

Some structures such as skid structures require the solid model to be converted to a shell model in order to reduce the size of the model. The same structural mechanics are kept especially to investigate the stress and strain responses accurately afterward. A good practice is to make a beam model first, in order to get the deformed shape plus the reactions load (shear and bending diagrams).

Fig. 10.16 Power and chemical engineering; the development of steam power during the Industrial Revolution depended to some extent on the construction of suitable boilers. These thin shells were constructed from suitable formed plates joined by riveting. More recently, the use of welding in pressure vessel construction has led to more efficient designs. Pressure vessels and associated pipework are key components in thermal and nuclear power plants and in all branches of the chemical and petroleum industries

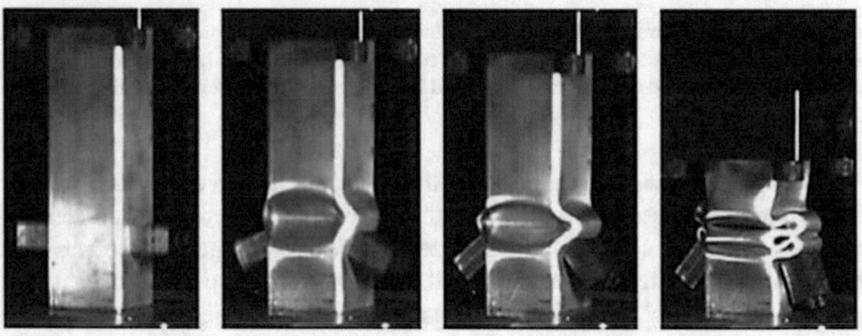

Fig. 10.17 Structural engineering; an important problem in the early development of steel for structural purposes was the need to design compression members against buckling. A striking advance was the use of tubular members in the construction of the Forth railway bridge in 1889: steel plates were riveted together to form reinforced tubes as large as 12 ft in diameter with a radius/thickness ratio of between 60 and 180

10.6.3 Step-by-Step Procedure to Convert Solid Model to Shell Model

A step-by-step procedure is described here, if the user is working on a model made with Abaqus then start from step #1 but skip steps #2 and #3.

1. Simplify the model geometry with Abaqus or a CAD software in order to remove unnecessary gaps and merge some of the solid bodies.

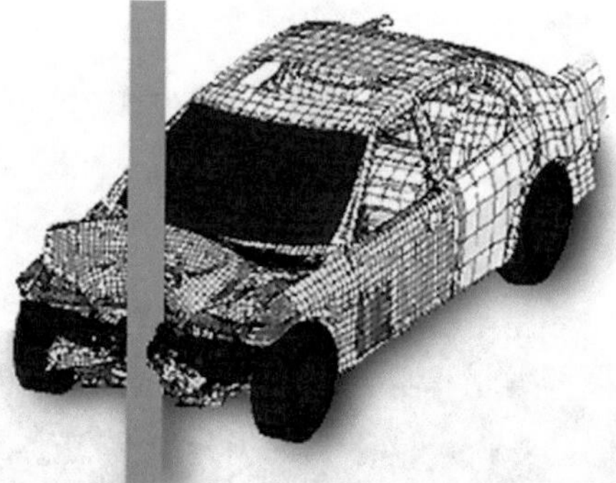

Fig. 10.18 Vehicle body structures; the construction of vehicle bodies for road transport involved a system of structural ribs and nonstructural paneling or sheeting. In the modern form of vehicle construction, the skin plays an important structural part, followed by the introduction of sheet–metal components preformed into thin doubly curved shells. The use of the curved skin of vehicles as a load-bearing member has similarly revolutionized the construction of railway carriages and aircraft. In the construction of all kinds of spacecraft, the idea of a thin but strong skin has been used from the beginning

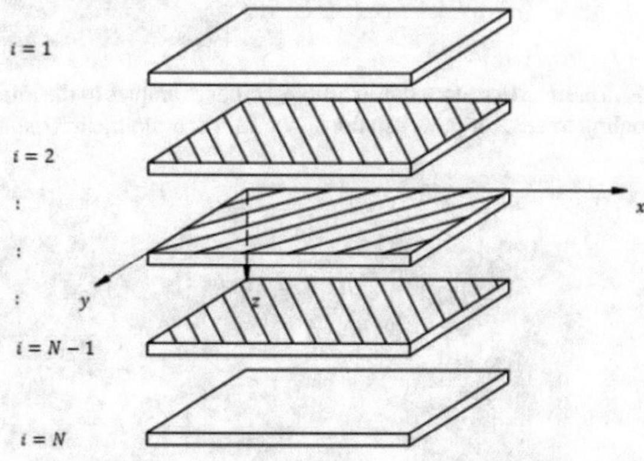

Fig. 10.19 Composite construction; the introduction of fiberglass and similar lightweight composite materials has impacted the construction of vehicles ranging from boats and racing cars to fighter and stealth aircraft, and so on. The exterior skin can be used as a strong structural shell

2. If the analyst has used CAD software to simplify the model, then it is recommended that the file be exported with the step file extension (.stp) so that it can be imported to Abaqus afterward.
3. If the user has a model step file (.stp) to be imported to Abaqus, then go to file menu, **File** → **Import** → Select step file (.stp) to open it.
4. Once the model is imported, open it in the part module and then assign the mid-surface region for all of the cells in each parts as shown in Fig. 10.21.
5. As a result, the user will get the transparent representation of the part, which the analyst can use as the reference surface for shell modeling. Several options are available to obtain offset faces for shell modeling shown in Fig. 10.22. By default, the shell model will represent the middle surface of the part shown in Fig. 10.23, with the different option parameters as shown in Fig. 10.24.

Fig. 10.20 Miscellaneous examples; other examples of the impact of shell structures include water cooling towers for power stations, grain silos, armour, arch dams, tunnels, submarines, and so forth

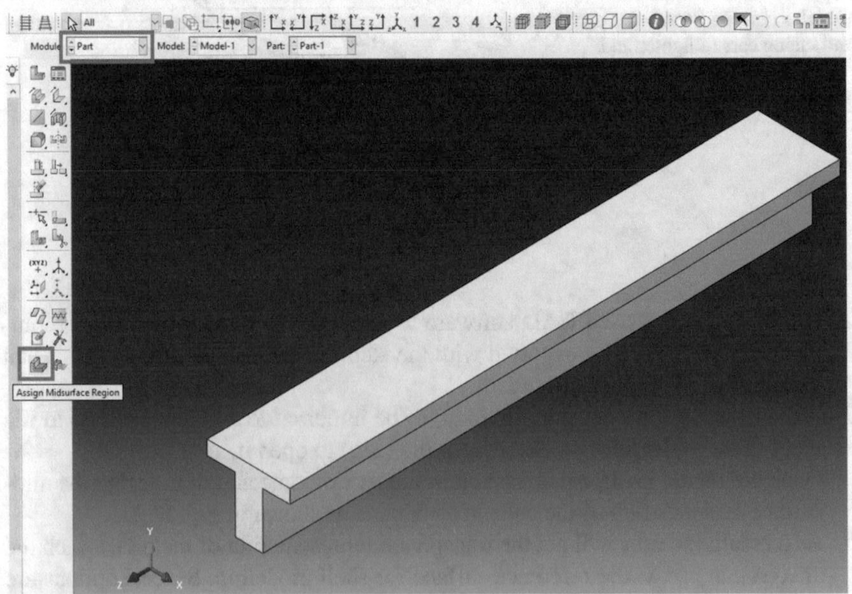

Fig. 10.21 Assign middle surface region

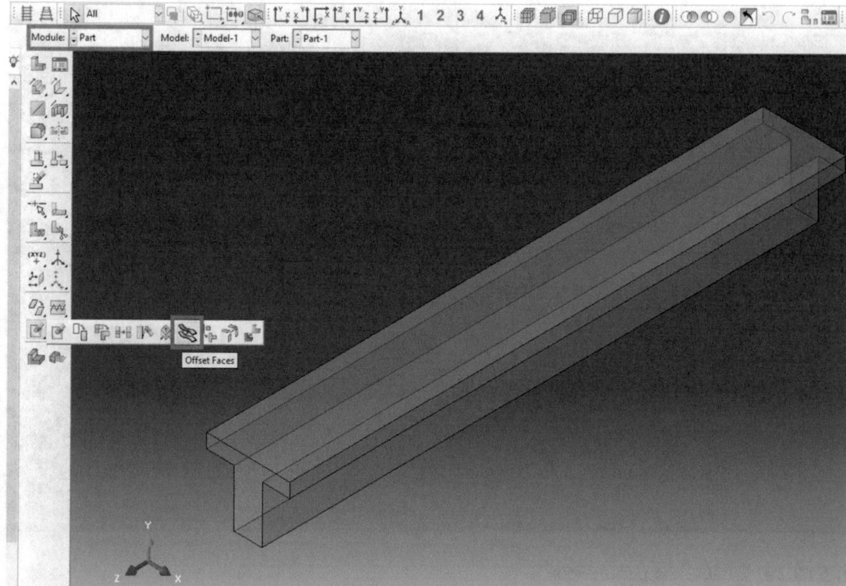

Fig. 10.22 Offset faces

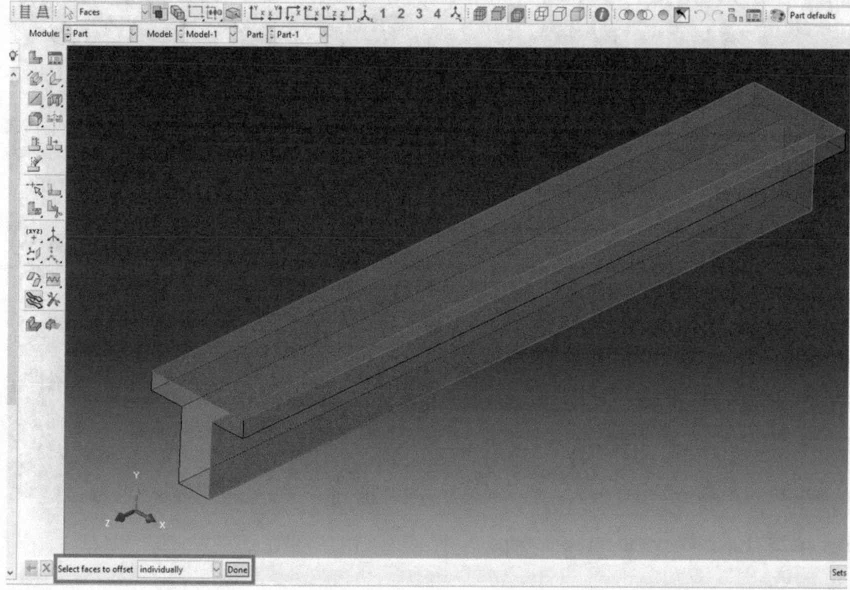

Fig. 10.23 Selection of faces to offset, here individually

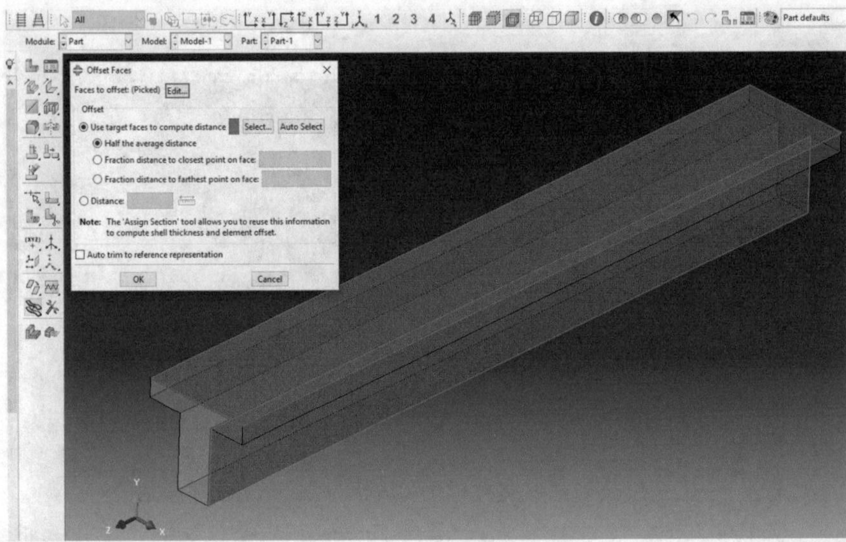

Fig. 10.24 Offset face parameters

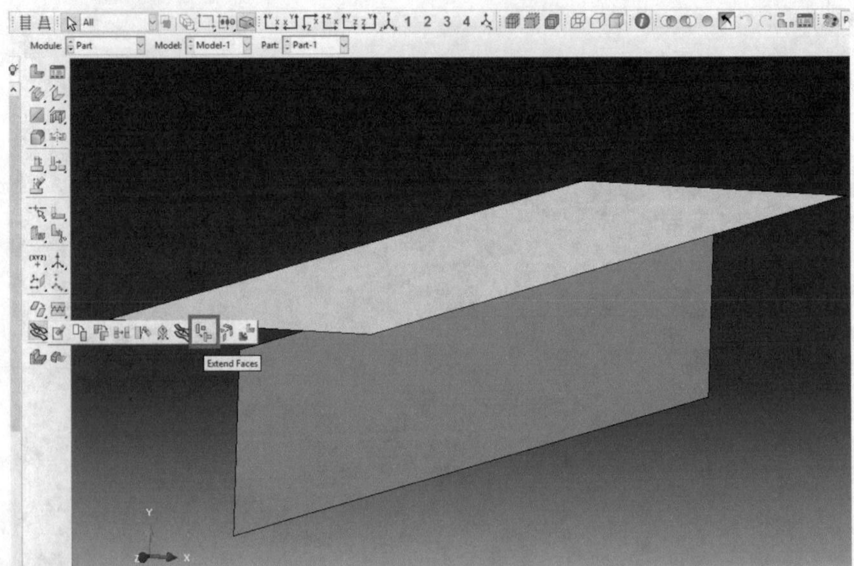

Fig. 10.25 Extend faces to fill some gaps

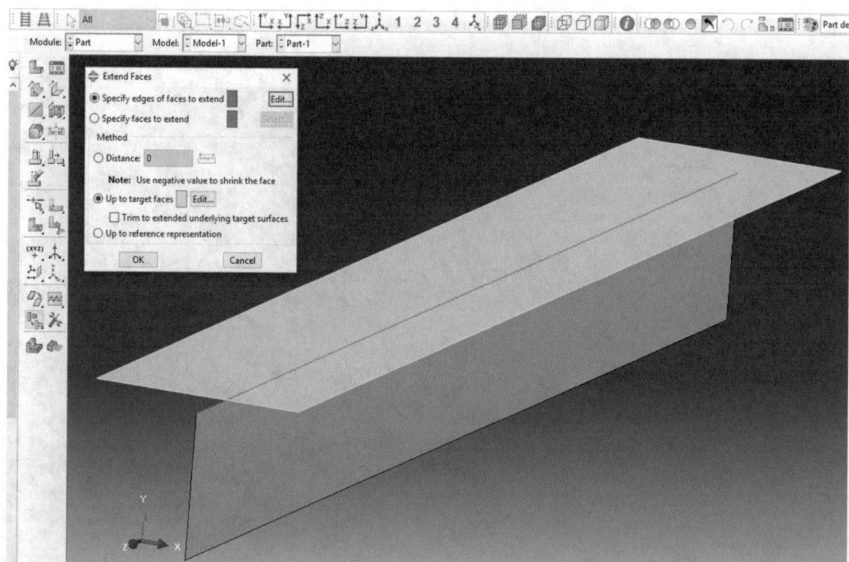

Fig. 10.26 Extend face parameters

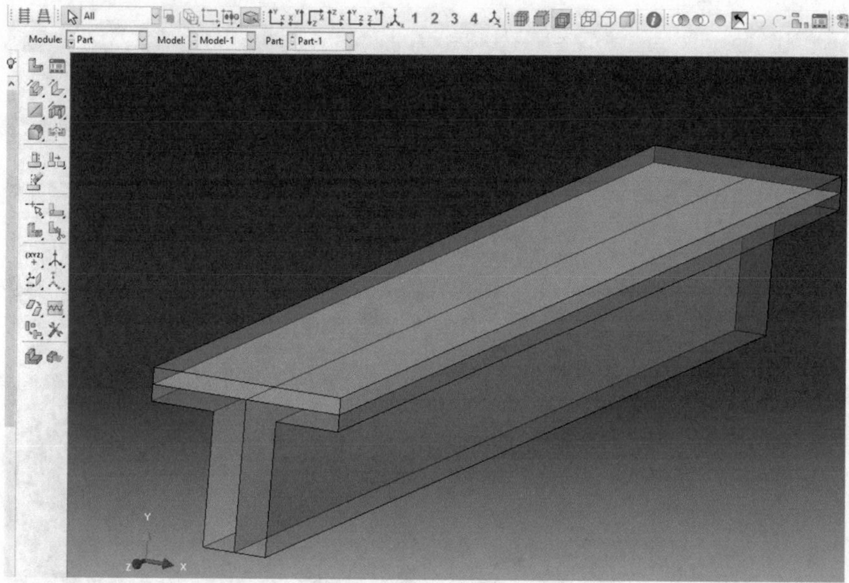

Fig. 10.27 Solid structure is now converted to a shell structure

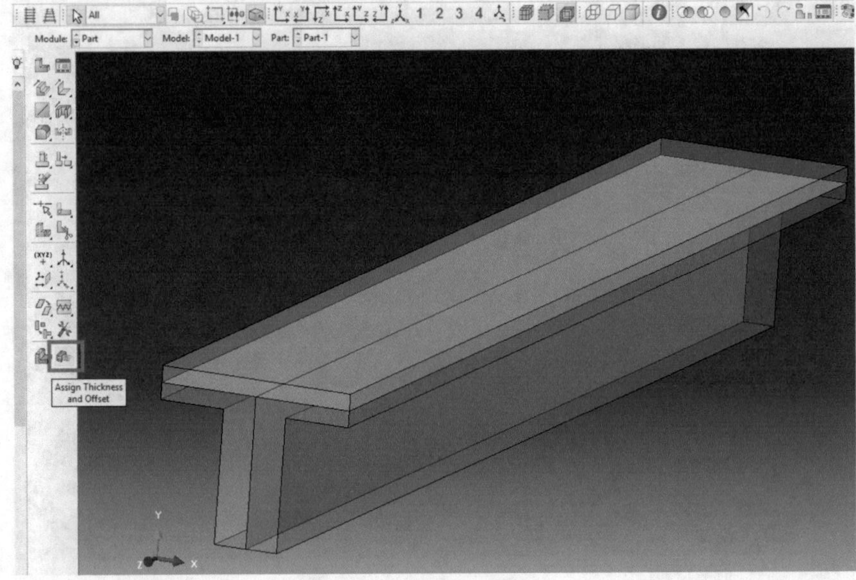

Fig. 10.28 Assign thickness and offset

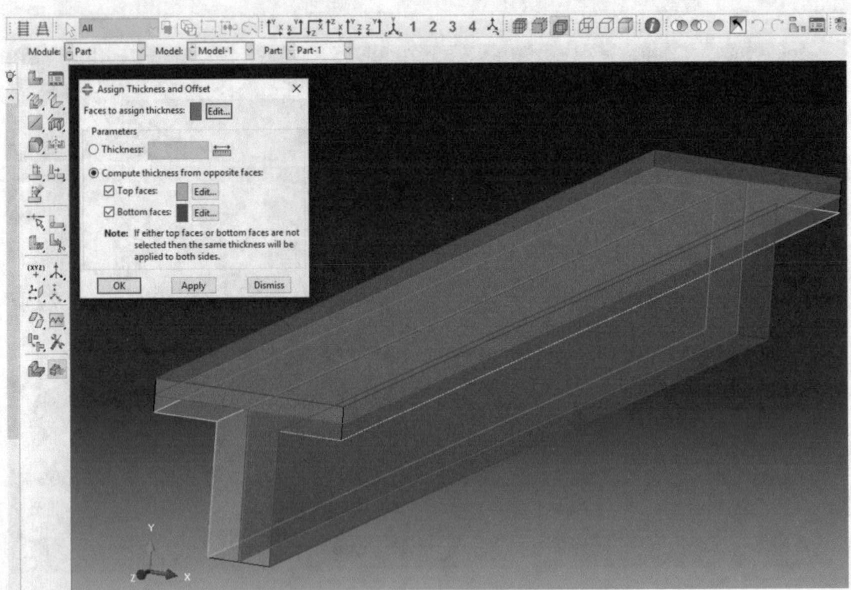

Fig. 10.29 Thickness and offset parameters

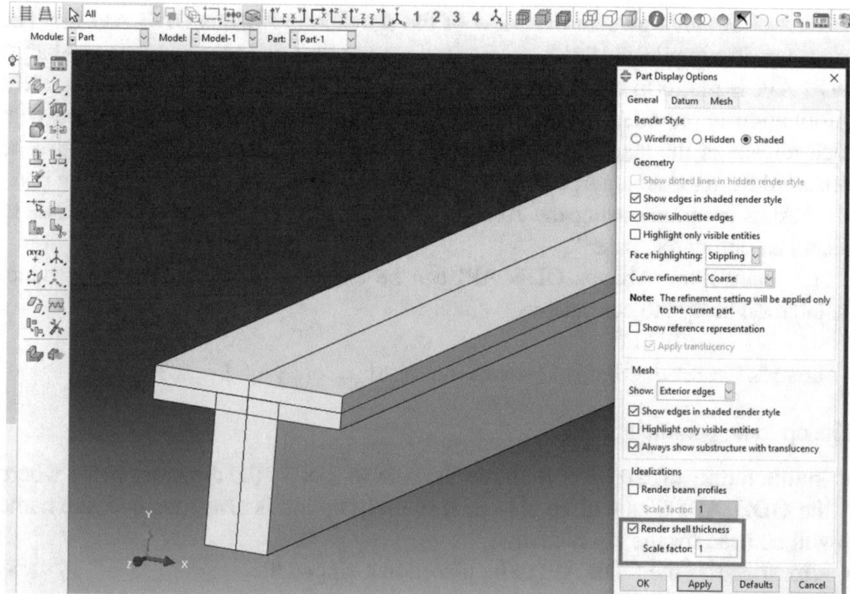

Fig. 10.30 Render shell thickness with a scale factor equal to one

6. If some gaps occurred in the model due to the middle surfaces, as shown in Fig. 10.25, the user will be able to remove these gaps by using the extend and/or blend faces options in Fig. 10.26. To extend the face, the user can use the option called: Up to target face option, which is quite useful for covering the gaps. The blend faces option is a little bit different, but still has a user-friendly interface to deal with surface issues.
7. The Solid model should be now converted to a Shell model, as shown in Fig. 10.27.
8. Regarding the thickness assignment in Fig. 10.28, there are two ways of proceeding. The first option is just to create the sections with the thickness made in the solid model to keep the control of the section shell defined in the model; the second option is to use the assign thickness and offset features, as shown in Fig. 10.29.
9. Finally, Fig. 10.30 shows the Shell model with the thickness assignments which are equivalent to the Solid model.

10.7 Compile and Link a Post-processing Program Using the Standalone Abaqus ODB API

This section explains how to use the standalone Abaqus ODB API software library kit to compile and link a program.

First of all, let's give an overview of the standalone Abaqus ODB API kit. The procedure to compile and link post-processing programs using the standalone Abaqus ODB API is platform and version dependent. The platform and version-dependent information is provided in platform-specific archives available for download with each release of the standalone Abaqus ODB API. The ODB API kit may also be extracted from the Abaqus product installation. For information on developing post-processing programs using the Abaqus ODB API, please refer to "Using C++ to access an output database[3]".

The standalone Abaqus ODB API can be extracted from an Abaqus product installation using the command:

```
abaqus extractOdbApi [-name name of archive] [-zip]
```

The optional parameters are

- **-name name_of_archive**. It allows the user to specify the directory name where the **ODB_API** files will be placed. If the **-zip** option is also specified, the name will be used for the zip filename.
- **-zip**. It causes the ODB API to be placed in a zipped file.

The Abaqus **extractOdbApi** utility requires the Microsoft Visual Studio C++ compiler to be installed on Windows.

The directory structure for the standalone ODB API is given below:

- **version_odb_api/lib**
- **version_odb_api/include**
- **version_odb_api/testOdbApi.[bat|csh]**
- **version_odb_api/odbDump.C**
- **version_odb_api/viewer_tutorial.odb**

The **lib** directory contains shared libraries that are required during linking and runtime. Runtime execution of ODB API programs requires the library search path to include the path to the lib directory. The include directory contains headers that are required during the compilation.

The user will have to test the ODB API library kit. A sample build script **version_odb_api/testOdbApi.[bat|csh]** is included in the standalone ODB API kit. This script contains the set of compile and link commands to build and run the **odbDump.C** sample program. Run this script and verify that the code does not return any build errors. When completed, examine the **odbDump.txt** file for ODB data content taken from the **viewer_tutorial.odb** file.

Building post-processing programs need to use the sample build script as a guide for the proper compile and link commands to build the program coded by user and using the Abaqus **ODB_API**. The example **odbDump.C** source file should also be examined for required **#include** files along with **main()** program definition and format. The **main()** function is defined at the bottom of the file.

---

[3]See Abaqus scripting user's guide v6.14 in Chap. 10 Using C++ to access an output database.

In order to set the program execution environment, the user needs to perform a runtime execution of ODB API programs which requires the library search path to include the path to the **odb_api/lib** directory. The user needs also to pay attention that proper setting of the library path must occur before the program can execute. Use the setting shown in the example build script to set the user runtime library search path. For example,

```
set path=6.14-1_odb_api\lib;%PATH%
setenv LD_LIBRARY_PATH lib
```

The full path to the libraries is recommended, when changing directories might be needed.

Regarding the execution of the post-processing programs, the user needs to run the application by entering the name of the application on the command line. Unix/Linux users may be required to enter a "**./**" before the command to execute programs from the current working directory. For example the command:

```
./odbDump.exe viewer_tutorial.odb
```

It opens the **viewer_Tutorial** output database and calls functions that print all the parts, part instances, the root assembly, connectors, and so on. It can be noted that the output takes several minutes unless redirected to file.

The output database must be created by or upgraded to the same version of Abaqus as that of the ODB API used to compile and link the application.

10.8 Create Executables Using the C++ ODB API Libraries Outside of Abaqus/Make

If the user wants to compile and link a C++ code (accessing the Abaqus ODB API libraries) outside of Abaqus/Make, this can be done by following the procedure explain below.

The Python programming language is used below to demonstrate the procedure to compile and link post-processing programs outside of Abaqus/Make. The following example was developed on an SGI platform. The user will have to customize it on the machine with a specific platform. Abaqus does not recommend or support this approach because it is highly platform and (Abaqus) version specific. However, this example may be used as a generic guideline for those people that want/need to do this.

The user can use a Python makefile to access the output database (.odb) file through a standalone C++ program. To create it, do the following:

1. Write a standalone C++ API code in an environment chosen by user (for example, Xemacs or Microsoft Visual Studio). For this example, this code is **myODBAPI-code.c**.

2. Use Abaqus/Make to compile and link it in the **verbose=3 mode**. The user will have to change the function **main**() inside the code temporarily to **ABQmain**(). Run Abaqus/Make using the command:

```
abaqus make job=your_code.c verbose=3
```

where **abaqus** is the command the user shall use to run Abaqus and **user_code.c** is the name of the file the user are compiling. This will dump a lot of information to standard output.

3. From this dump extract information that looks like the example given next:

```
username@trieste(10:23am)-> abq62 make j=myODBAPIcode v=3
platform: sgi
env: OS_MAJ_MIN = _6_5
Job name: myODBAPIcode
Input file: myODBAPIcode.c
Input file base name: myODBAPIcode
Input file extension: c
Input file without extension: myODBAPIcode
Is the file a source file: ON
Is the file a C++ file: ON
Abaqus JOB myODBAPIcode
Main function declaration: int ABQmain(int argc, char** argv);
Main function call: status = ABQmain(argc, argv);
Begin Compiling User Post-Processing Program
Fri Nov 2 10:29:57 2001
Compiling: /usr/username/myODBAPIcode.c
Compile command: CC
Compile arguments: ['-n32', '-mips3', '-DSGIn32', '-c', '-G', '0',
'-xansi', '-ptused', '-no_prelink', '-DSGI3000', '-DSGI', '-DSGI_ARCH',
'-D_SGI_MP_SOURCE', '-D_BSD_TYPES', '-DHKS_OPEN_GL',
'-DEMULATE_EXCEPTIONS=0', '-DHAS_BOOL', '-diag_error', '1201',
'-DTYPENAME=', '-D_POSIX_SOURCE', '-D_XOPEN_SOURCE', '-DFOR_TRAIL',
'-DSPECIALIZE', '-O0', '-I/amd/cyclone/b/abaqus60/releaseNoDist
/sgi4000/6.2-1/cae/include', '/usr/username/myODBAPIcode.c'] ...
End Compiling User Post-Processing Program
Fri Nov 2 10:30:09 2001
Begin Linking User Post-Processing Program
Fri Nov 2 10:30:09 2001
Executable name: /usr/username/myODBAPIcode.x
....
Linking: /usr/username/myODBAPIcode.x
Link command: CC
Link arguments: ['-n32', '-mips3', '-DSGIn32', '-no_prelink', '-Wl,-woff
,84,-woff,47', '-Wl,-woff,133,-woff,138,-woff,129', '-o', '/usr/username
/myODBAPIcode.x', '/usr/username/myODBAPIcode.o', '/usr/username
/main_153389.o', '/amd/cyclone/b/abaqus60/releaseNoDist/sgi4000/6.2-1/cae
/exec/lbr/standardB.sl', '/amd/cyclone/b/abaqus60/releaseNoDist/sgi4000
/6.2-1/cae/exec/lbr/HKSodb.sl', '/amd/cyclone/b/abaqus60/releaseNoDist
/sgi4000/6.2-1/cae/exec/lbr/HKSddb.sl', '/amd/cyclone/b/abaqus60
/releaseNoDist/sgi4000/6.2-1/cae/exec/lbr/HKSodiC.sl', '/amd/cyclone/b
/abaqus60/releaseNoDist/sgi4000/6.2-1/cae/exec/lbr/HKSnex.sl', '/amd
/cyclone/b/abaqus60/releaseNoDist/sgi4000/6.2-1/cae/exec/lbr/HKSwip.sl',
'-lftn', '-lm']
```

4. Create a Python file, **make.py**, like the one shown below using the above information. The various variables used should be self-explanatory.

```
import os
JOB = 'Job-1'
MAIN = 'myODBAPIcode'
SOURCE = MAIN+'.c '
OBJECT = MAIN+'.o '
EXE =JOB+'.exe '
COMPILE_CMD = 'CC '
COMPILE_OPT = ' -n32 -mips3 -DSGIn32 -c -G 0 -xansi -ptused -no_prelink
-DSGI3000 -DSGI -DSGI_ARCH -D_SGI_MP_SOURCE -D_BSD_TYPES -DHKS_OPEN_GL
-DEMULATE_EXCEPTIONS=0 -DHAS_BOOL -diag_error 1201
-DTYPENAME=-D_POSIX_SOURCE -D_XOPEN_SOURCE -DFOR_TRAIL -DSPECIALIZE -O0'
LINK_CMD = 'CC'
LINK_OPT = ' -n32 -mips3 -DSGIn32 -no_prelink -Wl,-woff,84,-woff,47
-Wl,-woff,133,-woff,138,-woff,129 '
FTN_LNK = ' -lftn -lm'
Abaqus_DIR = '/amd/cyclone/b/abaqus60/releaseNoDist/sgi4000/6.2-1/cae
/exec/lbr/'
HEADER_FILES='/amd/cyclone/b/abaqus60/releaseNoDist/sgi4000/6.2-1/cae
/include '
compile = COMPILE_CMD+COMPILE_OPT+' -I'+HEADER_FILES+SOURCE
os.system(compile)
print 'finished compiling'
link = LINK_CMD + LINK_OPT + ' -o ' +EXE + OBJECT
+Abaqus_DIR+'standardB.sl '+Abaqus_DIR+'HKSodb.sl '+Abaqus_DIR+'HKSddb.sl
'+Abaqus_DIR+'HKSodiC.sl '+Abaqus_DIR+'HKSnex.sl '+Abaqus_DIR+'HKSwip.sl
'+FTN_LNK
os.system(link)
print 'finished linking and building executable'
```

5. Change the source file, **myODBAPIcode.c**, to use the function **main()** instead of **ABQmain()**.
6. The source code must call **odb_initializeAPI()** to initialize the interface. This call is generated automatically when the Abaqus/Make procedure is run, but it must be included in any application that is not compiled and linked using Abaqus/Make. After all calls to the C++ interface have been completed, the interface may be deactivated by including a call to **odb_finalizeAPI()**; if this call is not made explicitly, it will be called automatically when the application exits.
7. Run the Python script, make.py, using the Abaqus driver:

```
abaqus python make.py
```

The output will look like

```
username@trieste(11:08am)-> abq62 python make.py
finished compiling
finished linking and building executable
```

This will create an executable, **Job-1.exe**.

8. The user can now run the executable from within the Abaqus driver using

```
abaqus Job-1.exe
```

Alternatively, the user may run the executable outside the Abaqus driver. This option is not supported, and the following issues apply:

- The correct path to the Abaqus runtime libraries must be specified prior to starting the user application. The **HKSodb** library and several utility libraries resolve all the functions available in the interface to the output database.
- The runtime library path is typically set using the system environment variable **LD_LIBRARY_PATH**, but the method used to set the path may vary depending on your operating system configuration. This is described in section Accessing the C++ interface from an existing application.[4]

[4]See Abaqus scripting user's guide v6.14 in Sect. 10.7 Accessing the C++ interface from an existing application.

Chapter 11
Hardware or Software Issues

11.1 Solving File System Error 1073741819

If the executable "standard.exe" is aborted with a system error code 1073741819, please check the (.dat), (.msg), and (.sta) files for error messages if the files exist. If there are no error messages and the problem cannot be resolved, the command **abaqus job=support information=support** should be run to report and save your system information. If there is an error message about system error code 1073741819 do this:

rename the **mkl_avx2.dll** in $(C:\backslash SIMULIA\backslash Abaqus\backslash 6.14-5\backslash code\backslash bin)$ to **mkl_avx2.dll.11.0.0.1**

If this solution does not work the user must contact the Abaqus support to get a workaround on the problem because an error code 1073741819 may be caused by damage to Windows system files. The corrupted system files entries can be a real threat to the well-being of your computer, which is not directly related to Abaqus itself.

11.2 Interpreting Error Codes

In case of abnormal terminations, it is also possible to obtain a traceback indicating the Abaqus routine where the failure occurred. Communication of this information to Abaqus technical support personnel can help in speedier resolution of failures.

- On occasion, an Abaqus analysis will terminate abnormally. While the code is designed to be as stable and robust as possible, it is impossible to anticipate every potential error that may occur, especially those that are related to the operating system of computer. Abnormal termination is one for which a proper error message cannot be generated. In the event that an abnormal termination does occur, the user may see an error message in the (.log) file such as

R. J. Boulbes, *Troubleshooting Finite-Element Modeling with Abaqus*,
https://doi.org/10.1007/978-3-030-26740-7_11

```
Abaqus Error: The executable /opt/Abaqus/6.5-1/exec
/standard.exe aborted with the system error "Illegal
floating point operation" (signal 8).

Please check the .dat, .msg, and .sta files for
error messages if the files exist. If there are
no error messages and you cannot resolve the
problem, please run the command
"abaqus job=support information=support" to report
and save your system information. Use the same
command to run Abaqus that you used when the
problem occurred.

Please contact your local Simulia support office
and send them the input file, the file support.log
which you just created, the executable name, and
the error code.

Abaqus /Analysis exited with errors
```

In an attempt to provide improved diagnostics in such situations, changes were made in Version 6.3-1 to provide error codes that correspond to the specific system errors that caused Abaqus to abort. It must be noted that the output of error codes in Versions 6.3-1 and higher is not causing the suppression of an otherwise meaningful error message. The error code information is strictly a supplement to any messages that would already be printed.

- Unix and Linux Systems
  The error codes depend on the system on which Abaqus is being executed and are standardized for each specific system by the system provider. Unix and Linux error codes are defined in file **signal.h** under the heading "Signal Numbers". This file is located in **/usr/include/sys** or in **/usr/include/asm** on Linux systems. There is not a direct correlation among all error codes on all Unix platforms; however, the following error codes typically have identical meanings across most platforms:
  - 2: **SIGINT**, The application received a keyboard interrupt. This signal is usually fixed by terminating.
  - 4: **SIGILL**, An illegal instruction was made.
  - 6: **SIGABORT**, An unexpected error condition has caused the program to terminate itself.

    On rare occasions, this error code may be thrown when the computer has run out of disk space. In this situation, it is more often the case that a proper error message will be delivered.
  - 8: **SIGFPE**, Floating point exception. The code has attempted an invalid floating-point operation.

- 9: **SIGKILL**, External job kill. The operating system or the user has killed the job. This signal cannot be handled by the application, and no core dump will occur.
- 7 or 10: **SIGBUS** Bus error. This is sometimes caused by writing past the end of a local array in a user subroutine.
- 11: **SIGSEGV**, Segmentation fault or illegal memory reference an error in which the program is attempting to access memory not allocated to it.
- 13: **SIGPIPE**, An error caused by an attempt to write to a pipe with no-one to read it.
- 15: **SIGTERM**, The application received an external request to terminate.

The main purpose of the error code is to assist Simulia support and development personnel in determining the cause of the analysis failure. Abaqus users are not expected to be familiar with these error codes and are encouraged to report them to the assisting support engineer.

Apart from error codes 2, 9, and 15, in most cases the only time these errors are caused by the user is when a user subroutine is part of the analysis. Good programming practices will reduce the number of occasions that may lead to the abovementioned error codes.

- Windows Systems

On Windows platforms, the error message shown in the log file will be different. For more specific information on Windows error codes, please visit the Microsoft web pages for the most recent content. A text string corresponding to the error code is normally printed. For example,

```
Abaqus Error: The executable C:\Simulia\Abaqus\6.8-1
\exec\explicit.exe aborted with system error "Access
is denied." (error code 5)
```

which usually means that a memory access violation caused the abort.

11.3 Obtaining a Traceback from a UNIX/Linux Core Dump

When a program aborts with a core dump, this almost always indicates a bug. A debugger can be used to obtain a traceback naming the function that was executing when the program aborted. Information obtained from a traceback can be useful to you and Simulia support engineers. To obtain a traceback on a UNIX/Linux platform, follow the steps below.

On Linux systems, the default user environment may prevent the creation of a core file. The following command may be required in the $HOME startup script to allow the core file to be written out.

- In the C-Shell startup script /.cshrc:

```
limit coredumpsize unlimited
```

- In the Bash/Korn Shell startup script /.bashrc, .profile:

```
ulimit -c unlimited
```

Some procedures are now given in order to obtain a traceback.

1. Obtaining a traceback from an Abaqus/CAE/Viewer abort, run the following command immediately after the abort occurrence:

   ```
   abaqus -tb
   ```

   If an abort occurs, the traceback will be shown in your command window. Linux machines may be configured to dump core files using the core.pid file naming. If this is the case for your machine, the user will need to rename the core.pid file(s) **before** running the above command.

   ```
   mv core.pid core
   mv gui_tmp/core.pid gui_tmp/core
   ```

2. To obtain a traceback from an Abaqus analysis abort, Abaqus provides the capability to obtain a traceback by making a single environment variable setting. The function assumes that a qualified debugger can be located in the system PATH. Add the following line to the abaqus_v6.env environment file:

   ```
   traceback_generator=ON
   ```

   If an abort occurs, Abaqus will run use the debugger to get a traceback from any core file located in the temporary scratch directory. In Windows, the application.dmp will be printed. The traceback will be recorded in the job log file.
3. To obtain a traceback from an Abaqus MPI based parallel analysis abort, an Abaqus MPI based parallel analysis may have processes that run on multiple hosts and the core file may be dumped in a local temporary scratch directory. The attached environment file is designed to check each local storage location and produce a traceback from all core files found. The function assumes that a qualified debugger can be located in the system PATH. If an abort occurs, the traceback(s) will be recorded in the job log file.
   Copy the content of "tbmpi.env" code given below to your abaqus_v6.env file.

Listing 11.1 tbmpi.env

```
import os, driverUtils
if not os.name == 'nt':
    import resource
    resource.setrlimit(resource.RLIMIT_CORE, (-1,-1))

traceback_generator=ON
verbose=2
os.environ["ABA_VERBOSE"]=str(verbose)
os.environ["MPI_ERROR_LEVEL"]='2'
mp_environment_export += ("MPI_ERROR_LEVEL",)

def onJobStartup():
    print 'Starting Job... ', id
    import driverInformation
    info=driverInformation.DriverInformation({},{})
    info.informationGetUserLimits()

def mp_mpiCommand():
    import os
    os.environ['ABA_PROGRAM'] = program
    return command, {}

def onJobCompletion():
    import os, driverTraceback, glob, socket
    from driverConstants import LOCAL
    from driverEnv import driverEnv
    from mpi import Mpi
    verbose = int(os.environ.get('ABA_VERBOSE', "0"))
    # Only look for cores when the execution
    # directory is local
    if os.environ.has_key('ABA_PROGRAM') and \
    file_system[1] == LOCAL:
        exe = os.environ['ABA_PROGRAM']
        if verbose:
            print host_list, local_host, exe
            os.environ['verbose'] = '1'
        # Create script
        if os.name == 'nt':
            tbFile = scrdir + os.sep + 'tb.bat'
            cmd = tbFile
            f=open(tbFile, 'w')
            f.write('@echo off\n')
            f.write('cd /d %s\n' % scrdir)
```

```
        if verbose:
            f.write('dir\n')
        f.write('if␣exist␣"*.dmp"␣')
        f.write('type␣*.dmp\n')
        f.close()
    else:
        os.umask(0o077)
        tbFile = scrdir + os.sep + 'tb.sh'
        cmd = 'bash␣' + tbFile
        f=open(tbFile, 'w')
        f.write('#!/bin/bash\n')
        f.write('export␣verbose=1\n')
        f.write('cd␣%s\n' % scrdir)
        if verbose:
            f.write('ls␣-al\n')
        f.write('for␣core␣in␣core*\ndo\n')
        f.write('%s␣python␣-c␣' % \
        os.environ['ABA_COMMAND'])
        f.write('"import␣driverTraceback;␣')
        f.write('driverTraceback.\
␣␣␣␣␣␣␣␣␣␣␣␣␣␣␣␣␣␣␣␣␣␣generateTraceback(\'%s\')"\n' \
                 % exe)
        f.write('done\n')
        f.write('cat␣*.dmp␣2>/dev/null\n')
        f.close()
    # uncomment to force ssh usage instead of MPI
    #os.environ['ABA_USE_MPI'] = '0'
    env = driverEnv().read()
    options = {'verbose':verbose}
    options['mp_host_list'] = host_list
    options['mp_head_node'] = local_host
    options['mp_file_system'] = file_system
    options['mp_rsh_command'] = rsh_command
    mpiImpl = env.get('mp_mpi_implementation')
    options['mp_mpi_implementation'] = mpiImpl
    options['mp_mpirun_path'] = \
    env.get('mp_mpirun_path')[mpiImpl]
    options['mp_mpirun_options'] = \
    env.get('mp_mpirun_options','')
    # Workaround for driverUtils bug
    class FixMpi(Mpi):
        def __init__(self, options, env):
            self.env = env
            self.options = options
            Mpi.__init__(self, options, env)
```

```
        globals()['Mpi'] = Mpi
        try:
            m = FixMpi(options, os.environ)
        except:
            print "Error in Mpi instantiate"
            import traceback
            traceback.print_exc()

        cleanupHosts = list(local_host)
        for entry in host_list:
            host = entry[0]
            if host not in cleanupHosts:
                if verbose:
                    print "Checking scratch on ", \
                    host
                m.rcp([tbFile], scrdir, \
                socket.gethostname(), host)
                m.rsh(host, cmd)
                cleanupHosts.append(host)

    if os.name == 'nt':
        if glob.glob('*.dmp'):
                driverTraceback.generateTraceback(exe)
    else:
            for i in range(len(glob.glob('core*'))):
                print "TB: ", i
                driverTraceback.generateTraceback(exe)
            if glob.glob('*.dmp'):
                os.system('cat *.dmp')
```

4. To obtain a traceback from an Abaqus post-processing program, run the debugger manually in the directory where the abort occurred, specifying the name of the executable followed by the core dump file name.

   ```
   gdb myprog.exe core.pid
   ```

   At the debugger prompt, enter the command **"where"**.

11.4 Windows HPC Compute Clusters

The Distributed Memory Parallel (DMP) Analysis capability for the Windows x86_64 platform is supported using Windows HPC Compute Clusters [1] only. In order to execute the Abaqus using Distributed Memory Based parallelization on a Windows cluster, the user must have the following software.

- Microsoft Windows HPC Server.
- Microsoft HPC pack installed on each compute node.
- Microsoft HPC pack installed on each workstation used to submit Abaqus jobs.
- Abaqus C++ runtime prerequisites installed on each machine: headnode, compute node, and workstation.

The installation is simply done in two main steps:

1. Install Abaqus into a file share on the head node of the cluster. For simplicity reasons, this is the recommended method. During installation, make sure to enter the UNC path as installation target directory.
2. Install the Microsoft Visual C++ Runtime levels on all the compute nodes. When using a shared installation of Abaqus, the Abaqus C++ runtime pre-requisite software may not be installed on the client machines that are only accessing the shared installation. This software must be manually installed on each compute node and workstation that will use the shared installation. In Abaqus 6.14 and earlier, these prerequisites can be located on the SIMULIA Abaqus DVD media in the: 1\win86_64 directory.

   ```
   2005_SP1_vcredist_x64.exe
   2005_SP1_vcredist_x86.exe
   2008_SP1_vcredist_x64.exe
   2008_SP1_vcredist_x86.exe
   2010_SP1_vcredist_x64.exe
   ```

   For Abaqus 2016 and later, use the following installers located in the win_b64 \ code \ bin \ directory:

   ```
   InstallDSSoftwarePrerequisites_x86_x64.msi
   InstallDSSoftwareVC9Prerequisites_x86_x64.msi
   InstallDSSoftwareVC10Prerequisites_x86_x64.msi
   InstallDSSoftwareVC11Prerequisites_x86_x64.msi
   InstallDSSoftwareVC12Prerequisites_x86_x64.msi
   ```

   Abaqus comes pre-configured to detect the Windows HPC environment and define a default set of batch queue definitions; therefore, there is nothing to do at the time for command-line execution.

At this point, it is a good practice to make a test about the cluster environment. Set, from the C:\Analysis directory, the fetch job c1 and submit it the shared and local queue using verbose=2, with the following lines of code:

```
abaqus job c1 fetch verbose 2
abaqus input c1 cpus 2 job c1c2share queue share
abaqus input c1 cpus 2 job c1c2local queue local
```

Then examine the job logs for errors, the **hostlist** defined and the execution host where the jobs ran.

The configuration on Windows client machines depends about how to submit Abaqus jobs from client workstations to the HPC compute cluster. The Microsoft HPC Pack client utilities must be installed on the client workstation. The same package used to install the HPC pack on the head node can be used for the workstation installation. If the analyst used Windows Deployment Services to provision and install the operating systems on the compute nodes, then the user can also install the HPC pack from the **REMINST** share on the head node of the cluster (\\HPC-HN1\REMINST\setup.exe).

The analyst using the HPC Pack configuration has to configure the **CCP SCHEDULER** as described below. It is an environment variable which tells the client workstation, what is the HPC Cluster to submit jobs.

1. Open **Control Panel → System → Advanced System Settings → Environment Variables**
2. Under **System Variables**, click **New...**
3. Enter **CCP_SCHEDULER** as the variable name and **HPC-HN1** as the value.

Now to run Abaqus jobs, the user can then submit jobs using one of the following selections:

```
abaqus job jobid cpus 8 -queue local
abaqus job jobid cpus 16 -queue share
abaqus job jobid cpus 24 -queue genxmllocal
abaqus job jobid cpus 32 -queue genxmlshare
```

In the above lines **jobid** is the name of the user's job and the number of cpus is the desired number of cores to be used for a job. All other available and appropriate Abaqus command line arguments can be used. The meaning of the parameters **local**, **shared**, **genxmllocal**, and **genxmlshare** are described below.

To understand Windows HPC job submission file management, consider the following relatively typical configuration of a Windows cluster as shown in Fig. 11.1. The user must submit Abaqus jobs from a directory on the head node that is shared out to all of the compute nodes. This is because the compute node must have access to the input files and must be able to write output files to the directory.

Figure 11.1 shows the logic about the cluster file management, such as the disk 1 is a local drive on the head node, physically connected to the head node, then the disk 2 is local to the first compute node, and the disk 3 is local the second compute node and so on. Finally, the directory **c:\share** is created on the local disk for the head node but is shared out to all of the local nodes.

There are two ways an Abaqus job can be executed.

1. By using the option **share**, in this case, a job can be executed using the directory from which the job was submitted as the output directory. This option is selected by specifying **-queue share** on the command line. As an example, if a job is

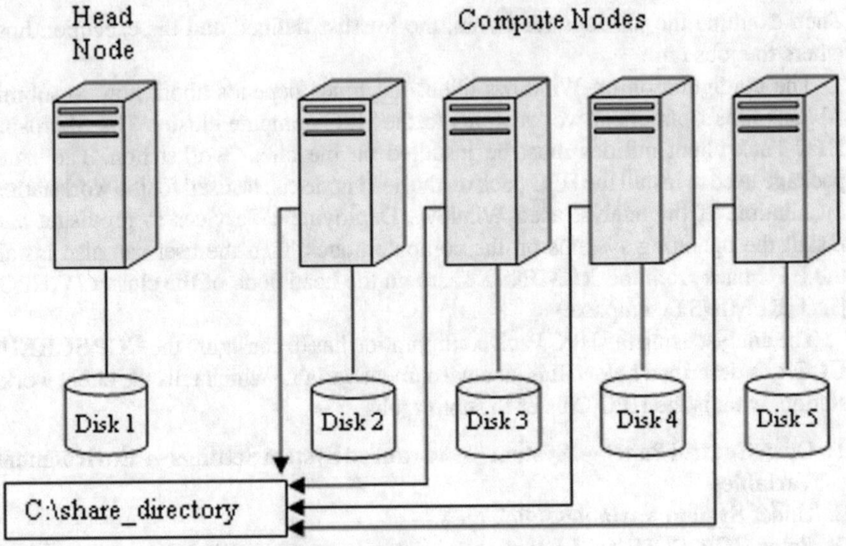

Fig. 11.1 File management with a Windows HPC cluster solution

submitted that executes on the second compute node, the Abaqus scratch directory will be created on disk 3 and the Abaqus scratch files will be written to that local drive. The Abaqus output files (.odb, .sta, .msg, etc); however, will be written to the share directory which physically means to disk 1 on the head node. This method of submitting jobs can be very convenient, but can be problematic for performance because if the jobs are executing on all the compute nodes, all of the compute nodes will be constantly writing data to disk 1 which will become a bottleneck.

2. Or by using the option **local**, in this case the alternative to use the share directory on the head node as the output directory for an Abaqus job is to run using the command **-queue local**. If this option is specified, Abaqus will create an output directory on the local disk of the compute node in the directory specified by the scratch parameter in the **abaqus_v6.env** file. Abaqus will then copy the necessary input files from the share directory on the head node to the output directory and run the job with both output and scratch directories on the local disk. When the job is complete, Abaqus will copy the contents of the output directory back to the directory on the head node from which the user submitted the job. This method of submitting jobs provides good performance, but is less convenient because the user must log in to the compute node or nodes in use in order to look at the job files while the job is executing. Nonetheless, this method of submitting jobs is recommended because share directories typically do not provide adequate performance.

There are two queue submission options that have not been discussed yet. The **-queue genxml** options where it is either local or share. If an Abaqus job is submitted using either **genxmllocal** or **genxmlshare**, Abaqus generates a (.xml) file that can be used to submit the job to the HPC job scheduler, but Abaqus does not actually submit the job to the job scheduler (as it is done with the non-genxml local and share options). The local and share specifiers are the same as described above. This technique is convenient if the user wants to use more advanced job scheduling techniques than what is supplied with the local and share queue options.

1. The job submission from Windows client workstations can be performed by installing the HPC Pack software and setting the **CCP SCHEDULER** variable to point to the HPC server, as described earlier. When a user submits a job to the Microsoft scheduler for the first time or from a different computer, they are prompted for their AD credentials. They are also asked whether they want to have these remembered for future jobs. It is recommended that they choose to have their credentials remembered for subsequent seamless job submission.
2. The job submission from Abaqus/CAE which defines the following Abaqus/CAE queue inside the Abaqus environment file. It can be outlined that the hostname parameter is not defined, since the job is actually submitted locally. Jobs can be submitted from the Abaqus/CAE Job Manager if the appropriate **onCaeStartup()** function is defined inside the Abaqus environment file. This function is given below. This function adds the **queue** definitions in the **Edit Job** window when the user creates a new job definition in the Abaqus/CAE **Job Manager**:

    a. Right Click on **Jobs** in the Model Tree and click **Manager**,
    b. In the **Job Manager** click **Create**,
    c. In the **Create** window select the model and click **Continue**,
    d. In the **Edit Job** windows, under **Run Mode** select the queue (HPClocal),
    e. Define any other desired job parameters and click Ok,
    f. The new job will appear in the **Job Manager** windows. Click **Submit** to send the job to the cluster.

    Before submitting jobs from the Abaqus CAE interface, it is recommended that users run a sample job from the command line first and save their AD credentials for future use.

Listing 11.2 onCaeStartup.py

```
def onCaeStartup():
    def makeQueues(*args):
        import driverUtils
        session.Queue(name='HPClocal',
                      queueName='local',
                      driver=driverName)
        session.Queue(name='HPCshare',
                      queueName='share',
                      driver=driverName)
    addImportCallback('job', makeQueues)
```

11.4.1 Classics Troubleshooting with HPC Cluster

Here are some things the user can do in order to diagnose problems with Abaqus and to operate on the cluster in general. By using the built-in Microsoft HPC diagnostic tools, the HPC Cluster manager has a large number of built-in cluster diagnostics that can be used to troubleshoot issues with the cluster. After any deployment, all of these tests should be run to identify any problem areas with the cluster. If any problems are identified these should be resolved before attempting to run Abaqus jobs. To use the built-in diagnostic tools:

- Open the **HPC Cluster Manager** → **Go** → **Diagnostics** or press **Crtl + 4** from anywhere within the **HPC Cluster Manager** interface.

When the user submitted a job from the command line, the user gets this error:

```
Abaqus Error: Could not convert submission directory
to UNC
```

This indicates that the directory the user is submitting the job from is not shared. As a solution, move the job files to an existing shared directory or share the directory the user is in. Then make sure the directory is shared so that the user has read/write privileges at both the share and file system (NTFS) levels.

11.4.1.1 Submit Job on a Workstation But Not Seen on HPC Job Manager

When the user submits a job from Abaqus/CAE on his workstation the Abaqus/CAE job Manager shows that the job has been submitted but the user does not see a corresponding job in the HPC Job Manager. In such case, something is wrong. The first time a user submits a job to the Microsoft job scheduler they are queried for their credentials and then prompted whether they want these remembered. If this is the first time a user is submitting a job to a Windows HPC cluster, Abaqus is likely waiting for them to enter their AD credentials before the job can actually be submitted. In Abaqus/CAE, this prompt is hidden in a command prompt window behind the main Abaqus/CAE interface. The solution is to switch to this window and then enter the correct credentials or run a small job from the command line and save the user credentials before using Abaqus/CAE for the first time.

11.4.1.2 Cannot Find the Result Files

The user submitted a job from his laptop before the user went home. When the user came back in the next day, the HPC job manager reported the job was complete but the user cannot find the results. In this case, the user needs to find the computed result files, if the user submits a job from a shared directory on a laptop and then take the laptop off the network, when the job completes, it is not able to copy the results back to the submission directory. In this case, the complete job files are still in the scratch directory on the first compute node where the job was running. If a user had a job called **myjob** running with JobID 123, in our example cluster, files would be in **C:\scratch\ user_myjob_123_exec.** The regular user will likely need the help of an administrator to access these files unless the user is using a topology model that provides access to the nodes from the enterprise network. To avoid this scenario, laptop users should always submit from a shared directory that is not on their machine.

11.4.1.3 Multi-nodes are Not Performing Correctly

If the user does not think that the multi-node jobs are performing as they should, how can the user check it? There are several things that can be done to troubleshoot suspected performance issues:

1. Run the built-in **mpipplatency** and **mpippthroughput** diagnostic tests on all nodes. Each test can be configured to run on a specific interface if desired. Comparing the latency and throughput of each interface can help to determine if there is a problem with the application/Infiniband network.
2. If the user is using a network topology that provides the compute nodes with 2 network interfaces, the user can have Abaqus using a different network interface for MPI traffic by setting this variable in the Abaqus environment file. The line below instructs MPI to use the **private** network interface:

   ```
   mp_mpirun_options = '-env MPICH_NETMASK 10.10.10.1
   /255.255.255.0'
   ```

   This can help identify whether the problem is with a specific interface or something else more general. For example, if an Abaqus job fails to run on the application network but runs properly on the private network, the user, therefore, knows that there is an issue with the application network.
3. In the user is using an Infiniband interconnect for the application network; verify that the **OpenFabrics Network Direct Provider** is installed on each node. Open command prompt with elevated privileges (Run as Administrator) and enter this command:

   ```
   clusrun ndinstall 1
   ```

   The command will run on each node. If the user sees a line in the output similar to the line below on each node then Network Direct is installed:

   ```
   0000001013 - OpenFabrics Network Direct Provider
   ```

4. The user can verify that MPI is using Network Direct by requesting the MPI connectivity table to be printed inside the Abaqus log. Add the following variable to the Abaqus environment file:

```
mp_mpirun_options = '-env MPICH_CONNECTIVITY_TABLE 1
```

The output printed in the Abaqus job log will show the mode of communication used between each rank in the job. If the user has an Infiniband interconnected on the application network, the user should see Network Direct as the communication mode between each rank. The example output below shows that Network Direct was used between NODE-01 and NODE03:

```
MPI Connectivity Table
Rank:Node Listing
------------------------------------------------
0: (NODE-01)
1: (NODE-03)
Connectivity
------------------------------------------------
SourceRank:[Indicators for all TargetRanks]
Where +=Shared Memory, @=Network Direct, S=Socket,
.=Not Connected
0
0:.@
1:@.
Summary
------------------------------------------------
Total Ranks:          2
Total Connections:    1
Shared Memory         0
Network Direct:       1
Socket:               0
```

5. Additional MPI trace and debug information can be requested by adding/modifying the **mp_mpirun_options** variable in the Abaqus environment file.

```
mp_mpirun_options = '-exitcodes -1 d 2
```

6. The user will need to add the **verbose=** command-line option when submitting jobs. This will generate more output in the Abaqus log files which can help to identify exactly where a problem is happening. Valid options are 1, 2 and 3 with increasing levels of output.

```
abaqus job=myjob cpus= 32 verbose=2
```

7. If the user needs to run an Anti-Virus software on the compute nodes, the user will have first to disable or remove this software from the compute nodes and retry the jobs. Please note that sometimes disabling Anti-Virus software services is not enough. Some Anti-Virus applications install drivers and hooks into the TCP/IP stack which can still have an effect on network latency. This is more prevalent if the analyst is using a network interface that uses TCP/IP for application traffic (Private or Enterprise interfaces).

Reference

1. Windows HPC Server 2008 R2 - Microsoft Technical Reference. https://docs.microsoft.com/en-us/previous-versions/windows/it-pro/windows-hpc-server-2008R2/ee783547(v=ws.10)

Appendix
Guidelines and Good Practices Examples

All you need to know about Abaqus, without daring to ask for it

The Appendix gives the Abaqus input file example about some models to help the user to understand or to carry out some standalone analysis tests.

A.1 Using *COUPLING to Simulate Pure Bending of Thin Walled Pipes

When a thin-walled circular cross-section pipe is loaded in pure bending, the cross-section will be ovalized. To induce pure bending in a shell model of such a structure, the proper definition of the kinematic conditions at both ends of the pipe section is required.

The in-plane deformation necessary to capture ovalization must remain free, but the out-of-plane deformation associated with warping must be suppressed. The ***COUPLING** and ***KINEMATIC** options can be used to load a thin-walled pipe in pure bending with an applied rotation.

The input model files[1] demonstrate the use of these options for full- and half-symmetry pipe models. In each model, one end of the pipe has symmetrically boundary conditions while the other end is loaded by applying a rotation to the reference node of the ***COUPLING** definition.

Note that once a combination of displacement degrees of freedom at a coupling node has been constrained, additional displacement constraints such as the boundary conditions cannot be applied to that node.

[1]See *Abaqus Example Problems Guide*, Sect. 1.1.2, for the example file named "Elastic–plastic collapse of a thin-walled elbow under in-plane bending and internal pressure".

R. J. Boulbes, *Troubleshooting Finite-Element Modeling with Abaqus*,
https://doi.org/10.1007/978-3-030-26740-7

To properly define the coupling and symmetry conditions in the half-symmetry model, two ***COUPLING** definitions are used. One couples the nodes on the symmetry plane, while the other couples the remaining nodes. Both ***COUPLING** definitions use the same reference node.

A.2 Available Degrees of Freedom with Kinematic Relation at Coupled Nodes

Once any combination of degrees of freedom (DOF) of displacement at a coupling node is constrained, additional displacement constraints cannot be applied to that node.

```
*ERROR: DEGREE OF FREEDOM 5 DOES NOT EXIST FOR NODE 1. IT HAS
ALREADY BEEN ELIMINATED BY ANOTHER EQUATION, MPC, RIGID BODY,
KINEMATIC COUPLING CONSTRAINT, TIE CONSTRAINT OR EMBEDDED ELEMENT
CONSTRAINT. THE REQUIRED EQUATION CANNOT BE FORMED.
```

This includes boundary conditions, multi-point constraints, equations, or other kinematic coupling definitions. The same limitation applies to rotational degrees of freedom. For example, if we have the following lines in the input file:

```
** Define the surface to be used with *COUPLING
*NSET, NSET=NSET_2
1,14
*SURFACE, TYPE=NODE, NAME=S2
NSET_2, 1.
** Define the kinematic coupling constraint so that
** degrees of freedom 1, 2, 3, 4 and 6 are constrained
**
*COUPLING, CONSTRAINT NAME=RBE2,REF NODE=28,SURFACE=S2
*KINEMATIC
1, 4
6, 6
** Define the following equation:
**
** 0.5*udof_5Node1 + udof_5Node14 - 2*udof_5Node28 = 0
**
*EQUATION
3,
1,  5, 0.5,
14, 5, 1.0,
28, 5, -2.,
```

The equation to constraint with a kinematic coupling is

$$\frac{1}{2} \times uDOF_5^{Node1} + uDOF_5^{Node14} - 2 \times uDOF_5^{Node28} = 0 \quad \text{(A.1)}$$

The user will receive the error message shown above because even though the fifth degree of freedom at the first node is not included in the kinematic coupling definition, it is not available for other constraint definitions. The reason why nodes involved in kinematic couplings are handled in this way is that for all but the simplest cases, participation of the unconstrained degrees of freedom in other constraints would result in an over-constraint.

A.3 Stability and Accuracy of the Trapezoidal Rule

The unidimensional system shown in Fig. A.1 is loaded with an applied initial speed v_0 on the mass M, giving the initial energy to the spring–mass system, which will be dissipated by oscillation. The spring, mass, and speed are consistent with a system of units according to one of the consistent system of units listed in Table 3.1. The solution of this system is given by Eq. A.2, with $\omega = \sqrt{\frac{K}{M}}$.

$$x\left(t\right) = \frac{v_0}{\omega} sin\left(\omega t\right) = sin\left(t\right) \quad \text{(A.2)}$$

Here, the solution of the system in displacement (x) is a pure sinusoidal function with a time period equal to 2π s. The solutions regarding the speed ($\dot{x}$) and the acceleration ($\ddot{x}$) are therefore the first and second derivatives of Eq. A.2, respectively.

The input data of the spring–mass model modeled according to the loading and boundary conditions shown in Fig. A.1 are given below. The code is made with four different values of the time increments $\Delta t = 0.25$, $\Delta t = 0.5$, $\Delta t = 1$, and $\Delta t = 2$ in order to check how the solution will be computed with different time increment as a function of the implicit scheme driven by the integration parameter α, which is initially set equal to zero.

It will be possible to compare the computed solution with the exact solution determined in Eq. A.2 afterward, to observe the deviation behavior of the implicit

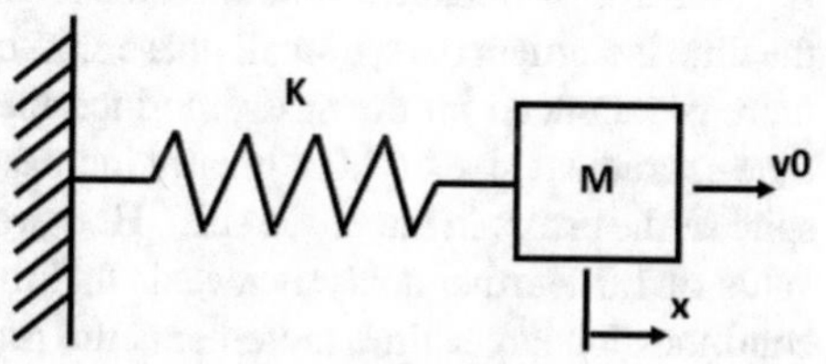

Fig. A.1 Implicit dynamic model of a system with a spring K and a mass M

numerical scheme as a function of the α parameter and then to come to a conclusion regarding the stability and accuracy of the numerical settings used.

```
*HEADING
Implicit Integration Demonstration
*NODE, NSET=NALL
1, 0., 0., 0.
2, 2., 0., 0.
*ELEMENT, TYPE=SPRINGA, ELSET=SPRING
1, 1, 2
*SPRING, ELSET=SPRING

1.,
*ELEMENT, TYPE=MASS, ELSET=MASS
2, 2
*MASS, ELSET=MASS
1.,
*BOUNDARY
1, 1, 3
2, 2, 3
*INITIAL CONDITIONS, TYPE=VELOCITY
2, 1, 1.0
*STEP, INC=40
*DYNAMIC, ALPHA=0.0, NOHAF
.25, 10.
*NODE FILE, NSET = NALL
U, V, A
*Restart, write, frequency=0
*Output, field, variable=PRESELECT, frequency=1
*Output, history, frequency=1
*Energy Output ALLIE, ALLKE, ALLWK, ETOTAL
*END STEP
```

The effects of the time-increment size with implicit integration for an integration parameter **ALPHA** set equal to zero as a function of the time increment Δt are shown in Figs. A.2, A.3, A.4 and A.5.

In conclusion, it can be observed that compared to the exact solution given in Eq. A.2, for the smallest time increment (0.25 s) the results are virtually exact. For the time increment 0.5 s, a small phase shift can be detected. This phase shift becomes more pronounced for the next time-increment size (1.0 s) and is quite strong for the time-increment size of 2.0 s, greatly increasing the apparent period of oscillation, in spite of the errors in the phase shift. However, the amplitude remains at the expected value of 1.0. Further analysis reveals that this trend (displacement amplitude of 1.0) continues for larger time increments the procedure becomes unconditionally stable and exhibits no artificial damping.

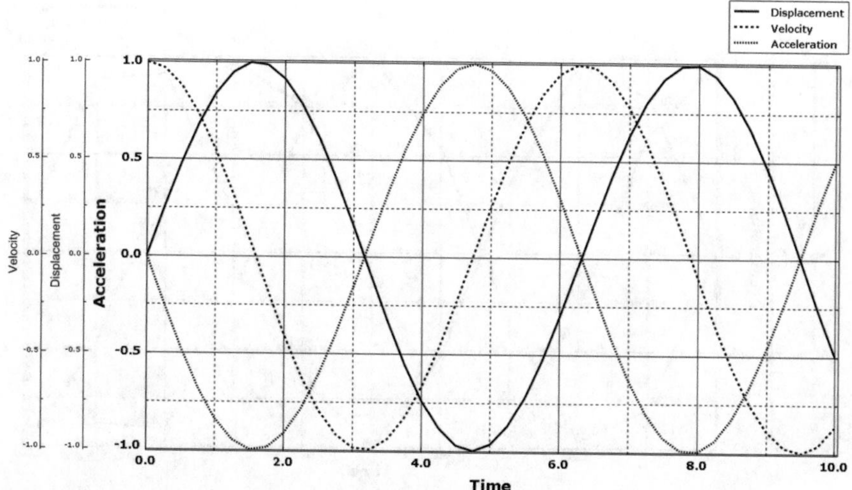

Fig. A.2 Computed solution for $\alpha = 0$ and $\Delta t = 0.25$

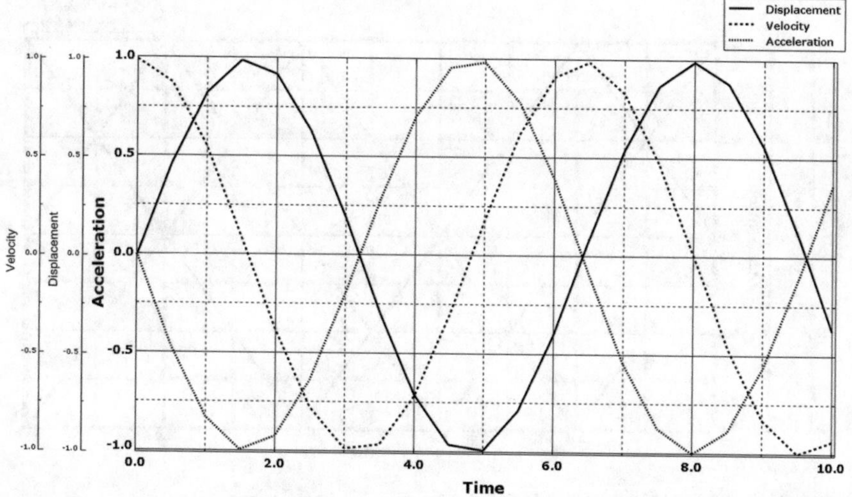

Fig. A.3 Computed solution for $\alpha = 0$ and $\Delta t = 0.5$

Hence, the time incrementation for this procedure can be chosen to fit the physical response of the system without regard to the stability, and relatively large time increments can be used. Of course, the method only has second-order accuracy. Consequently, the accuracy of the solution is lost as the size of the time increment increases. This loss of accuracy is manifested by the phase shift.

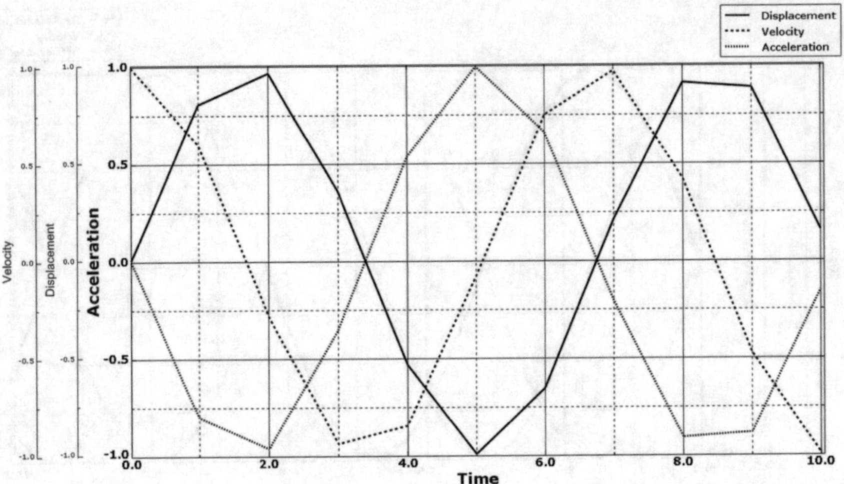

Fig. A.4 Computed solution for $\alpha = 0$ and $\Delta t = 1$

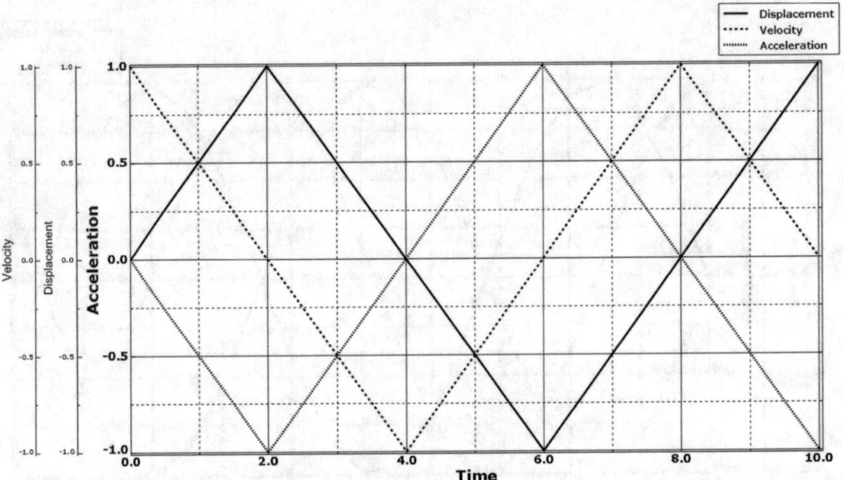

Fig. A.5 Computed solution for $\alpha = 0$ and $\Delta t = 2$

The stabilization effects due to numerical damping into the solution can be achieved after playing with the value of the integration range set by the parameter **ALPHA**, which is nonzero and as a function of the same time-increment intervals. Indeed, previously in the above simulations, the numerical damping was removed. The HHT method allows the introduction of numerical dissipation through the choice of the parameter α. A careful analysis of the algorithm shows that the method very selectively dampens the vibrations with a period of less than or the same order of magnitude as the time increment. Of course, the same spring–mass model is used to investigate the effects of numerical damping to compare the effects of numerical integration parameter α on the solutions afterward.

The analysis is, therefore, run again to the same time increments, except that the total final time is 50 s instead of 10 s, to give the finest stabilized behavior of the computed solution. Two values of damping are shown: first, the default value of $\alpha = -0.05$ is shown in Figs. A.6, A.7, A.8, and A.9, and second, the maximum damping value of $\alpha = -1/3$ is shown in Figs. A.10, A.11, A.12, and A.13.

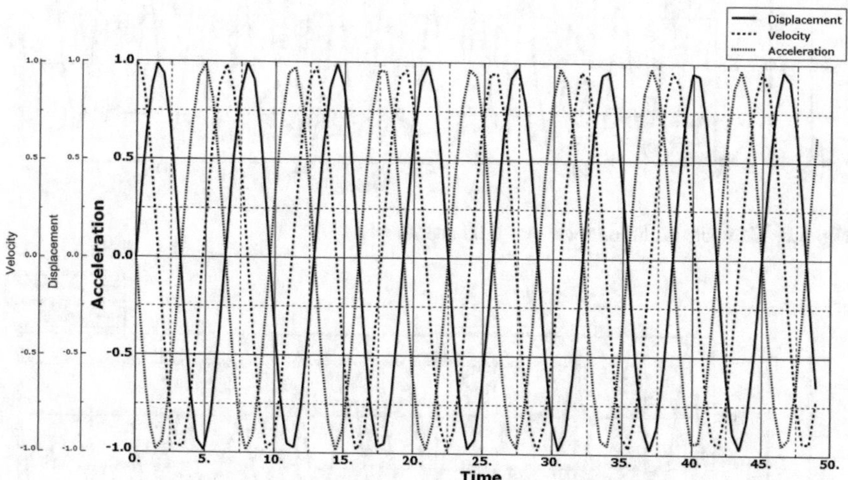

Fig. A.6 Computed solution for $\alpha = -0.05$ and $\Delta t = 0.25$

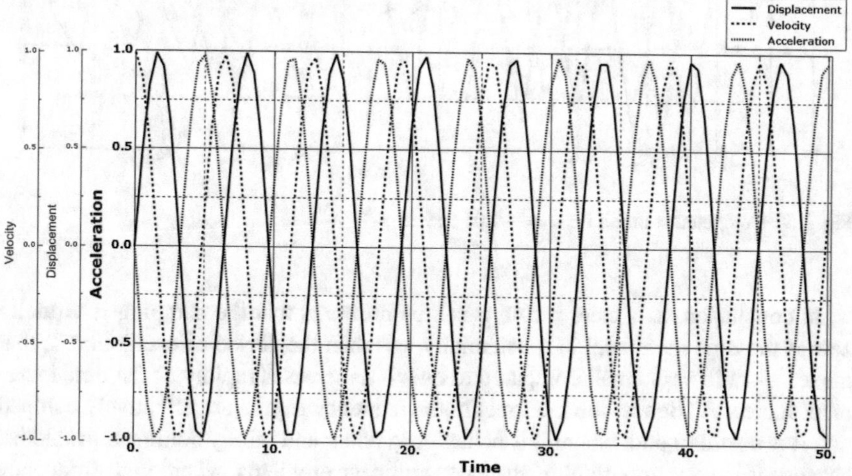

Fig. A.7 Computed solution for $\alpha = -0.05$ and $\Delta t = 0.5$

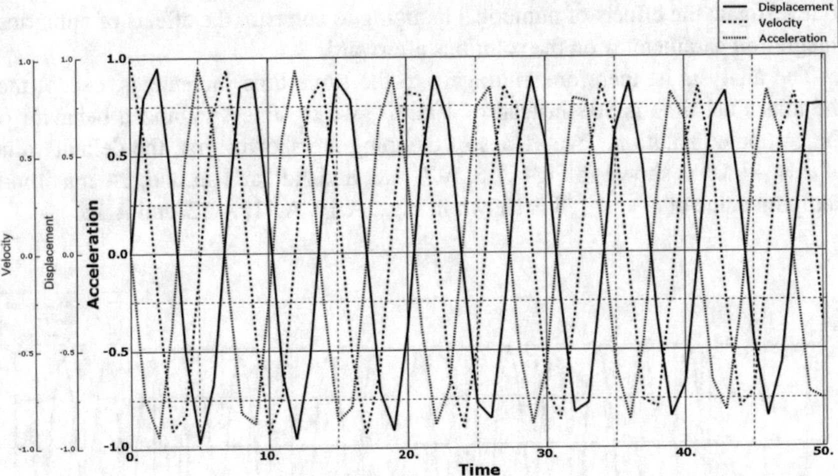

Fig. A.8 Computed solution for $\alpha = -0.05$ and $\Delta t = 1$

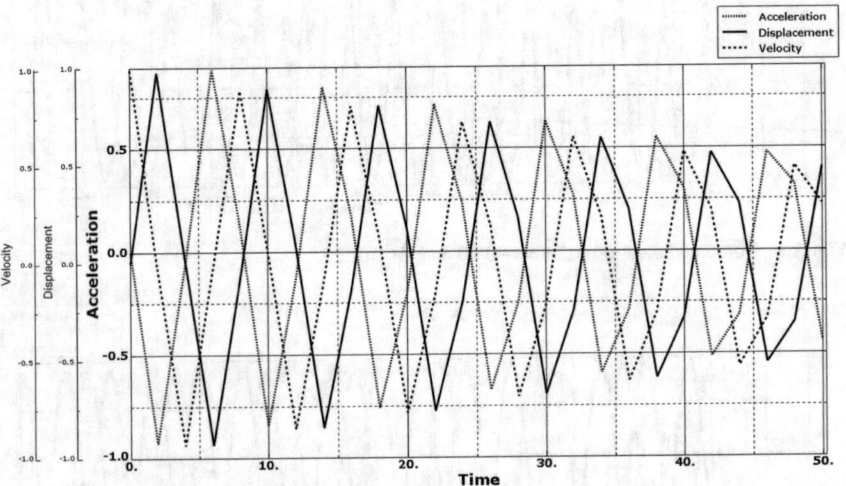

Fig. A.9 Computed solution for $\alpha = -0.05$ and $\Delta t = 2$

In conclusion, as a clear result, it can be observed that the damping is virtually zero if the time increment is substantially less than the period of oscillation T, that makes $\frac{\Delta t}{T}$ ratio very small compared to one; but increases rapidly as the time increment increases. Hence, the incorrect high-frequency results are effectively damped out. It is useful to understand this behavior in linear and mildly nonlinear problems, because it is very important in strongly nonlinear problems, where high-frequency noise can cause convergence problems. This is particularly true in impact problems, where a high-frequency response is always excited. Moreover, with the default value

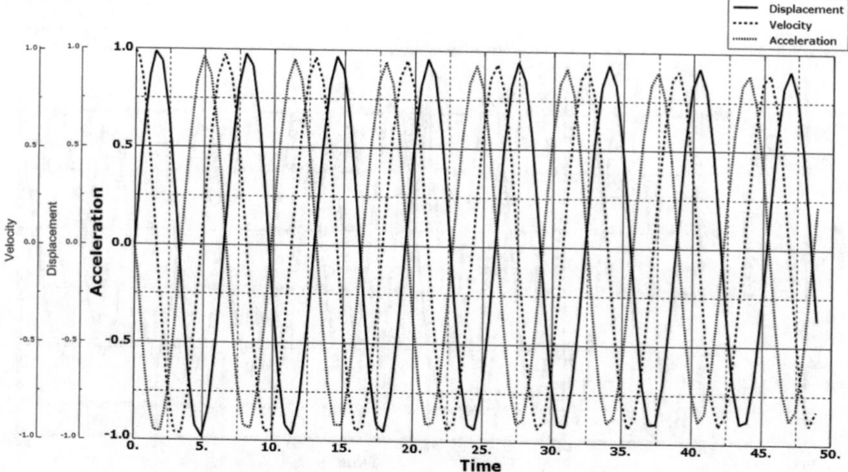

Fig. A.10 Computed solution for $\alpha = -1/3$ and $\Delta t = 0.25$

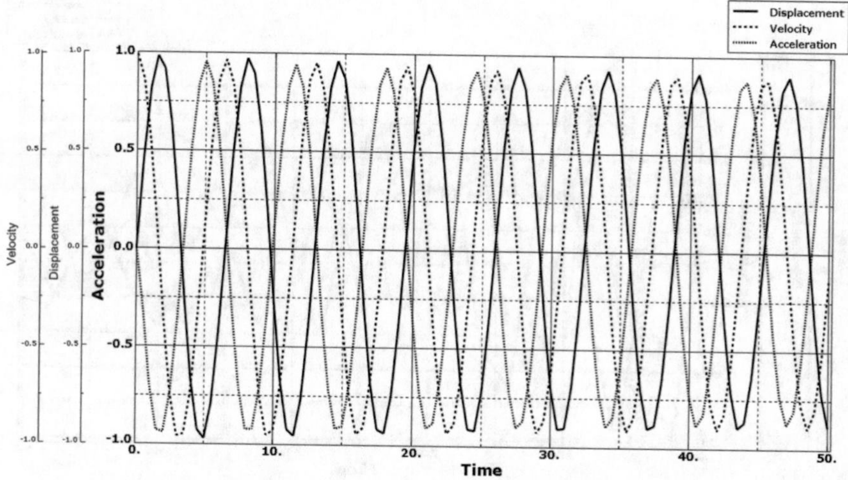

Fig. A.11 Computed solution for $\alpha = -1/3$ and $\Delta t = 0.5$

of $\alpha = -0.05$, the total numerical dissipation is very small for most problems. For the problems included in the *Abaqus Example Problems and Benchmarks* manuals, the dissipation is always less than 0.5% of the total energy in the system. An energy control check is always strongly recommended to evaluate the dissipation contribution in a system or structure.

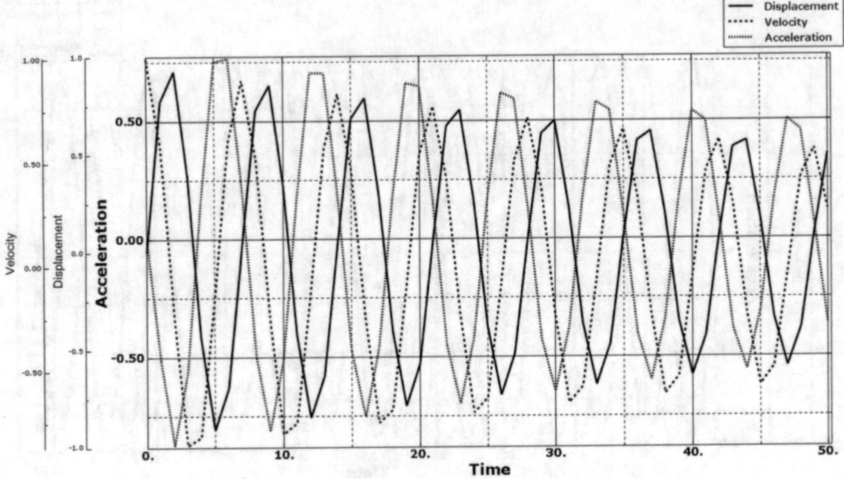

Fig. A.12 Computed solution for $\alpha = -1/3$ and $\Delta t = 1$

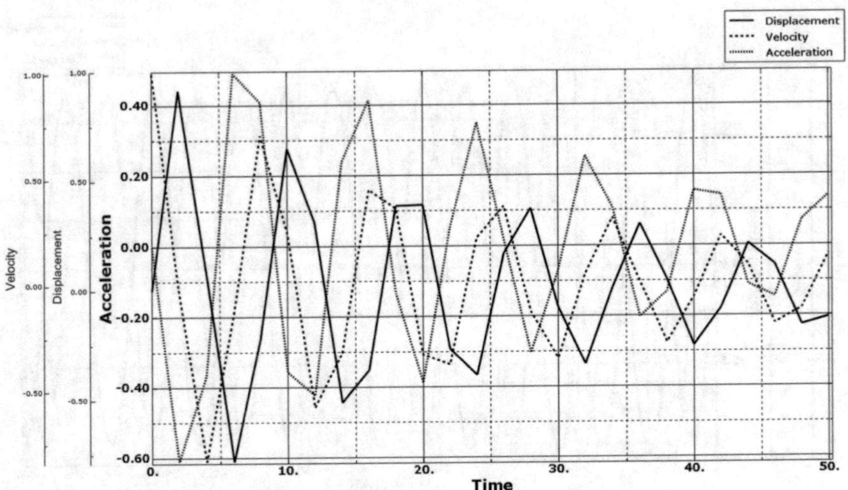

Fig. A.13 Computed solution for $\alpha = -1/3$ and $\Delta t = 2$

A.4 Accuracy Control in Highly Nonlinear Problems with a Half-Increment Residual Tolerance

As was noted earlier, for highly nonlinear problems, an accurate choice of half-increment tolerance is less important. In these problems, the time-increment size is usually dictated by the convergence requirements due to the nonlinearity of the problem. Accuracy of the time integration is then a fortunate by-product of the strong

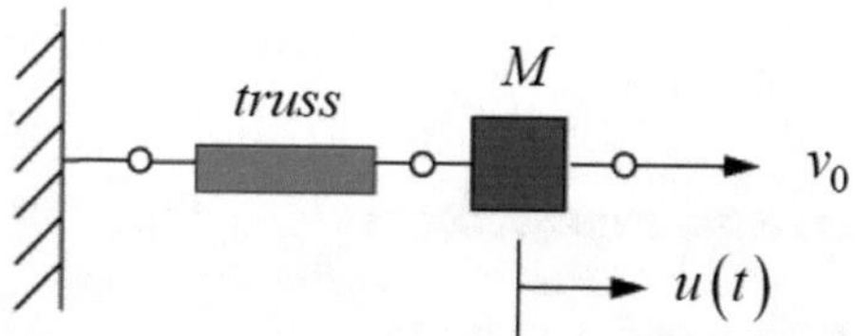

Fig. A.14 Elastic–plastic truss element connected to a mass point. Whatever the consistent system of units used, the element length $l = 1$, the cross-sectional area $A = 10^{-4}$, the Young's modulus $E = 10{,}000$, the yield stress $\sigma_Y = 100$, the mass $M = 1$ and the initial velocity given is $v_0 = 1$

nonlinearity. To demonstrate, consider replacing the spring in the spring–mass system shown in A.3 with an elastic–plastic truss element as shown in Fig. A.14.

Assuming that the post-yield response is piecewise linear and taking the density of the material to be zero, this is simply a spring–mass system with a nonlinear spring. The use of the geometric nonlinearity **NLGEOM** with the truss element type **T2D2** allows the cross-sectional area to change with the level of strain.

In order to exaggerate the effects of the nonlinearity, a very tight equilibrium tolerance has been used by employing the ***CONTROLS** option. The input file is as follows with **ALPHA** equal to zero because the stiffness dampens low-frequency modes for four different values of the half-increment residual parameter **HAFTOL** to check the effects of the half-increment residual tolerance in solution.

```
*HEADING
Nonlinear effects: elastic-plastic truss and mass
*NODE, NSET=NALL
1, 0., 0., 0.
2, 1., 0., 0.
*ELEMENT, TYPE=T2D2, ELSET=TRUSS
1, 1, 2
*SOLID SECTION, ELSET=TRUSS, MATERIAL=MAT
0.001,
*MATERIAL, NAME=MAT
*ELASTIC
10000.0, 0.
*PLASTIC
100., 0.0
200., 0.1
300., 0.5
400., 1.0
500., 10.
*ELEMENT, TYPE=MASS, ELSET=MASS
2, 2
*MASS, ELSET=MASS
1.,
```

```
*BOUNDARY
1, 1, 2
2, 2,
*INITIAL CONDITIONS, TYPE=VELOCITY
2, 1, 1.0
*STEP, INC=500, NLGEOM
*DYNAMIC, ALPHA=0., HAFTOL=0.01
**DYNAMIC, ALPHA=0., HAFTOL=0.1
**DYNAMIC, ALPHA=0., HAFTOL=1.0
**DYNAMIC, ALPHA=0., HAFTOL=1.E20
.25, 100.
*NODE FILE, NSET=NALL
U, V, A
*CONTROLS, PARAMETER=FIELD,
FIELD=DISPLACEMENT
1.E-8,
*Restart, write, frequency=0
*Output, field, variable=PRESELECT, frequency=1
*Output, history, frequency=1
*Energy Output ALLIE, ALLKE, ALLWK, ETOTAL
*END STEP
```

The effects of the half-increment residual parameter **HAFTOL** on the solution of the model shown in Fig. A.14 are plotted in Fig. A.15.

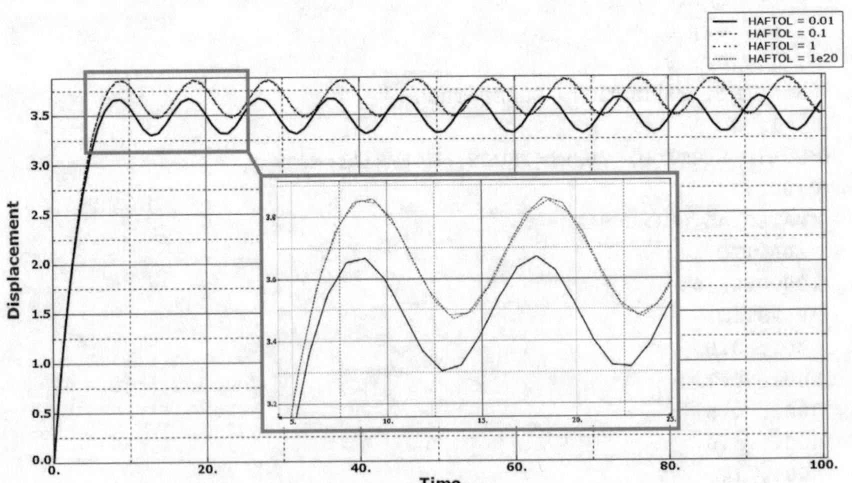

Fig. A.15 Solutions of truss mass system for different values of the parameter **HAFTOL**, the half-increment residual tolerance parameter

It is clear from the results of this problem that the response is shown in Fig. A.15 in the early, dissipative phase of the problem does not depend on the accuracy control **HAFTOL**. The results diverge significantly only after the plastic deformation is completed and the mass starts to vibrate elastically.

In many heavily nonlinear problems (impact, forming, etc.), the interest is in the amount of deformation during the dissipative phase. In these cases, the choice of half-increment tolerance is not critical. Therefore, very large values of **HAFTOL** are generally used in such cases. Moreover, these half-increment tolerances provide "reasonable" accuracy so long as the system has enough natural dissipation for any high-frequency content to disappear quite rapidly relative to the time scale of the event being modeled.

A.5 The Art of Meshing

In this example, the different mesh control options will be investigated on a 2D structure shape like such as padeye geometry. The main aim of this section is to understand the algorithm performances applicable to a 2D structure and to extend the overview ideas for a 3D structure too. In order to focus only on the mesh control performances, a global seed element size has been set on this test case model and fixed for all combinations of mesh control parameters, equal to 5 mm.

Figure A.16 shows a geometry model used to explain and understand how an Abaqus mesher works with the different options shown in Fig. 6.1 or Fig. 6.5 for a 2D or 3D structure, respectively.

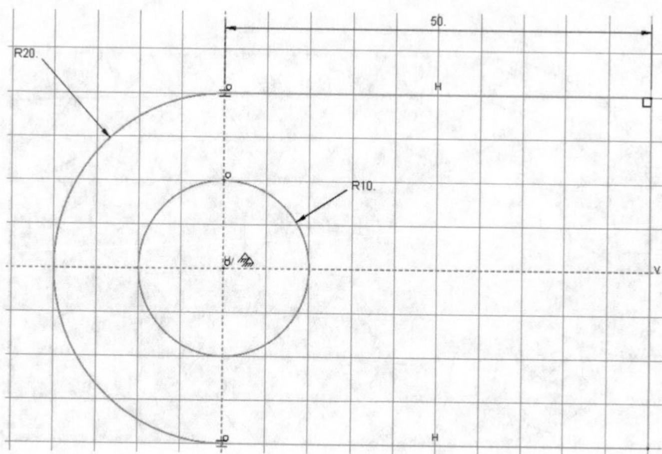

Fig. A.16 Padeye geometry

A.5.1 Free Meshing Technique

It is time to see the performance of the Abaqus mesher with different configurations regarding the element types and algorithm used. Figure A.17 shows a typical free meshing part with triangular elements. Table 6.1 shows that unstructured triangulation with triangular elements can fit a complex structure and is very robust, but on the other hand it increases the computational time because more elements are required to mesh the structure. It can also be difficult to deal with the mesh sizes in specific cases. Moreover, the triangular elements are stiffer than quadrilaterals and can, therefore, be inadequate for some certain types of loading conditions (e.g., bending moment).

Figure A.18 shows the model meshed with quadrilateral-dominated elements. By comparing this figure with Fig. A.17, it can be observed that a structured grid is more beneficial for model computation than unstructured triangulation. On the other hand, this type of mesh is not capable of performing well with a complex geometry model without using some partitioning strategies in the model.

Figure A.19 shows the structure meshed using the same type of element used to mesh Fig. A.18 but with a different algorithm; here the algorithm used is the medial axis described in Sect. 6.1.5.6. It can be observed that the mesh pattern is more regular than the mesh shown in Fig. A.18, which allows initial element distortion to be avoided.

Figure A.20 shows the model meshed with quadrilateral elements using the same algorithm as was used in Fig. A.18 and described in Sect. 6.1.5.1. It can be observed that the mesh pattern is very similar to the mesh made in Fig. A.18, and therefore an obvious conclusion here should be that there is no major difference between using quadrilateral elements and quadrilateral-dominated element types to mesh a structure set with an advanced front algorithm.

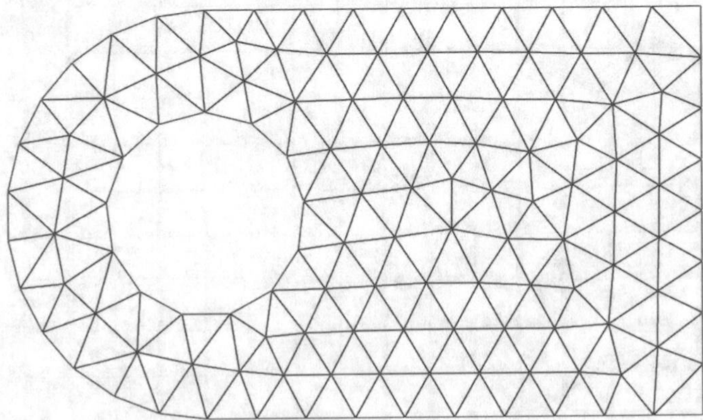

Fig. A.17 Padeye meshed with triangular element

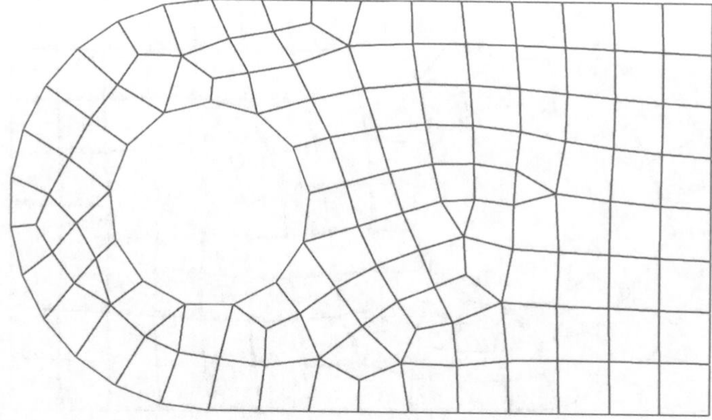

Fig. A.18 Padeye meshed with quadrilateral-dominated elements and use of advanced front algorithm

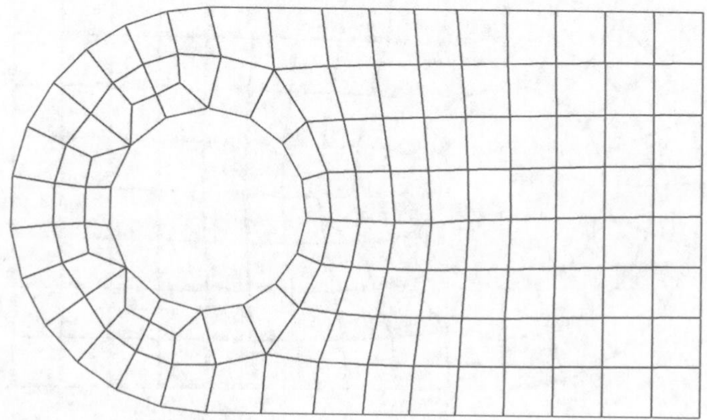

Fig. A.19 Padeye meshed with quadrilateral-dominated elements and use of medial axis algorithm

Figure A.21 shows the model meshed with the same element type as used in Fig. A.19 but with a different algorithm, now meshed using a medial axis. Again, it can be observed that no major difference exists in the mesh patterns made in Fig. A.19 and Fig. A.21. However, the structure meshed in Fig. A.21 looks more uniform in terms of the mesh pattern around the hole than the mesh made in Fig. A.19, meaning that the quadrilateral element using the medial axis algorithm should have a better mesh performance and can preferably be used with design transition in the model.

In conclusion, when using the free technique mesh in Abaqus, it is not necessary to think about any partitioning strategies in the model and it is directly applicable to any simple structure to be meshed. Moreover, it gives very good mesh performance in a structure model if the right element type has been set with the proper algorithm.

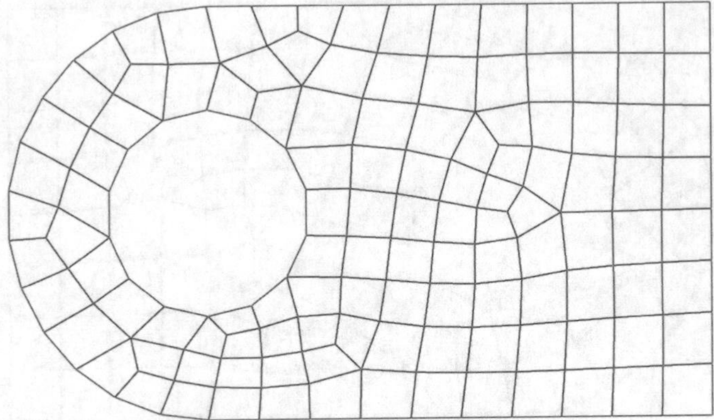

Fig. A.20 Padeye meshed with quadrilateral elements and use of an advanced front algorithm

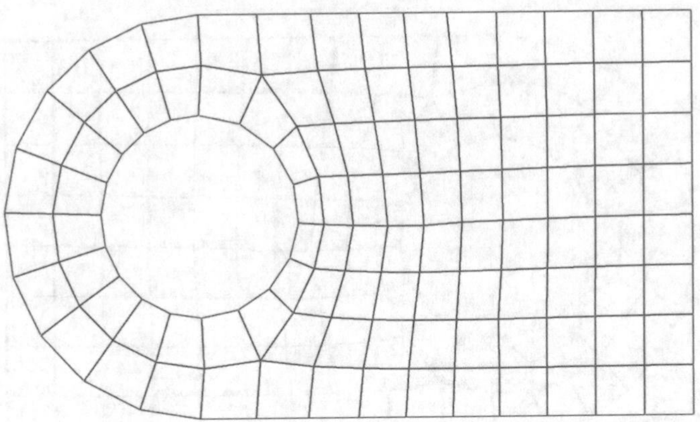

Fig. A.21 Padeye meshed with quadrilateral elements and use of medial axis algorithm

A.5.2 Model Partitioning with a Strategy Based on Design Symmetry

A complex model design would most likely need to be meshed by using model partitioning to obtain, as far as possible, a uniform meshed structure to avoid any initial element distortion and to ensure a good and smooth compatibility between nodal fields in the solution. There is no specific rule for deciding on appropriate model partitioning strategies that work as a function of the model designed, which is why here it can be seen like an Art of mesh more than a technical guideline. This section will investigate a type of model partitioning based on symmetry or having kinds of pseudo-symmetry lines in the model, as shown in Fig. A.22. Thanks to these partition lines, it will be easier to mesh the model.

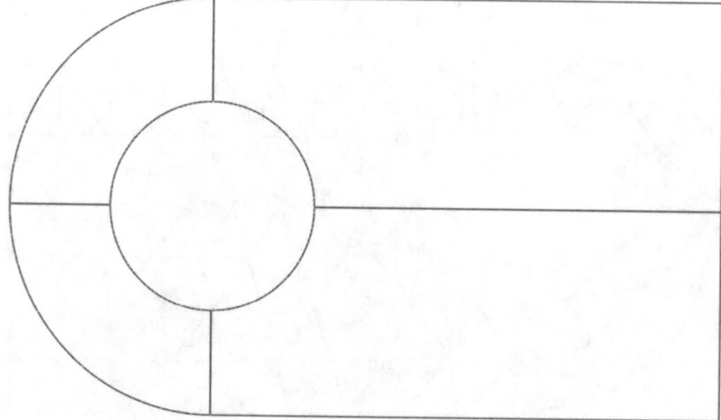

Fig. A.22 Symmetric partition lines used in the padeye design model in Fig. A.16

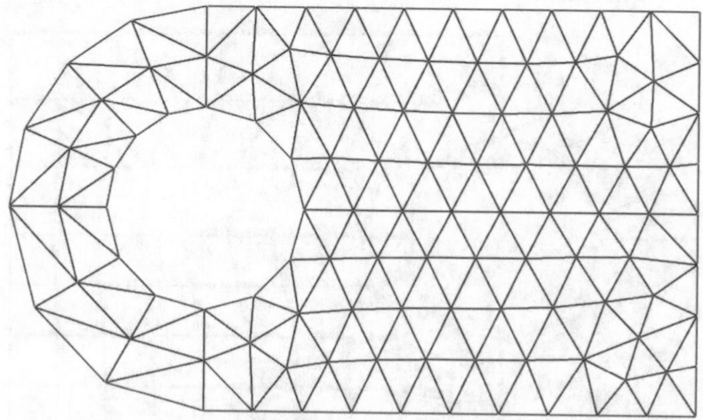

Fig. A.23 Padeye meshed with a free technique for meshing a triangular element using the model partitioning strategy based on design symmetry

Figure A.23 shows an unstructured triangulation mesh set with the same mesh control parameters as were used in Fig. A.17. It can be observed that partition lines have a positive effect on the meshed structure because the triangulation pattern looks more uniform as it intersects with these partition lines. Therefore, the unstructured triangulation mesh looks more regular than a kind of structured triangulation thanks to the partitioning strategies used on the model.

Figure A.24 shows a structured triangulation mesh that looks like it has a kind of quadrilateral-dominated mesh. Indeed, here the mesh control has been set with a triangular element using a structured meshing technique. By comparing it with Fig. A.25, a type of quadrilateral mesh pattern using triangular element can be clearly observed. These combined settings can be powerful for meshing models with complex designs in an easy way.

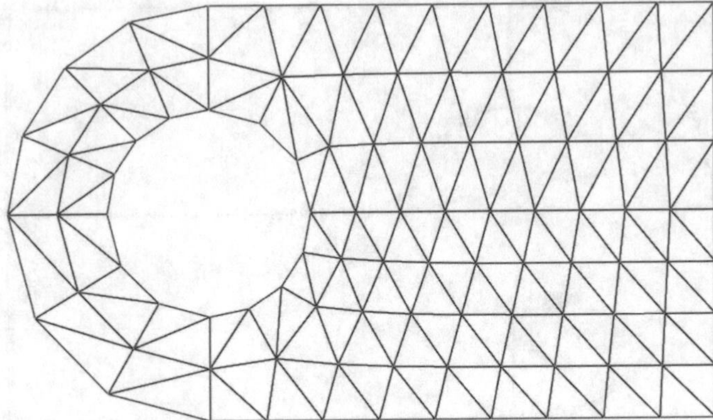

Fig. A.24 Padeye meshed with a structured meshing technique of triangular element using the model partitioning strategy based on design symmetry

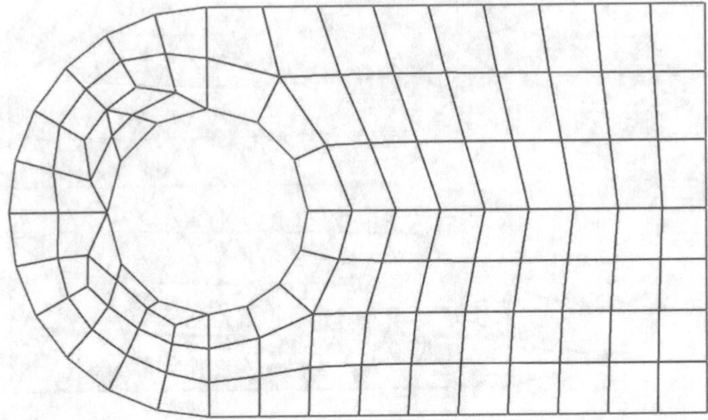

Fig. A.25 Padeye meshed with quadrilateral-dominated elements and the use of the medial axis algorithm and model partitioning strategy based on design symmetry

Figure A.25 now uses different settings for the mesh control parameter with a structured technique of quadrilateral-dominated elements and the medial axis algorithm, as was used in Fig. A.19. A good uniform mesh pattern can be observed with some deviation around the hole transition shape because the mesh now intersects with partition lines.

Figure A.26 looks almost identical to the mesh pattern made in Fig. A.21 with a little improvement in accordance with the design symmetry thanks to the partition lines. It, therefore, proves that having a smart partitioning strategy in both complex and simple designs can help to make a mesh of better quality with regard to the element geometry ratio.

In conclusion, model partitioning can be used to improve the quality of the mesh pattern in both simple and complex design models. The symmetry of the design

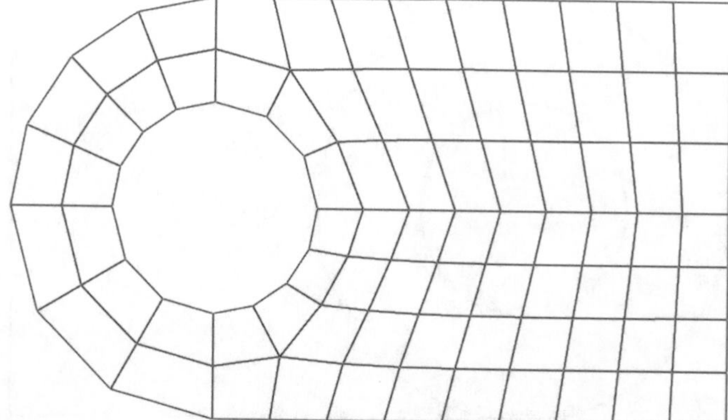

Fig. A.26 Padeye meshed with quadrilateral element and the use of the medial axis algorithm and model partitioning strategy based on design symmetry

geometry can be used to think about how to draw the partition lines without caring about the boundary and loading conditions applied to the model. Therefore, the way in which a model is meshed, is less dependent of the boundary and loading conditions applied to the model, as it is mainly related to the structure shape.

A.5.3 *Model Partitioning with a Strategy Based on the Dominant Geometry*

This section will present an investigation of an alternative model partitioning strategy that has been developed by considering not the symmetry but the geometry shape. The difference between the two is shown in Fig. A.27. Now, instead of having symmetry lines in 2D or 3D, it is possible to consider the model partitioning by looking at the dominant shape in the meshing region of the model. Here, for instance, the dominant geometry is a circle, and therefore the partition must have a circular shape. It can be assumed that the most difficult area to mesh should be close to the top and bottom edges of the partitioned circle with the rectangular shape. These areas should have the maximum initial element distortion in the mesh. The question is how the Abaqus mesher will handle that.

Figure A.28 shows a quite similar mesh pattern as was made in Fig. A.17, with the main difference is that because of the circle partition, the mesh is more uniform around the hole. Of course, as expected, the areas close to the edge between the partition circle and the rectangle show a warning message due to an initial element distortion as highlighted in Fig. A.29.

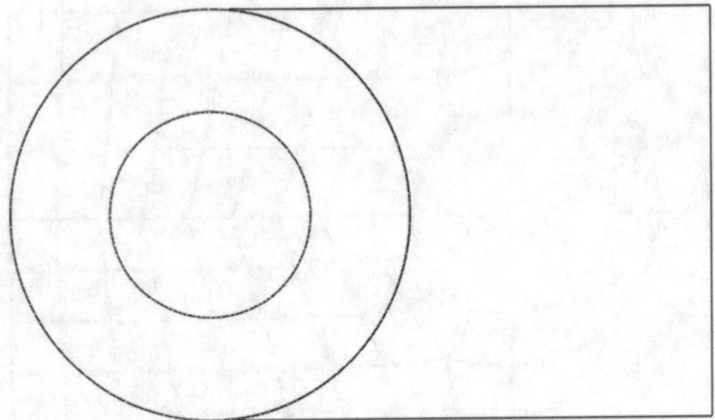

Fig. A.27 Geometric partition lines used on the Padeye design model in Fig. A.16

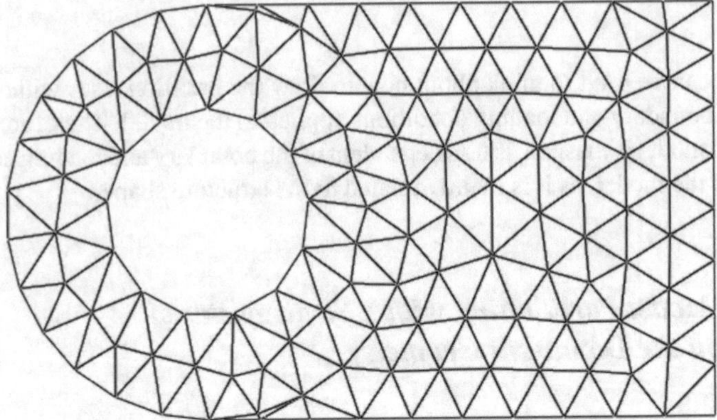

Fig. A.28 Padeye meshed with a free meshing technique of triangular elements using the model partitioning strategy based on the dominant geometry

Figure A.29 shows the detection of element distortion determined with Abaqus thanks to the Verify Mesh[2] feature in the mesh module. This is an check of the analysis mesh done by Abaqus with default parameters regarding the element ratio geometry tolerance.[3] These parameter values can be changed but modification of the default criteria is not recommended; otherwise, the solution should be handled with care.

Figure A.30 shows a mesh pattern obtained by using a quadrilateral-dominated element with an advanced front algorithm as was used in Fig. A.18. It can be observed first that the element distortion warning has been removed in both areas due to a change in the element geometry ratio preprogrammed in Abaqus after performing

[2]See *Abaqus CAE User's Guide*, Sect. 17.19.1, "Verifying element quality for more information".

[3]See *Abaqus CAE User's Guide*, Sect. 17.6.1, "Verifying your mesh for more information".

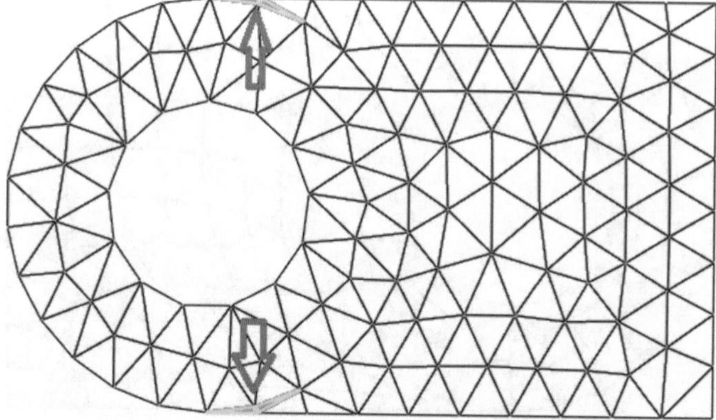

Fig. A.29 Initial element distortion warning detected from the meshed model made in Fig. A.28

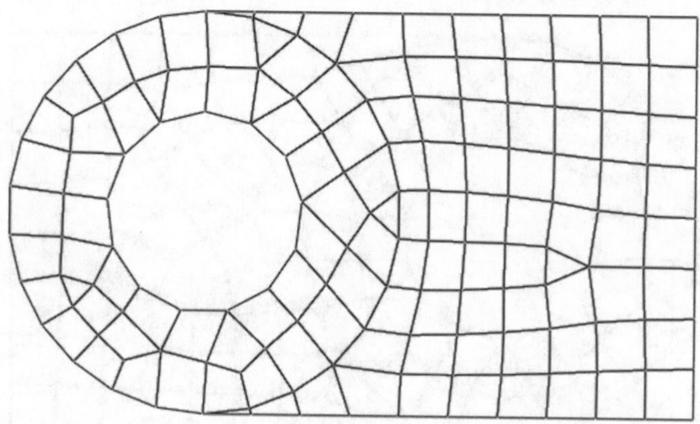

Fig. A.30 Padeye meshed with quadrilateral-dominated element and the use of the advanced front algorithm with a geometric partitioning strategy

the analytical check of the mesh thanks to the mesh verification feature. Second, the mesh looks more uniform around the hole than in Fig. A.18. By comparing it with the previous free triangle technique, it seems a better choice here to select a quadrilateral-dominated element to get rid of the element distortion warning.

Figure A.31 shows what happened when the algorithm was changed from the advanced front option used in Fig. A.30 to a medial axis option as is used in Fig. A.19. In Fig. A.19, the medial axis algorithm showed a significant improvement in the mesh pattern compared to Fig. A.18 with regard to getting a uniform meshed structure. But here, the opposite case occurs: there is no improvement in Fig. A.30 because the mesh pattern looks less uniform in the rectangular zone than the previous mesh. On the other hand, the mesh pattern around the hole designed has a better distribution with quadrilateral element meshed. There is a balanced situation here, where the rectangular zone is less uniformly meshed but the hole has better uniformity in the

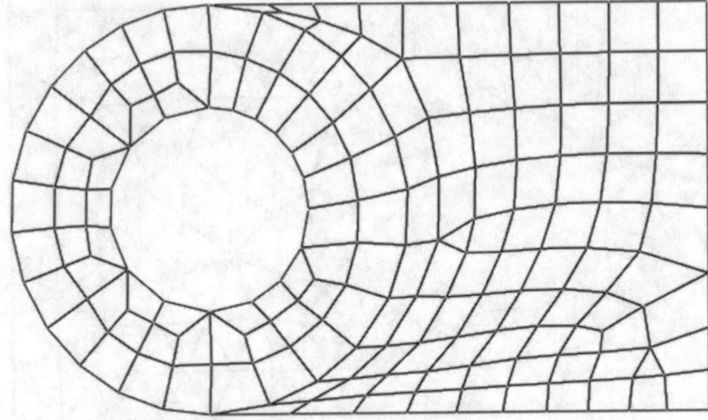

Fig. A.31 Padeye meshed with quadrilateral-dominated element using the medial axis algorithm and a geometric partitioning strategy

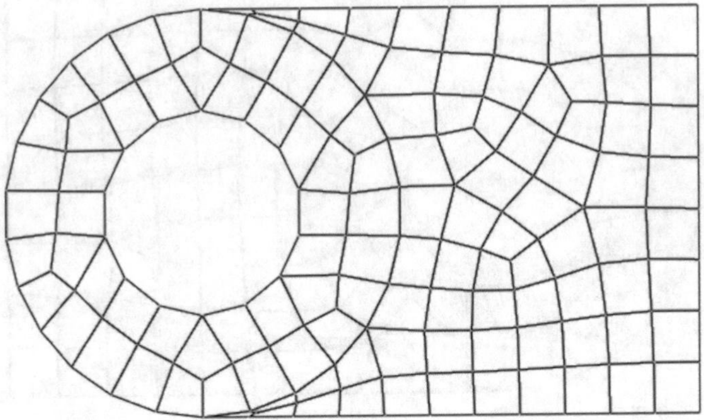

Fig. A.32 Padeye meshed with quadrilateral element using an advanced front algorithm and a geometric partitioning strategy

mesh. The analyst can decide which zone is of most interest for the analysis in order to have a better control of the mesh quality. As long as there is no element distortion warning, the compatibility of the solution versus the mesh passes all the numerical criteria and can be accepted by users.

Figure A.32 shows the model meshed with quadrilateral elements using an advanced front algorithm; the result looks similar to the mesh pattern obtained in Fig. A.20 and there is still an improvement regarding the quality of the mesh around the hole. Even though the mesh is an unstructured quadrilateral pattern, it seems more uniformly distributed through the whole model than previously observed in Fig. A.31.

Finally Fig. A.33 shows the best selected option with the mesh control feature to get a very good uniform mesh pattern of the whole structure with a quadrilateral element using the medial axis algorithm. Even though the two areas on the top and

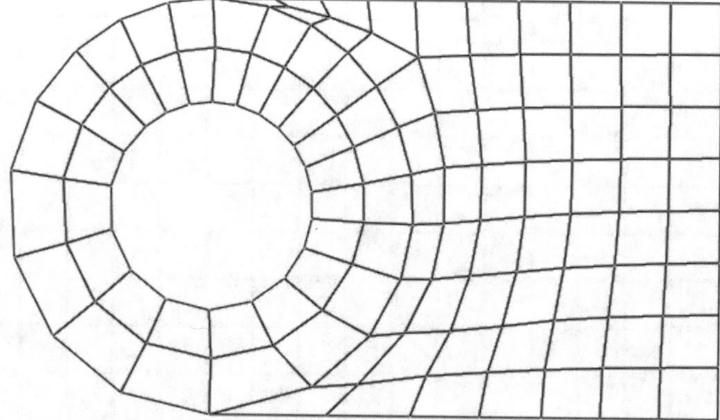

Fig. A.33 Padeye meshed with quadrilateral element using medial axis algorithm and a geometric partitioning strategy

bottom of the edge transition between the partition circle and the rectangular shape are not meshed identically, the compatibility of the solution is ensured because no element distortion has been detected by the tool for checking and verifying the mesh analysis.

In conclusion, it was already stated that the mesh is independent of the boundary and loading conditions but is highly dependent on the model partitioning strategy used. Indeed, model partitioning can be done in order to focus on a specific region in the model connected with the solution scheme which is somehow related to the boundary and loading conditions. As a function of the partitioning strategy used, the different options in the mesh control feature can affect the accuracy of the solution and act differently from model to model. It is, therefore, important that the analyst should be concerned with making a smart choice regarding the partitioning of model and selecting the proper mesh options by trying different mesh mappings on the model, and comparing the advantages and disadvantages for every mesh mapping to draw a conclusion about the correct settings to use afterward. All modifications in the model partitioning strategy will change the mesh pattern in the whole model. Thinking wisely about the model partitioning, carrying out a visual inspection with different mesh options, and controlling the mesh with the mesh verification module feature to remove element distortion are the smartest operations to carry out prior to submitting the model job.

A.5.4 Small Edges and Consequences for the Mesher

Sometimes there are some small edges, faces, and vertexes made in the design which create some difficulties in achieving a proper quality mesh. To understand how this can happen, users have to keep in mind that an imported model in Abaqus made by

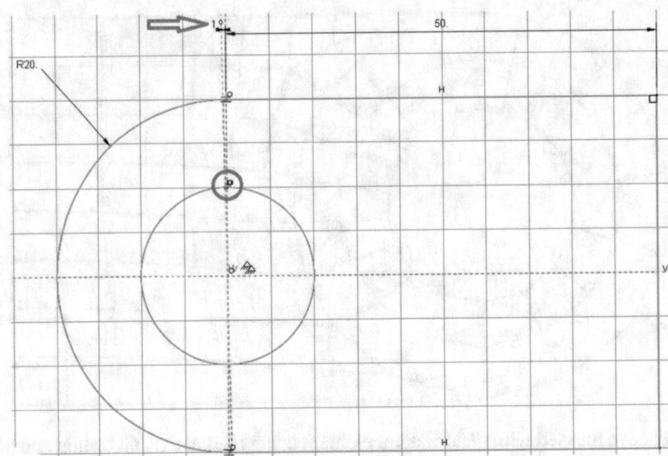

Fig. A.34 Padeye with a mismatch in design construction lines showing a very small edge arc length that should not be present in the model

CAD software needs to respect certain rules to avoid major difficulties in analytical tasks such as in the meshing part. Indeed, FEA software reads the imported model file as a list of instructions to generate the model according to all points, construction lines, edges, faces, and so on, made with CAD software. It is, therefore, important to understand that as CAD modeling obeys different rules from FEA modeling, a mismatch between them can lead to a huge waste of time cleaning the model. The analyst must pay attention to any details in order to remove all unnecessary small edges, and small faces in the model that are not structurally significant for the model response.

Figure A.34 shows an example of a mismatch between a model made with CAD software, and imported into Abaqus, showing a very small edge in the internal circle construction lines that has about one degree of deviation from the vertical line used to make the external circle. This means that instead of having two circles constructed using a picked point aligned with a vertical axis, the model shows that the internal circle is made with two construction lines, one of which is a very short arc. This section investigates the consequence of this type of mismatch with the mesher options.

Figure A.35 shows the meshed model using a free technique triangle option. It can be observed that the mesh pattern shows some difficulties in properly meshing the area, where the design mismatch occurred and shows an initial element distortion warning.

Figure A.36 shows a different mesh option using a free technique with a quadrilateral-dominated element and a medial axis algorithm. In this case, the element distortion warning has been removed because of differences in the geometry criteria for element ratio tolerance.

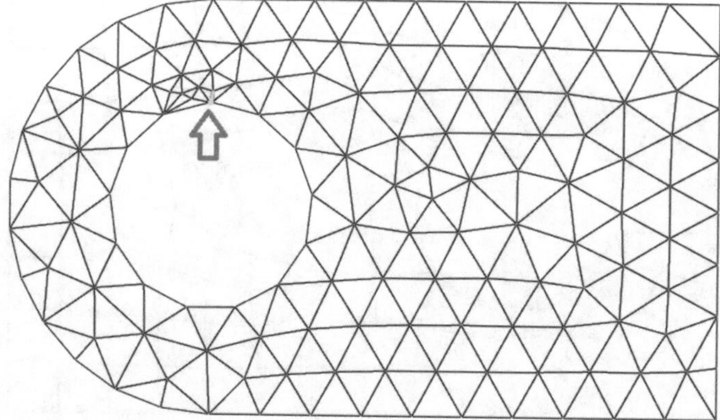

Fig. A.35 Padeye meshed with a design mismatch using a free technique triangle option mesh

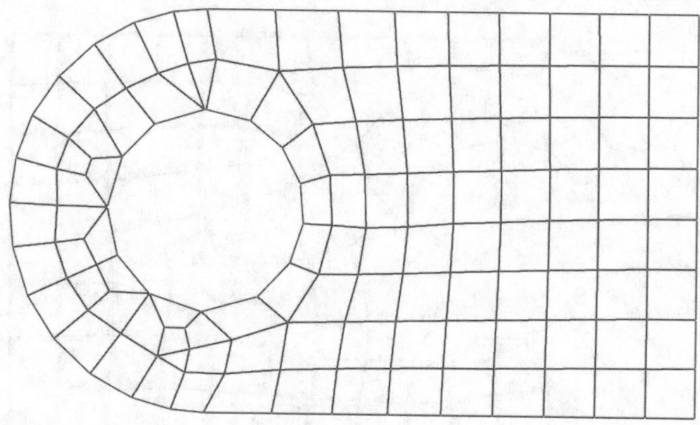

Fig. A.36 Padeye meshed with a design mismatch using a free technique with a quadrilateral-dominated element and a medial axis algorithm

Figure A.37 switches to a quadrilateral element with an advanced front algorithm to mesh the structure using a free meshing technique, then the element distortion warning comes back.

Figure A.38 uses a quadrilateral element instead to check the difference from the mesh pattern shown in Fig. A.37, after meshing the part it appears unfortunately that there is no difference because the element distortion warning is still there. It can, therefore, be observed that the change in the element type does not remove the element distortion warning because the distortion is dependent on the algorithm used to mesh the model. In addition to the element geometry ratio tolerance, a wrong choice of algorithm can lead to numerical difficulties in meshing the model properly.

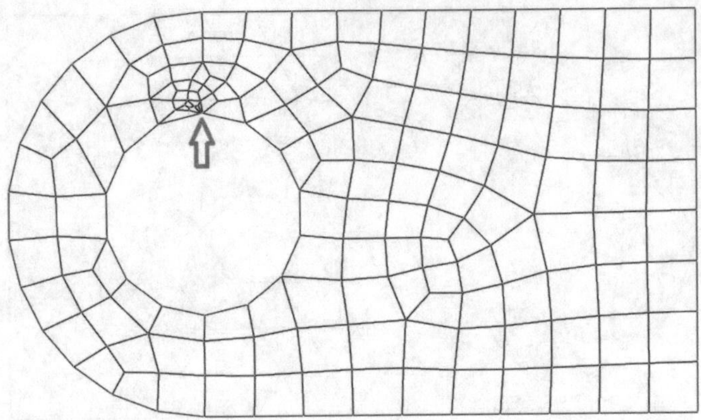

Fig. A.37 Padeye meshed with a design mismatch using a free technique with a quadrilateral element and an advanced front algorithm

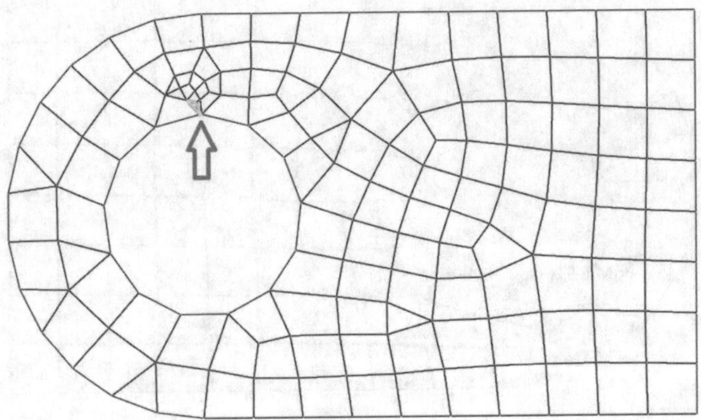

Fig. A.38 Padeye meshed with a design mismatch using a free technique with a quadrilateral-dominated element and an advanced front algorithm

In that case, it seems that the smartest choice for the combination of mesh control parameters is to use a free technique with a quadrilateral element and a medial axis algorithm to keep dealing with such small edges, as shown in Fig. A.39. Of course, in many cases including small edges and similar, small design features can be removed by using the Virtual Topology toolset[4] in the mesh module in order to combine edges, faces, and so on.

[4]See *Abaqus CAE User's Guide*, Sect. 75, "The Virtual Topology Toolset", for more information.

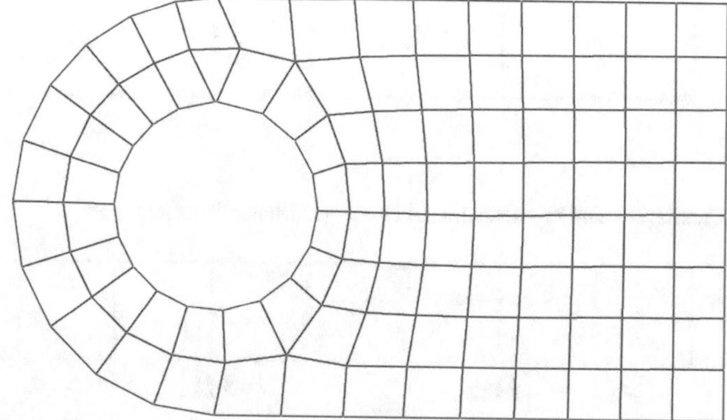

Fig. A.39 Padeye meshed with a design mismatch using a free technique with a quadrilateral element and a medial axis algorithm

A.5.5 Incompatible Mesh

An incompatible mesh is obtained when the analyst is using different element types to mesh the same part, for instance, by mixing quadrilateral and triangular element types together. In such cases, Abaqus will return a message about the incompatible mesh to notify the user and then if the analyst still wants to use an incompatible mesh, the transition region lines or surfaces between the two element types will be defined as **TIE** lines or surfaces. The use of different element types is not recommended to mesh the same part or to define contact interaction in an assembly, because this may result in a deviation in the solution caused by the change in the stiffness region due to the different shape functions associated with different element types.

To understand how this work, just take an example with a simple fixed-free cantilever beam loaded with a vertical single concentrated load force F at the beam end and equal to 10 N. The total axial length is equal to 45 mm and the second moment of the area is calculated from a square-shaped cross-sectional beam with a width equal to 5 mm. The material properties used are purely elastic with a Young's modulus equal to 200 000 MPa and a Poisson's ratio equal to 0.3. Only small deformations are considered to solve the static equilibrium.

To determine what kind of deviation can occur in mixed element types in a meshed structure for model solution, two configurations will be investigated as shown in Figs. A.40 and A.41, respectively. Both meshes are 2D structural plane stress analyses and it would be easier to figure out the change in solution if, instead of considering a plane stress analysis, both structures could be seen as truss structures. In that case, instead of talking about the transition in the shape function, it would be more appropriate to talk about the element connections, which differ from one configuration structure to another. Indeed, the elements connectivity in Fig. A.40 will lead to a different arrangement of the stiffness matrix from the second configuration shown

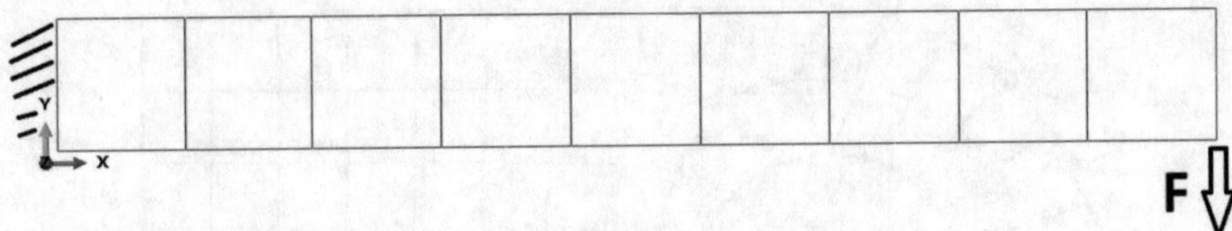

Fig. A.40 Cantilever beam meshed with structured quadrilateral element type

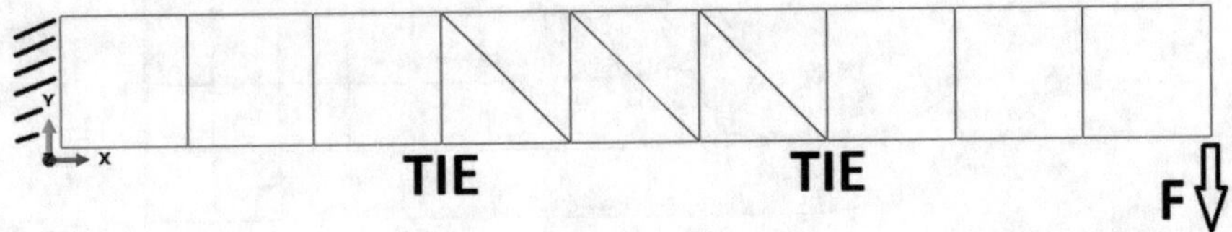

Fig. A.41 Cantilever beam meshed with a structured mix of quadrilateral and triangular element types

in Fig. A.41. Therefore if a stiffness-matrix arrangement K is different, the nodal vector solution u with the static equilibrium equation $Ku = f$ (f as the nodal force vector) will be different. The user will get two different solutions from the two configurations shown in Fig. A.40 and in Fig. A.41, respectively. The question here is how this deviation in the solution can be evaluated.

In a complex model mesh, it is almost impossible to determine the solution deviation calculated with a mixed element type mesh. But to give a rough idea about the solution deviation, Fig. A.41 shows one-third of the structure meshed with triangles. As the cantilever beam is a trivial solution, it will be possible to evaluate such a solution deviation by comparing with the theoretical and computed solutions.

First, let us write the theoretical solution using the Euler–Bernoulli beam theory, where the equilibrium of a beam structure is

$$EI\frac{d^2y}{dx^2} = M_z(x) \tag{A.3}$$

where E is the Young's modulus, I the second moment of area $I = \frac{a^4}{12}$, y the deflection and M_z the bending moment on the axis z along the beam length x.

The equilibrium equation, Eq. A.3, gives the solution in both structure cases applicable to Figs. A.40 and A.41.

$$\begin{cases} EI\frac{d^2y}{dx^2} = -F(L-x) \\ EI\theta(x) = -FLx \ + \ F\frac{x^2}{2} \ + \ K_1 \\ EIy(x) = -FL\frac{x^2}{2} + \frac{F}{2}\frac{x^3}{3} + K_1x + K_2 \end{cases} \tag{A.4}$$

The fixed boundary condition gives $\theta(x=0) = 0$ and $y(x=0) = 0$ to determine the integration constants $K_1 = 0$ and $K_2 = 0$. At the end, the deflection equation can

Table A.1 Maximum deflection solution of a cantilever beam

Models	Element types[a]	Solutions u_y (mm)	Deviations (%)
Theory	Not applicable	−0.02916	0.00
Beam	Beam	−0.02944	0.96
2D plane stress	Quadrilateral	−0.02915	0.03
2D plane stress	Quadrilateral and triangle	−0.02830	2.95

[a]To ensure the accuracy of the results all mesh types were made with a quadratic element formulation

be written as

$$EIy(x) = \frac{F}{2}x^2\left(\frac{x}{3} - L\right) \tag{A.5}$$

The theoretical solution of the maximum deflection is, therefore, given by the equation:

$$y(x = L) = u_y = -\frac{FL^3}{3EI} = -4\frac{FL^3}{Ea^4} \tag{A.6}$$

The conclusion from the results in Table A.1 is clear: the solution deviation is higher than that of any other structure configuration compared with the theoretical result. This confirms a behavior that was completely expected, because the mixture of quadrilateral and triangular element types cannot produce a solution that is closer than the theoretical result. Indeed, the triangular shape is stiffer than the quadrilateral shape, which leads to an interesting observation, that when different element types are used to mesh the same structure, the stiffness no longer represents only the state of the material properties but also represents the mesh configuration pattern. Therefore, a change of stiffness due to mesh mapping is really an unphysical behavior because material properties are assumed to be uniformly distributed along the structure and a local change of the stiffness due to mesh mapping will calculate a unrealistic solution. Mixed element types meshed on the same part should be avoided as far as possible or minimized in complex regions to avoid the computation of unrealistic solutions.

Index

R. J. Boulbes, *Troubleshooting Finite-Element Modeling with Abaqus*,
https://doi.org/10.1007/978-3-030-26740-7

Printed in the United States
By Bookmasters